Lecture Notes in Civil Engineering

Volume 220

Lecture Notes in Civil Engineering (LNCE) publishes the latest developments in Civil Engineering - quickly, informally and in top quality. Though original research reported in proceedings and post-proceedings represents the core of LNCE, edited volumes of exceptionally high quality and interest may also be considered for publication. Volumes published in LNCE embrace all aspects and subfields of, as well as new challenges in, Civil Engineering. Topics in the series include:

- Construction and Structural Mechanics
- Building Materials
- Concrete, Steel and Timber Structures
- Geotechnical Engineering
- Earthquake Engineering
- Coastal Engineering
- Ocean and Offshore Engineering; Ships and Floating Structures
- Hydraulics, Hydrology and Water Resources Engineering
- Environmental Engineering and Sustainability
- Structural Health and Monitoring
- Surveying and Geographical Information Systems
- Indoor Environments
- Transportation and Traffic
- Risk Analysis
- Safety and Security

To submit a proposal or request further information, please contact the appropriate Springer Editor:

- Pierpaolo Riva at pierpaolo.riva@springer.com (Europe and Americas);
- Swati Meherishi at swati.meherishi@springer.com (Asia - except China, and Australia, New Zealand);
- Wayne Hu at wayne.hu@springer.com (China).

All books in the series now indexed by Scopus and EI Compendex database!

More information about this series at https://link.springer.com/bookseries/15087

Manoranjan Parida · Avijit Maji · S. Velmurugan ·
Animesh Das
Editors

Proceedings of the Fifth International Conference of Transportation Research Group of India

5th CTRG Volume 3

Springer

Editors
Manoranjan Parida
Department of Civil Engineering
Indian Institute of Technology Roorkee
Roorkee, Uttarakhand, India

Avijit Maji
Department of Civil Engineering
Indian Institute of Technology Bombay
Mumbai, Maharashtra, India

S. Velmurugan
CSIR—Central Road Research Institute
New Delhi, Delhi, India

Animesh Das
Department of Civil Engineering
Indian Institute of Technology Kanpur
Kanpur, Uttar Pradesh, India

ISSN 2366-2557 ISSN 2366-2565 (electronic)
Lecture Notes in Civil Engineering
ISBN 978-981-16-9924-5 ISBN 978-981-16-9925-2 (eBook)
https://doi.org/10.1007/978-981-16-9925-2

This Springer imprint is published by the registered company Springer Nature Singapore Pte Ltd.
The registered company address is: 152 Beach Road, #21-01/04 Gateway East, Singapore 189721, Singapore

Preface

Transportation Research Group of India (TRG) has brought out this edited book titled *Proceedings of the Fifth International Conference of Transportation Research Group of India* in three volumes. Volume I of the book includes the papers on themes TCT-A01: Pavements and materials, TCT-D01: Travel behavior and transport demand, and TCT-H01: Emerging travel technology. Volume II includes the papers on the themes TCT-B01:Traffic flow theory, operations and facilities, TCT-C01: Transport planning, policy, economics and project finance, and TCT-I01: Other transportation modes (including NMT) and pedestrian. Volume III includes the papers on the themes TCT-E01: Environment (including energy) and sustainability in transportation, TCT-F01: Transportation safety and security, and TCT-G01: Transport and mobility networks (including public transportation, freight and logistics). We are pleased to write the preface for Volume III of this edited book and acknowledge the support rendered by the co-chairs of the respective Technical Committee of TRG (TCT), namely Dr. Madhu Errampalli, Prof. P. Vedagiri, and Prof. Prasanta Kumar Sahu in handling the review process.

The Volume III of the fifth conference of TRG consists of selected 24 papers on the themes of TCT-E01, TCT-F01, and TCT-G01. Out of these, seven papers are on the theme of TCT-E01, another eight papers are on the theme of TCT-F01, and the remaining nine papers are on the theme of TCT-G01. These papers cover wide variety of topics such as sustainable transportation system, renewal energy, transit-oriented development, risk assessment, safety management, driver behaviour, public transportation, freight etc. All the papers have undergone a double-blind review process for selection.

This year TRG is celebrating 10 years of its successful existence and has been providing a platform to all academicians and professionals working on transportation-related problems and their solutions.

This book on conference proceedings of the fifth conference of Transportation Research Group of India held at Bhopal, India, is a compilation of good works of transportation professionals from India and abroad. We hope that this volume and the other two volumes will intrigue the readers with new research ideas and to look for their possible solutions, which can take forward the transportation systems into a new era.

Roorkee, India Manoranjan Parida
Mumbai, India Avijit Maji
New Delhi, India S. Velmurugan
Kanpur, India Animesh Das

About TRG and CTRG

Transportation Research Group of India (TRG) is a not-for-profit registered society with the mission to aid India's overall growth through focused transportation research, education, and policies in the country. It was formally registered on 28 May 2011, and has completed 10 years of its journey this year. The following are the vision and objectives of TRG.

Vision

- To provide a unique forum within India for the interchange of ideas among transportation researchers, educators, managers, policy-makers from India and all over the world, with the intention of covering all modes and sectors of transport (road, rail, air, and water; public and private; motorized and non-motorized) as well as all levels (urban, regional, intercity, and rural transport) and for both passenger and

freight movement, in India, and at the same time to also address the transportation-related issues of safety, efficiency, economic and social development, local and global environmental impact, energy, land use, equity and access for the widest range of travellers with special needs, etc.
- To serve as a platform to guide and focus transportation research, education, and policies in India towards satisfying the country's needs and to assist in its overall growth.

Objectives

- To conduct a regular peer-reviewed conference in India so as to provide a dedicated platform for the exchange of ideas and knowledge among transportation researchers, educators, managers, and policy-makers from India and all over the world, from a perspective which is multi-modal, multidisciplinary, multi-level, and multi-sectoral, but with an India-centric focus. Initially, this conference will be held every two years; however, the frequency may change as per the decision of the society from time to time.
- To publish a peer-reviewed journal of good international standard that considers and recognizes quality research work done for Indian conditions, but which also encourages quality research focused on other developing and developed countries that can potentially provide useful learning lessons to address Indian issues.
- To conduct other activities such as seminars, training and research programmes, meetings, discussions as decided by the society from time to time, towards fulfilling the mission and vision of the society.
- To identify pertinent issues of national importance, related to transportation research, education, and policy through various activities of the society and promote transportation researchers, educators, managers, and policy-makers in an appropriate manner to address the same.
- To collaborate with other international societies and organizations like WCTRS, ASCE, TRB, etc., in a manner that works towards fulfilling the mission and vision of the society.

The Conference of Transportation Research Group of India (CTRG) is the premier event of TRG. It is held every two years and traditionally moves around India. In the past, CTRG has been organized in Bangalore (December 2011), Agra (December 2013), Kolkata (December 2015), Mumbai (December 2017), Bhopal (December 2019), and Trichy (upcoming in December 2021 jointly with NIT Trichy, in association with IISc, Bangalore; IIT Madras; IIT Palakkad; and NATPAC). CTRG has been getting wide scale recognition from reputed Indian and international institutions/organizations like IIT Kanpur, IIT Kharagpur, IIT Guwahati, IIT Bombay (Mumbai), SVNIT Surat, MANIT Bhopal, NIT Trichy, TRB, WCTRS, CSIR-CRRI, ATPIO, T&DI-ASCE, EASTS, to name a few. CTRG is a large conference typically attended by around 400–500 participants, usually from 12 to 15 countries, with about

200 double-blind peer-reviewed technical papers being presented. The conference provides a wide range of executive courses, tutorials, workshops, technical tours, keynote sessions, and special sessions.

Transportation in Developing Economies (TiDE) is the official journal of TRG and is published by Springer. TiDE was formally launched in 2014 and has so far published seven volumes.

Prof. Ashish Verma
Indian Institute of Science Bangalore
Founding and Immediate Past President, TRG

Prof. Akhilesh Kumar Maurya
Indian Institute of Technology Guwahati
Current President, TRG

List of Reviewers

Ramesh A., IIT, Ropar
Rao A. Mohan, CRRI
Ganesh A. R., IIT Madras
Norhidayah Abdul Hassan, UTM Universiti Teknologi, Malaysia
Mukti Advani, CRRI
Amit Agarwal, IIT Roorkee
P. K. Agarwal, MANIT Bhopal
Ruchika Agarwala, IIT Bombay
Juraidah Ahmad, Universiti Teknologi MARA
Notani Mohammad Ali, Purdue University
D. S. N. V. Amarkumar, Shell Bitumen, India
M. V. L. R. Anjaneyulu, NIT, Calicut
Sabreena Anowar, University of Missouri
Uma Arepalli, Maine Department of Transportation
Shriniwas Arkatkar, NIT Surat
Ashutosh Arun, CSIR
Anil Kumar Bachu, IIT Patna
Rajneesh Bahari, Indian Army
Ranja Bandyopadhyaya, NIT, Patna
Arunabha Banerjee, IIT Guwahati
Ipsita Banerji, IISc, Bangalore
Manik Barman, University of Minnesota, Duluth
Munwar Basha, IIT Hyderabad
Debasis Basu, IIT, Bhubaneswar
Ambika Behl, CRRI
Ashik Bellary, KLS, VDIT
Reema Bera, IIT Kharagpur
P. R. Bhanumurthy, JNTU, Andhra Pradesh
Gottumukkala Bharath, CRRI
Ashish Bhaskar, Queensland University of Technology
Prithvi Bhat, Queensland University of Technology

Sudip Bhattacharjee, Alabama A&M University
Tanmoy Bhowmik, University of Central Florida
Sasanka Bhushan, VIT University, Vellore
Prasanta Bhuyan, NIT, Rourkela
Mehek Biswas, IISc, Bangalore
Nomesh Bolia, IIT Delhi
Bhaswati Bora, IIT Kanpur
Anuj Budhkar, IIEST, Shibpur, West Bengal
Anush Chandrappa, IIT, Bhubaneswar
Arti Chaudhary, CRRI
Akhilesh Chepuri, Sreenidhi Institute of Science and Technology, Ph.D. from SVNIT
Munavar Fairooz Cheranchery, TKM College of Engineering, Kerala
Bhargava Rama Chilukuri, IIT Madras
Tanuj Chopra, Thapar Institute of Engineering and Technology
Pushpa Choudhary, IIT Roorkee
Rajan Choudhary, IIT Guwahati
Venkaiah Chowdary, NIT Warangal
Saurabh Dandapat, IIT Kharagpur
Atasi Das, GR Infratech
Bibhutibhusan Das, NIT Surathkal
Pritikana Das, CSIR, CRRI
Sanhita Das, IIT Guwahati
Subasish Das, Texas A&M University
Vivek R. Das, Dayanand Sagar College of Engineering, Kerala
Venkata Santosh Kumar Delhi, IIT Bombay
Ravi Deshpande, Indian Army
Partha Pratim Dey, IIT, Bhubaneswar
P. V. Divya, IIT Palakkad
Venkata Ramana Duddu, University of North Carolina, Charlotte
Madhu Errampalli, CRRI
Galkin Andrii, O. M. Beketov National University of Urban Economy in Kharkiv, Kharkiv, Ukraine
Tejash Gandhi, Kao Corporation
Venu Garikapati, National Renewable Energy Laboratory, NREL
G. S. Ghataora, University of Birmingham, UK
Indrajit Ghosh, IIT Roorkee
Kaniska Ghosh, IIT Kharagpur
Ripunjoy Gogoi, Amity University
S. S. S. V. Gopalaraju, MVGR College of Engineering
Sutapa Goswami, University of Calcutta
Bharath Gottumukkala, CRRI
Shiva Kumar Govindaraju, University of Pisa UNIPI, Italy
Bivina G. R., IIT Roorkee

Pradip Gundaliya, L. D. College of Engineering, Gujarat
Ankit Gupta, IIT BHU
Sanjay Gupta, School of Planning and Architecture, Delhi
Ayyanna Habal, IIT Bombay
Mohd. Rosli Hainin, UTM Universiti Teknologi, Malaysia
Md. Bashirul Haque, University of Leeds, UK
Shimul Haque, Queensland University of Technology
Gayathri Harihan, IISc, Bangalore
Vasant Havangi, CRRI
Simon Hesp, Queen's University, Canada
Praveen Idara, University of Missouri, USA
Suresh Immanuel, University of Evansville
Norfarah Nadia Ismail, International Islamic University, Malaysia
Abhijith Iyengar, IIT Madras
Nur Izzi, UKM National University of Malaysia
Bhavesh Jain, IIT Roorkee
Subham Jain, IIT Kanpur
Udit Jain, IIT Roorkee
Nataraju Jakkula, CSIR
Lalita Jangpangi, CSIR
Ashok Julaganti, NIT Warangal
Raghuram Kadali, Sri Vasavi Engineering College, Andhra Pradesh
Nurul Hidayah Kamaruddin, Universiti Tun Hussein Onn Malaysia, UTHM
Venkatesan Kanagaraj, IIT Kanpur
Siksha Kar, CRRI
Gunasekaran Karuppanan, Anna University, Tamil Nadu
Aniket Kataware, IIT Bombay
Ankit Kathuria, IIT, Jammu
Ravinder Kayitha, CRRI
Ahmad Khan, Queen's University, UK
Durgarani Korlapati, Andhra University, Andhra Pradesh
Suhana Koting, University of Malaysia
Rajesh Krishnan, Imperial College of London, UK
Ambika Kuity, NIT, Silchar
Venu Madhav Kukkapalli, University of North Carolina, Charlotte
Aditya Kumar, IIT Bombay
Anjan Kumar, IIT Guwahati
Anupam Kumar, Delft University
Binod Kumar, CRRI
Prabin Kumar, IIT Bombay
Rajiv Kumar, NIT Jalandhar
Ravindra Kumar, CRRI
Vasantha Kumar, VIT, Vellore

Molugaram Kumar, UCE, Osmania University, Hyderabad
Phani Kumar Patnala, Echelon Institute of Technology, Haryana
Kranti Kumar Kuna, IIT Kharagpur
Ganeshbabu K. V., Amaravati Development Corporation, Andhra Pradesh
Deepa L., TKM College of Engineering, Kerala
Sam Labi, Purdue University, USA
K. M. Laxmanrao, JNTUH College of Engineering, Hyderabad
Harikrishna M., NIT, Calicut
Shashi M., Indian School of Mines
Manoj M., IIT Delhi
Nivedya M. K., Worcester Polytechnic Institute WPI, USA
Nagabhushana M. N., CRRI
Geetimukta Mahapatra, IIT Guwahati
Smruti Sourava Mohapatra, IIT-ISM, Dhanbad
Saurabh Maheshwari, University of California, Davis
Ramesh Chandra Majhi, CRRI
Avijit Maji, IIT Bombay
Bandhan Majumdar, BITS, Pilani, Hyderabad
Haritha Malladi, Lafayette College, NYC
C. Mallikarjuna, IIT Guwahati
Ajinkya Mane, University of North Carolina, Charlotte
Bappaditya Manna, IIT Delhi
Samson Mathew, NIT Trichy
Gabriella Mazzulla, University of Calabria, Italy
Arpan Mehar, NIT Warangal
Prakash Mehta, Tawata Technology, Gujarat
Swati Mitra, IIT Kharagpur
Abhishek Mittal, CRRI
Mithun Mohan, NIT Karnataka
Ranju Mohan, IISc, Bangalore
Abhisek Mudgal, IIT BHU
Caleb Munigety, IISc, Bangalore
Dasaka Murty, IIT Bombay
Sai Kiran Myakuntla, IISc, Bangalore
Lohit Kumar Nainegali, IIT, Dhanbad
Syam Nair, IIT Kanpur
Prakash Nanthagopalan, IIT Bombay
Binanda Narzary, Tezpur University
Akhandappagol Ningappa, NIT-K, Surathkal
Sangram Nirmale, IISc, Bangalore
Seckin Ozkul, University of South Florida
Rajesh Paleti, Pennsylvania State University
Pan Pan, Queen's University, Canada

Mahabir Panda, NIT, Rourkela
Anurag Pande, California Polytechnic State University
Manoranjan Parida, IIT Roorkee
Purnima Parida, CBRI
Bhaskar Paul, IIT Kharagpur
Digvijay Pawar, IIT Hyderabad
Praveena Penmetsa, University of Alabama
Vedagiri Perumal, IIT Bombay
C. S. R. K. Prasad, NIT Warangal
Prashant Prasad, IIT Kharagpur
M. Lakshmi Vara Prasad, NIT, Silchar
G. V. R. Prasadaraju, JNTU Kakinada, Andhra Pradesh
Sushma Prusty, IIT Bombay
Farzana Rahman, University of Asia-Pacific
Bharat Rajan, IIT Bombay
Sridhar Raju, BITS, Pilani, Hyderabad
Gitakrishnan Ramadurai, IIT Madras
Ayothiraman Ramanathan, IIT Delhi
T. V. Ramanayya, NIT Warangal
G. D. Ransinchung, IIT Roorkee
K. Ramachandra Rao, IIT Delhi
Sita Rami Reddy, VNIT, Nagpur
Amaranatha Reddy, IIT Kharagpur
K. S. Reddy, IIT Kharagpur
Padma Rekha, SRM Institute of Science and Technology
S. Rokade, MANIT Bhopal
Sandeepan Roy, IIT Bombay
Minal S. CRRI
Shankar Sabavath, NIT Warangal
Nikhil Saboo, IIT BHU
Romeil Sagwal, CSIR
Dulal Saha, University of Waterloo
Amarendra Kumar Sahoo, GMR Infrastructure Ltd.
Umesh Chandra Sahoo, IIT, Bhubaneswar
Prasant Sahu, BITS, Pilani, Hyderabad
Jagadish Prasad Sahu, IIT Roorkee
Subbarao Saladi, Mahindra Ecole College of Engineering, Hyderabad
Moses Santhakumar, NIT Trichy
Binu Sara, College of Engineering, Trivandrum
Goutham Sarang, VIT, Vellore
Sajitha Sasidharan, IISc, Bangalore
Shobhit Saxena, IISc, Bangalore

Robert Schneider, School of Architecture and Urban Planning, University of Wisconsin, Milwaukee
Padma Seetharaman, CSIR
Ravi Sekhar, CSIR
Sathish Kumar Selvaraj, University of Melbourne
Yogesh Shah, IITRM, Ahmedabad
K. V. R. Ravi Shankar, NIT Warangal
S. Shankar, NIT Warangal
Ravi Shankar S., CRRI
Niraj Sharma, CSIR
R. Shivashankar, NIT Surathkal
Burhan Showkat, IIT Bombay
Manoj Shukla, CRRI
Raj Sidharthan, University of Nevada
Gourab Sil, BITS, Pilani
Bhupendra Singh, NIT, Patna
Dharamveer Singh, IIT Bombay
Priyansh Singh, BITS, Pilani
Rina Singh, CRRI
Varun Singh, MNNIT, Allahabad
Anil Kumar Sinha, CRRI
Kalaanidhi Sivagnanasundaram, Technion Israel Institute of Technology
M. Sivakumar, NIT, Calicut
R. Sivananadan, IIT Madras
Anand Sreeram, Hong Kong Polytechnic University; Visiting Scholar at University of Texas
R. Srinivasakumar, Osmania University, Hyderabad
Karthik K. Srinivasan, IIT Madras
Y. Sudheerbabu, Nimra College of Engineering and Technology, Andhra Pradesh
Sanjeev Kr. Suman, NIT, Patna
V. Sunitha, NIT Trichy
S. N.Suresha, NIT Karnataka
Mansha Swami, IIT Roorkee
Aravind Krishna Swamy, IIT Delhi
Subba Rao Swarna, Survey of India, Dehradun
Deprizon Syamsunur, UCSI University, Malaysia
Rahul T. M., IIT, Ropar
Shams Tanvir, North Carolina State University
Manoj Tipnis, L&T
Devesh Tiwari, CRRI
Lelitha Vanajakshi, IIT Madras
Harsha Vajjarapu, IISc, Bangalore
Sudhir Varma, IIT Patna

Nagendra Velaga, IIT Bombay
K. Venkatareddy, NIT Warangal
Veena Venudharan, IIT Palakkad
A. Veraragavan, IIT Madras
Ashish Verma, IISc, Bangalore
Meghna Verma, Ramaiah Institute of Management, Bangalore
Durgesh Vikram, BITS, Pilani
V. Vinayaka Ram, BITS, Pilani, Hyderabad
Guru Vittal, CRRI
Haryati Yaacob, Universiti Teknologi Malaysia
Shamsunnahar Yasmin, Queensland University of Technology, Australia

Contents

About the Editors

Prof. Manoranjan Parida is currently Professor in the Department of Civil Engineering at IIT Roorkee. He is working on an imprint research project called "Propagation and Mitigation Model of Mixed Traffic Noise for Planning Mid-Sized Indian Cities." His present doctoral research includes walkability, transit-oriented development, service quality models of metro system, planning of P&R around MRTS, and traffic noise modelling for mixed traffic situation. He has provided sustainable solutions for more than 350 road infrastructure projects for urban areas, intercity corridors rural roads, and expressway. Professor Parida received the Pt. Jawaharlal Nehru Birth Centenary Award in 2004 from Indian Road Congress and Outstanding Teacher Award of IIT Roorkee. Professor Parida is also Independent Director on the Board of Jharkhand Urban Infrastructure Development Company Ltd. and Convener of Urban Roads and Streets Committee, Indian Roads Congress, New Delhi.

Dr. Avijit Maji is Associate Professor in the Department of Civil Engineering, IIT Bombay, India. He received Doctor of Engineering from Morgan State University, USA, in 2008; M.Tech. from IIT Kanpur in 2004; and B.E. from Bengal Engineering College (now Indian Institute of Engineering Science and Technology) in 2000. He has more than sixteen years of academia and industry experience working for organizations in India and USA. His area of expertise includes design, operations, and safety of transportation infrastructure. He has published more than sixty journal and conference papers. For his contribution, Maryland Society of Professional Engineers conferred him with the "Young Engineer of the Year Award" in 2011. He is Licensed Professional Engineer (PE) in Maryland, USA, and Transportation Professional Certification Board's certified Professional Traffic Operations Engineer (PTOE).

Dr. S. Velmurugan is presently working as Chief Scientist at CSIR-Central Road Research Institute (CRRI), New Delhi, and also the Former Head one of the vibrant divisions of CSIR—CRRI namely, Traffic Engineering and Safety (TES) Division. He completed his Ph.D. in Transportation Systems Engineering at IIT, Bombay, in 1995. Dr. Murugan had two stints of Post-Doctoral Research (PDR) in Belgium and Australia in the fields of his specialization i.e. traffic and transportation engineering. He was instrumental in the formulation/revision of 18 publications of Indian Road Congress (IRC) during the last 15 years. Dr. Murugan is serving as Project Coordinator for the prestigious Project: "iRASTE (Intelligent Solutions for Road Safety through Technology and Engineering) in Nagpur" being executed by CSIR—CRRI in collaboration with International Institute of Information Technology (IIIT), Hyderabad, INAI (Applied Research Center for AI at IIIT-Hyderabad), M/s. Intel India as well as M/s. Mahindra and Mahindra under the initiative of Sh. Nitin Gadkari Ji aimed at the reduction of 50% fatalities in Nagpur city. The maiden version of the Indian Highway Capacity Manual (Indo—HCM) was successfully completed by Dr. Murugan in collaboration with seven academic institutes in the country by assuming the role of Champion in 2018 which is extensively referred by the research fraternity, academicians and practicing engineers. Further, he has provided sustainable solutions for more than 160 road related projects covering the subject areas of road safety, road safety audit, public transport evaluation, and estimation of road user costs addressing the issues in the inter-urban and urban roads including expressways. Dr. Murugan was bestowed with the Pt. Jawaharlal Nehru Birth Centenary Award in 2008 from IRC and the Outstanding Scientist Medal by the Construction Industry Development Council (CIDC) in 2015. Dr. Murugan is presently serving as the Co-convener of the "Intelligent Transport System" i.e. G-7 Committee of IRC.

Dr. Animesh Das is a Professor in the Department of Civil Engineering, IIT Kanpur. Dr. Das's area of expertise is pavement material characterization, analysis, design, and evaluation. He is interested in both theoretical as well as experimental studies. He has worked on problems related to characterization of asphalt mix and aggregates, asphalt recycling, asphalt micromechanics, microstructure and image analysis of asphaltic material, pavement design optimization, pavement surface characterization etc. He has published several papers in various journals and conference proceedings of repute.

A Data Envelopment Analysis-Based Benchmarking of Indian City Bus Systems

Ravi Gadepalli and Siddartha Rayaprolu

1 Introduction

Public transport plays an important role in providing reliable, safe, and affordable access to mobility to users. Governments in developed countries often subsidize public transport services toward meeting these objectives. In the case of developing countries, the investments in public transport are limited, thereby making the systems reliant on their internal sources of revenue. As a result, these countries are witnessing a rapid shift in user preference from public transport to personal vehicles, leading to increased congestion, emissions, and traffic crashes [1]. Therefore, in the case of both developed and developing countries, it is important that public transport systems perform efficiently and deliver the maximum societal value for their resource inputs. We analyze the bus-based public transport systems of Indian cities to establish the key variables impacting their efficiency and quantify the potential of each of these variables in scaling up the services to meet user needs adequately. The learnings from efficiency improvements in high transit usage environments like India can offer learnings for other developed and developing countries looking to improve their public transport efficiencies.

Public bus services in India are commonly provided by State Road Transport Undertakings (STUs) owned and operated by the government. Despite carrying 68 million users per day traveling 1.48 billion passenger km, these services have been operating in losses ranging from 6 to 27% of their total costs [2, 3]. Their losses can be attributed to multiple reasons including increased costs caused by inefficiencies in planning and operations, reduced revenue caused by providing services to low

R. Gadepalli (✉)
Transit Intelligence LLP, Bengaluru, India
e-mail: ravi@transitintelligence.in

S. Rayaprolu
George Mason University, Fairfax, VA, USA

M. Parida et al. (eds.), *Proceedings of the Fifth International Conference of Transportation Research Group of India* , Lecture Notes in Civil Engineering 220,
https://doi.org/10.1007/978-981-16-9925-2_1

demand areas, and the subsidized fares for user groups like students and senior citizens. However, the specific contribution of various factors to the overall performance of the STUs has not been established in the literature, thereby limiting the knowledge on the initiatives needed to improve the bus systems.

The current article addresses this gap by presenting a methodology to establish the role of various internal and external factors impacting the operational and financial performance of bus systems. A data envelopment analysis (DEA)-based approach was adopted to benchmark the operational and financial efficiencies of Indian city bus systems over the past seven years. The relative contribution of various inputs to the system toward various categories of outputs, i.e., service supply, consumption, and revenue efficiency was established. Further, the impact of external factors like city development, travel demand patterns, and economic characteristics on the efficiency of the bus system was also derived through regression analysis of city bus performance efficiencies and the external variables. Even though the analysis was carried out for the Indian context, the methodologies and variables used to identify measures for improving bus system efficiencies are applicable even in other developed and developing country contexts.

2 Literature Review

To compare the performance of various bus systems and identify the variables impacting their efficiency, the literature review focused on methods for benchmarking of public transport performance efficiency and the key internal and external variables required for the analysis.

2.1 Methods for Benchmarking Public Transport Systems

Efficiency represents the performance of an organization or a service by the amount of output produced for a given amount of input (energy, time, money). Efficiency evaluation of a service enterprise depends on how well the service is managed, how well that is received, and how big the organization is [4]. Benchmarking involves the comparison of individual service enterprise efficiencies to the most efficient service of the sample under consideration. It is commonly carried out through frontier techniques which put the most efficient services on the frontier and evaluate the rest of the services relatively [5]. The methods used for benchmarking analysis can be classified into parametric and nonparametric techniques. Stochastic frontier analysis (SFA) is the most preferred technique for parametric benchmarking while for nonparametric benchmarking, data envelopment analysis (DEA) is the preferred technique [6, 7]. SFA does not consider more than one output variable and also includes external variables within the frontier analysis and hence does not establish their relationship with efficiencies explicitly [8]. Although DEA does not propose any mechanism to attain efficiency, it quantifies the changes needed for the inefficient unit to become efficient according to the outputs being sought [9]. It also determines the weightages of each input and output variables within the overall efficiency as an output, as against the predetermined weightages adopted by SFA [4].

DEA further has two key methods of benchmarking: Charnes, Cooper, and Rhodes (CCR Model) and Banker, Charnes, and Cooper (BCC Model). The CCR model is based on constant returns to scale (CRS) which assumes that an increase in the inputs by a factor to have a proportional increase in the output by the same factor and vice versa [4]. The efficiencies calculated for each decision-making unit (DMU) using this CCR-CRS model are known as overall technical efficiencies (OTE). BCC model is based on variable returns to scale (VRS) which assumes that an increase in all the inputs by a factor may not change the output by the same factor and vice versa [5]. The efficiencies calculated using BCC-VRS model are known as pure technical efficiencies (PTE). The ratio of these two efficiencies (OTE/PTE) is known as scale efficiencies (SE), which measures the impact of scale size on the efficiency of a DMU [6]. The outputs from scale efficiency will provide insights on increasing or diminishing returns to scale of the DMU, which can be used to understand if a firm needs to increase or decrease its size for maximizing its efficiency. Even though DEA is widely applied across various sectors, its application for the transport sector has been increasing only recently [7, 8].

In summary, DEA offers the benefit of being able to use multiple input and output variables to derive relative efficiency performance of bus systems. Being nonparametric, it can also be used to compare efficiency between entities of varying scales and sizes and further check for scale efficiency of the bus agencies. Another inherent advantage of DEA is its method of assigning coefficients to various input and output variables, which can be used to carry out slack analysis. Slack analysis estimates the change in value of each input and output of a DMU required to improve its efficiency to be equivalent to the most efficient DMU [9].

2.2 *Internal and External Variables Impacting Public Transport Efficiency*

The input and output variables used in public transport efficiency evaluation literature and their frequency of usage in recent articles were presented by [6, 10]. Majority of the literature on efficiency evaluation of transit focuses on the system's productivity, i.e., the services offered per unit input with limited focus on the effectiveness of these services in meeting the user and societal objectives [7, 11]. A further classification of these variables was presented by [10] where input variables were classified into physical measures, capital expenses (CAPEX), and operating expenses (OPEX), and output variables were classified into service supply, service consumption, and revenue. The current article adopts a similar classification system to derive efficiencies according to users', operators', and the city's perspective while maintaining the homogeneity in the selection of input variables.

In addition to the internal input and output variables, the efficiency of a transit system also depends on external variables that influence its performance. These include the city characteristics like population density, economic development, and mobility characteristics like trip lengths, population with access to services, etc. However, the literature analyzing the impact of external variables on public transport efficiency is limited.

2.3 Benchmarking of Indian Bus Systems

Applications of DEA to measure performance efficiency of Indian bus systems have been limited. Saxena and Saxena [9] used DEA to compare the performance of 35 STUs providing intra-city and inter-city services for the year 2004–05 using both CCR and BCC methods. The scale efficiency of each of the STUs and the percentage reduction potential for various input resources to reach the efficiency frontier were derived. Similarly, [12] applied CCR and BCC methods of DEA to study the performance of 25 Indian STUs between 2002–03 and 2004–05. They derive the relative efficiencies of STUs and identify their potential for improving their technical and scale efficiencies. However, both the articles combine intra-city and inter-city bus systems for their analysis, which was not an accurate comparison considering their varying operational characteristics, passenger requirements, etc., [13] classified measurement variables into three categories of operations, finance, and accident-based, and later analyzed the importance of each category based on the analytical hierarchy process (AHP) [13]. However, this study did not consider the variables related to service consumption and did not provide a detailed insight on how to utilize the calculated efficiency numbers toward overall improvement.

Segregation of supply, consumption, and revenue perspectives of public bus agencies are missing in the Indian literature. Further, none of the previous studies evaluate the impact of external variables on the STUs' efficiency. Hence, the findings from existing literature do not provide conclusive insights into the specific measures needed to improve city bus systems.

3 Methodology

We adopted the two-level analysis proposed by [10] to benchmark Indian city bus systems. A DEA-based benchmarking analysis was carried out initially based on the input–output variables with each State Transport Undertakings (STU) providing city bus services as a DMU. The impact of external variables on efficiency was derived based on regression analysis between the efficiencies from the DEA analysis and external variables representing the city development and mobility characteristics.

3.1 Efficiency Analysis Using DEA

DEA-based efficiency analysis was carried out separately for three categories of outputs, i.e., service supply, service consumption, and revenue efficiency to enable a more disaggregated understanding of their performance. Both CRS and VRS-based efficiencies were derived to measure the scale efficiencies of the DMUs and were used for the regression analysis with external variables. The potential for reducing various inputs toward improving DMU efficiencies was estimated using slack analysis. The variables considered for analysis, formulation of DEA, and the data collection are explained in Sect. 4.

3.2 Formulation of DEA

An input-oriented DEA-based benchmarking was adopted to establish the reduction potential for each of the inputs for various categories of outputs. Equations (1)–(4) provide the formulation adopted for DEA. The objective function of the DEA formulation is to maximize efficiency h of target DMU j_0 where a total of n DMUs operate with m inputs and s outputs; y_{rj} is the amount of rth output from entity j, and x_{ij} is the amount of ith input from the same entity j. The decision variables $u = (u_1, u_2, \ldots, u_r, \ldots, u_s)$ and $v = (v_1, v_2, \ldots, v_r, \ldots, v_m)$ are weights given to the s outputs and m inputs, respectively. Thus, the objective equation (Eq. 1) is iterated n times to calculate the relative efficiencies of one entity at a time. The weights are constrained such that they are positive, and the efficiency of any entity is not greater than one (Eq. 2). This is further ensured by assigning an infinitesimally small positive value for weightages (Eqs. 3 and 4).

$$\max h_{j_0}(u, v) = \frac{\sum_{r=1}^{s} u_r y_{rj_0}}{\sum_{i=1}^{m} v_i x_{ij_0}} \tag{1}$$

Subject to:

$$\frac{\sum_{r=1}^{s} u_r y_{rj}}{\sum_{i=1}^{m} v_i x_{ij}} \le 1, \quad j = 1, 2, \ldots, n, \tag{2}$$

$$\frac{u_{rj_0}}{\sum_{i=1}^{m} v_i x_{ij}} \ge \varepsilon, \quad r = 1, 2, \ldots, s \tag{3}$$

$$\frac{v_{ij_0}}{\sum_{i=1}^{m} v_i x_{ij}} \ge \varepsilon, \quad i = 1, 2, \ldots, m \tag{4}$$

MaxDEA 7.1 Basic, an open-source spreadsheet-based application, was used for the DEA [14]. The application forms a frontier based on efficient units within the reference data set and generates a "lambda" or "efficiency reference set" for the inefficient units [15]. The lambda values, when multiplied with the existing input resources of the inefficient DMU, form a composite hypothetical unit, whose difference with the original DMU will result in "proportionate movement", i.e., the potential for increase or reduction in resources being utilized for each input. If a DMU does not attain the efficiency frontier despite applying the "proportionate movement" to the resources, "slack" would be required to push the units toward efficiency frontier. Finally, "projection" represents the efficient target of input or output resource that should be used. For the input-oriented DEA model, Eq. (5) represents the calculation of projected values, based on the proportionate and slack values of an inefficient unit [9]. For a target DMU with an optimal efficiency h, x_{ij} is the amount of ith input from the same entity j, and $\overline{x_{ij}}$ represents the final projected amount of resource that is required.

$$\overline{x_{ij}} = h * x_{ij} - S_{ij}^{-} \tag{5}$$

3.3 *Regression Analysis for Impact of External Variables*

The performance efficiency of a city bus system is also impacted by the external environment in which it is operating. This includes overall city development characteristics like its population, geographic spread, and economic development along with its overall mobility characteristics like the average trip length, percentage of trips by various length categories (0–5, 5–10, 10–30 km), and mode share of buses in the city. Therefore, a linear regression-based analysis was carried out to establish the level of correlation between the efficiency of city bus systems calculated through DEA and various external variables that are likely to impact its efficiency. Only variables which could be collected through secondary data sources were considered for analysis. Variables like potential shift from personalized modes to public transport and change in level of service indicators for users which required primary data collection were not part of the scope of the current study.

4 Data Collection and Selection of Variables for Benchmarking

A total of eight city bus systems covering the largest Indian cities were benchmarked over the seven-year period between 2009–10 and 2015–16. Secondary data required for benchmarking of these city bus systems, i.e., the physical and financial performance, were collected for seven financial years between 2009–10 and 2015–2016 [3, 16–18]. Data for a total of 36 variables were initially collected, out of which only 16, i.e., nine input and seven output variables which were consistently reported by all DMUs for all the years, as presented in Table 1. Total buses held, total staff, and staff per each bus were used as the input physical measures of DMUs. OPEX was measured through staff cost, fuel and lubricant cost, total operational cost per bus, and the average fuel efficiency of all buses. Total capital expenditure and the sum of CAPEX and OPEX were considered as separate input variables. Among output variables, effective kilometres, i.e., the actual vehicle km of bus service provided were considered as the variable representing service supply. Service consumption was measured through passengers carried, total passenger km traveled on buses and load factor, i.e., the ratio of passenger occupancy to bus capacity. Revenue output of the bus system was measured using total revenue of the DMU, revenue per each bus, and traffic revenue, i.e., revenue made through tickets.

Table 2 provides a summary of eight DMUs considered for benchmarking analysis along with secondary data of external variables representing key city development and mobility characteristics. The data on external variables were only available for the year 2010–11. Therefore, the regression analysis between efficiency and external variables was carried out only for 2010–11, while the efficiency benchmarking was carried out for all the seven years.

Table 1 Input and output variables considered for sensitivity analysis

Input category	Input variables	Output category	Output variables
Physical measures	Buses held	Service supply	Effective kilometers
	Total staff	Service consumption	Passengers carried
	Staff per bus		Passenger kilometers traveled
OPEX	Staff cost		Load factor
	Fuel and lubricant cost	Revenue	Total revenue
	Total cost per bus per day		Total revenue per bus per day
	Fuel efficiency		Traffic revenue
CAPEX	Capital expenditure		
OPEX + CAPEX	Total cost		

Table 2 Cities, DMUs, and external variables considered for analysis

City	Name of STU/DMU	Population (million)+	Area (km^2)[a]	Density (Pop/km^2)	Average trip length (km)+	% trip length distribution (5–10 km)+	GDP per capita (in USD)	Economy index score[a]
Bangalore	Bangalore Metropolitan Transport Corporation (BMTC)	8.5	800	10,624	7.5	27	2960	44
Kolkata	Calcutta State Road Transport Corporation (CSTC)	14.11	186.23	75,780	5.4	27	1822	36
Chennai	Metropolitan Transport Corporation—Chennai (MTC)	8.67	175	49,691	7.0	30	2630	43
New Delhi	Delhi Transport Corporation (DTC)	16.31	1482.71	11,003	7.7	23	4057	36
Ahmedabad	Ahmedabad Municipal Transport Services (AMTS)	6.35	468.92	13,547	6.7	27	2959	32
Mumbai	Brihanmumbai Electric Supply and Transport Undertaking (BEST)	18.41	1400.4	13,149	4.9	25	3987	41
Chandigarh	Chandigarh Transport Undertaking (CHNTU)	1.02	105.68	9706	6.1	32	3629	39.6
Pune	Pune Mahanagar Parivahan Mahamandal Limited (PMPML)	5.05	276.4	18,271	7.3	26	2916	40

[a]Economic score index of city was estimated using employment levels, equity, and growth of new businesses

Given that the current study benchmarks eight cities, the total number of variables for each category of benchmarking, i.e., service supply, consumption, and revenue efficiency, was limited to four. Three input variables which are common across all categories of benchmarking and one output variable for each category of benchmarking were shortlisted based on the correlation between input and output variables. The DEA model was then applied for several combinations of strongly correlated variables, and the variables with 20% or more impact on efficiency were retained.

5 Analysis and Findings

Benchmarking of the eight DMUs was carried out for three categories of output variables, i.e., service supply, service consumption, and revenue efficiency. Such classification of efficiencies enables a disaggregated analysis of the efficiency in utilizing resources toward various types of outputs expected from the bus system. Correlation between input and output variables presented in Table 2 and sensitivity analysis for the impact of variables on the efficiency of DMU was carried out to derive three common input variables across three categories and one output variable for each efficiency category. Table 3 presents the final input and output variables selected for analysis and their data for the seven-year analysis period. Buses held, total staff, and total cost incurred by the DMU in a year were used as the input variables across benchmarking categories. Effective kilometers of services delivered were selected as the output variable for service supply while the total passenger kilometers traveled were selected as the output measure of service consumption. Total revenue earned by the DMU was identified as the output variable to benchmark revenue efficiency.

Table 4 presents the OTEs from the CRS analysis, PTEs from VRS analysis, and the scale efficiencies derived for the service supply, consumption, and revenue of the eight DMUs for the seven-year analysis period of 2009–10 to 2015–16. Bus services in Bangalore, Chandigarh, and Chennai were observed to perform the best across categories. Among the less efficient DMUs, Kolkata had the least efficiency across categories, followed by Ahmedabad and Delhi. Among the three categories of efficiency analyzed, cities performed best on revenue efficiency with a minimum average efficiency of 0.8 across cities over seven years. This was followed by supply efficiency with a minimum average efficiency of 0.74 while the consumption efficiency had the least value of 0.62. These values indicate the significant potential for improvement in consumption efficiency in Indian city bus services. The scale efficiencies were summarized in Fig. 1 as per the proportion of constant, increasing and diminishing returns to scale (RTS) observed for each efficiency category, i.e., supply, consumption and revenue efficiencies. Majority of supply and consumption efficiencies were increasing RTS while for revenue efficiencies had the highest proportion of diminishing RTS. This indicates the scope for improving supply and consumption efficiencies, while the revenue efficiencies have already peaked for the majority of the DMUs.

Table 3 Data of input and output variables used for DEA

DMU	Year	Input variables			Output variables (millions)		
		Buses held	Staff size	Total cost (millions ₹)	Effective km	Passenger km	Total revenue (₹)
BMTC	2009–2010	5715	30,996	10,666	442	17,853	11,317
	2010–2011	6110	32,953	12,772	458	20,102	13,275
	2011–2012	6091	32,300	14,817	466	21,148	15,031
	2012–2013	6330	34,273	18,080	464	1769	16,605
	2013–2014	6603	36,054	21,615	480	20,739	20,139
	2014–2015	6649	36,474	23,217	471	21,725	22,568
	2015–2016	6448	35,554	21,938	336	15,648	22,075
CSTC	2009–2010	978	6719	2246	40	1913	1881
	2010–2011	956	6102	2514	35	1211	654
	2011–2012	839	5813	2339	29	1200	616
	2012–2013	779	5485	2185	26	108	704
	2013–2014	718	5059	2305	22	1012	620
	2014–2015	813	4998	4324	28	765	3581
	2015–2016	782	4799	3992	21	631	2738
MTC	2009–2010	3210	23,000	9087	333	19,469	8110
	2010–2011	3414	23,500	11,516	347	21,796	9202
	2011–2012	3444	22,146	12,986	351	21,325	10,489
	2012–2013	3585	23,519	13,549	344	1754	12,568
	2013–2014	3666	23,982	15,314	360	19,217	13,605
	2014–2015	3787	25,219	17,236	351	18,996	13,309
	2015–2016	3832	24,930	18,045	354	19,087	13,048
DTC	2009–2010	3845	29,495	26,194	209	8492	5766
	2010–2011	5765	35,557	33,221	292	13,041	9870
	2011–2012	6078	40,721	37,119	374	9024	12,808
	2012–2013	5603	34,376	42,647	354	1075	13,503
	2013–2014	5341	35,503	25,962	317	15,543	12,325
	2014–2015	4977	32,864	51,013	287	11,019	11,099
	2015–2016	4564	30,527	57,009	267	10,579	10,050
AMTS	2009–2010	966	5592	2369	51	2115	1177
	2010–2011	942	5274	2301	53	2102	1089
	2011–2012	985	4715	2605	49	2129	1196
	2012–2013	1120	5428	3301	54	240	1433
	2013–2014	1036	5225	3739	54	1535	1537
	2014–2015	946	5503	3768	53	1585	1303

(continued)

Table 3 (continued)

DMU	Year	Input variables			Output variables (millions)		
		Buses held	Staff size	Total cost (millions ₹)	Effective km	Passenger km	Total revenue (₹)
	2015–2016	993	5498	4069	58	1790	1304
BEST	2009–2010	4078	29,750	14,316	252	12,757	9267
	2010–2011	4652	30,183	14,942	262	12,307	11,128
	2011–2012	4669	36,028	17,094	256	12,335	13,176
	2012–2013	4259	36,796	20,321	265	1410	13,986
	2013–2014	4314	36,610	22,158	255	9843	14,351
	2014–2015	4247	35,705	23,677	244	9072	15,097
	2015–2016	4094	34,174	25,157	234	7349	14,538
CHNTU	2009–2010	409	2096	1374	41	1882	983
	2010–2011	471	2136	1491	44	2022	1115
	2011–2012	493	2055	1620	43	1961	1182
	2012–2013	472	1921	1775	37	67	1143
	2013–2014	468	1826	1755	35	1542	1043
	2014–2015	432	2102	1807	34	1544	1111
	2015–2016	494	1967	2052	39	1598	1341
PMPML	2009–2010	1620	10,294	4272	108	4109	3865
	2010–2011	1562	9780	4324	103	3640	4185
	2011–2012	1634	9633	4704	100	3437	4255
	2012–2013	1832	11,385	5952	106	460	4941
	2013–2014	1841	10,466	7020	110	4477	6026
	2014–2015	2087	10,186	8751	115	4940	7074
	2015–2016	2075	9945	18,045	115	4950	7768

Table 5 presents the average reduction potential of resources to improve efficiencies of DMUs, derived from the slack analysis across seven years for both CRS and VRS methods. The potential for the reduction was lower under the VRS analysis. The input resources for consumption efficiency have the maximum potential for reduction across DMUs, which can be attributed to the lowest efficiency performance in this category. However, the reduction potential between input variables within these categories was relatively similar. Many previous studies highlighted the lack of adequate public transport in Indian cities [19–21]. Low consumption efficiencies even in cities with good supply efficiencies and high latent demand for public transport indicate that the bus systems need to identify other customer needs that affect their mode choice beyond service supply. This could include issues like demand responsiveness, safety, accessibility, affordability, comfort of travel, etc. Figure 2 presents the relationships between scale efficiencies and input variables. These charts indicate

Table 4 Efficiencies of all the DMUs based on CRS and VRS methods

DMU	Year	OTE			PTE			SE (OTE/PTE)		
		S[a]	C[a]	R[a]	S	C	R	S	C	R
BMTC	2010	1.00	0.78	1.00	1.00	0.79	1.00	1.00	0.99	1.00
	2011	1.00	0.83	1.00	1.00	0.83	1.00	1.00	1.00	1.00
	2012	1.00	0.87	1.00	1.00	0.87	1.00	1.00	1.00	1.00
	2013	1.00	0.84	0.99	1.00	1.00	1.00	1.00	0.84	0.99
	2014	0.94	0.76	1.00	1.00	1.00	1.00	0.94	0.76	1.00
	2015	0.99	0.85	1.00	1.00	1.00	1.00	0.99	0.85	1.00
	2016	0.78	0.67	1.00	0.78	0.68	1.00	1.00	0.99	1.00
	Average	*0.96*	*0.80*	*1.00*	*0.97*	*0.88*	*1.00*	*0.99*	*0.92*	*1.00*
CSTC	2010	0.46	0.40	0.86	0.61	0.62	0.97	0.75	0.64	0.89
	2011	0.42	0.25	0.27	0.59	0.59	0.59	0.71	0.43	0.45
	2012	0.42	0.31	0.28	0.69	0.69	0.69	0.61	0.45	0.40
	2013	0.46	0.40	0.35	0.81	0.81	0.81	0.57	0.49	0.43
	2014	0.40	0.35	0.29	0.76	0.76	0.76	0.52	0.46	0.38
	2015	0.39	0.20	1.00	0.53	0.53	1.00	0.74	0.38	1.00
	2016	0.29	0.17	0.99	0.63	0.63	1.00	0.46	0.27	0.99
	Average	*0.41*	*0.30*	*0.58*	*0.66*	*0.66*	*0.83*	*0.62*	*0.45*	*0.65*
MTC	2010	1.00	1.00	1.00	1.00	1.00	1.00	1.00	1.00	1.00
	2011	1.00	1.00	1.00	1.00	1.00	1.00	1.00	1.00	1.00
	2012	1.00	1.00	1.00	1.00	1.00	1.00	1.00	1.00	1.00
	2013	1.00	1.00	1.00	1.00	1.00	1.00	1.00	1.00	1.00
	2014	1.00	1.00	1.00	1.00	1.00	1.00	1.00	1.00	1.00
	2015	1.00	1.00	0.88	1.00	1.00	0.99	1.00	1.00	0.89
	2016	1.00	1.00	0.97	1.00	1.00	0.97	1.00	1.00	1.00
	Average	*1.00*	*1.00*	*0.98*	*1.00*	*1.00*	*1.00*	*1.00*	*1.00*	*0.98*
DTC	2010	0.52	0.36	0.59	0.53	0.38	0.60	1.00	0.96	1.00
	2011	0.51	0.39	0.64	0.55	0.39	0.70	0.93	1.00	0.92
	2012	0.60	0.24	0.69	0.65	0.26	0.76	0.92	0.93	0.91
	2013	0.68	0.68	0.73	0.71	0.68	0.76	0.96	0.99	0.96
	2014	0.60	0.56	0.62	0.61	0.56	0.63	1.00	0.99	0.99
	2015	0.62	0.45	0.51	0.62	0.45	0.62	1.00	0.98	0.81
	2016	0.63	0.47	0.59	0.64	0.48	0.60	0.99	0.96	0.97
	Average	*0.60*	*0.45*	*0.62*	*0.62*	*0.46*	*0.67*	*0.97*	*0.97*	*0.94*
AMTS	2010	0.59	0.44	0.55	0.68	0.62	0.65	0.87	0.71	0.84
	2011	0.68	0.48	0.49	0.75	0.67	0.65	0.91	0.73	0.75
	2012	0.64	0.50	0.51	0.70	0.66	0.63	0.92	0.75	0.81

(continued)

Table 4 (continued)

DMU	Year	OTE			PTE			SE (OTE/PTE)		
		S[a]	C[a]	R[a]	S	C	R	S	C	R
	2013	0.66	0.43	0.49	0.73	0.58	0.63	0.90	0.75	0.78
	2014	0.65	0.36	0.51	0.68	0.47	0.61	0.95	0.77	0.84
	2015	0.69	0.38	0.38	0.73	0.49	0.53	0.95	0.78	0.71
	2016	0.73	0.42	0.38	0.74	0.55	0.50	0.99	0.77	0.75
	Average	*0.66*	*0.43*	*0.47*	*0.71*	*0.58*	*0.60*	*0.93*	*0.75*	*0.78*
BEST	2010	0.60	0.52	0.90	0.60	0.53	1.00	1.00	0.98	0.90
	2011	0.57	0.44	0.89	0.58	0.45	1.00	0.99	0.98	0.89
	2012	0.55	0.44	0.94	0.55	0.45	1.00	1.00	0.97	0.94
	2013	0.65	0.51	0.94	0.65	0.53	1.00	0.99	0.97	0.94
	2014	0.60	0.44	0.90	0.61	0.46	0.93	0.99	0.95	0.97
	2015	0.62	0.43	0.81	0.62	0.44	1.00	0.99	0.96	0.81
	2016	0.62	0.36	0.99	0.63	0.39	1.00	0.99	0.93	0.99
	Average	*0.60*	*0.45*	*0.91*	*0.61*	*0.46*	*0.99*	*0.99*	*0.96*	*0.92*
CHNTU	2010	1.00	1.00	1.00	1.00	1.00	1.00	1.00	1.00	1.00
	2011	1.00	1.00	1.00	1.00	1.00	1.00	1.00	1.00	1.00
	2012	1.00	0.99	1.00	1.00	1.00	1.00	1.00	0.99	1.00
	2013	1.00	1.00	1.00	1.00	1.00	1.00	1.00	1.00	1.00
	2014	1.00	1.00	0.99	1.00	1.00	1.00	1.00	1.00	0.99
	2015	1.00	0.98	0.74	1.00	1.00	1.00	1.00	0.98	0.74
	2016	1.00	1.00	1.00	1.00	1.00	1.00	1.00	1.00	1.00
	Average	*1.00*	*1.00*	*0.96*	*1.00*	*1.00*	*1.00*	*1.00*	*1.00*	*0.96*
PMPML	2010	0.70	0.47	1.00	0.72	0.55	1.00	0.98	0.85	1.00
	2011	0.74	0.44	1.00	0.75	0.53	1.00	0.98	0.83	1.00
	2012	0.72	0.44	0.98	0.74	0.53	1.00	0.97	0.84	0.98
	2013	0.69	0.50	0.89	0.74	0.58	0.96	0.94	0.87	0.94
	2014	0.68	0.53	1.00	0.69	0.57	1.00	0.98	0.93	1.00
	2015	0.71	0.64	0.97	0.79	0.65	1.00	0.90	0.99	0.97
	2016	0.66	0.63	1.00	0.75	0.64	1.00	0.88	0.98	1.00
	Average	*0.70*	*0.52*	*0.98*	*0.74*	*0.58*	*0.99*	*0.95*	*0.90*	*0.98*
Average (yearly)	*2010*	*0.73*	*0.62*	*0.86*	*0.77*	*0.69*	*0.90*	*0.95*	*0.89*	*0.95*
	2011	*0.74*	*0.61*	*0.79*	*0.78*	*0.68*	*0.87*	*0.94*	*0.87*	*0.88*
	2012	*0.74*	*0.60*	*0.80*	*0.79*	*0.68*	*0.89*	*0.93*	*0.87*	*0.88*
	2013	*0.77*	*0.67*	*0.80*	*0.83*	*0.77*	*0.89*	*0.92*	*0.86*	*0.88*
	2014	*0.73*	*0.62*	*0.79*	*0.79*	*0.73*	*0.87*	*0.92*	*0.86*	*0.90*

(continued)

Table 4 (continued)

DMU	Year	OTE			PTE			SE (OTE/PTE)		
		S[a]	C[a]	R[a]	S	C	R	S	C	R
	2015	*0.75*	*0.62*	*0.79*	*0.79*	*0.70*	*0.89*	*0.95*	*0.87*	*0.87*
	2016	*0.74*	*0.62*	*0.81*	*0.79*	*0.70*	*0.88*	*0.93*	*0.87*	*0.90*
Cumulative average		*0.74*	*0.62*	*0.80*	*0.79*	*0.71*	*0.88*	*0.93*	*0.87*	*0.89*

[a]S—supply efficiency; C—consumption efficiency; R—revenue efficiency

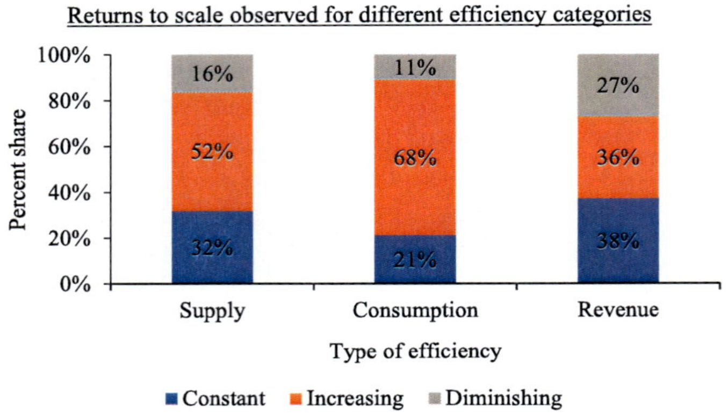

Fig. 1 Returns to scale observed for different efficiencies for all DMUs

the optimal range of input variables for maximum scale efficiency, beyond which DMUs exhibit an increasing or diminishing scale efficiency. DMUs with bus fleets between 2000 and 4000 buses, total staff between 10,000 and 30,000, and total annual cost between 10,000 and 30,000 million were observed to have the maximum scale efficiency.

5.1 Correlation Between Efficiencies and External Variables

The correlation between efficiencies and external variables representing city development and mobility characteristics was carried out for the few key data points available through [22]. Table 2 presents the variables and data collected for the year 2010–11, i.e., city population, area, urban population density, average trip length, percentage of trips by length category (0–5, 5–10, 10–30 km), mode share of buses in the city, gross domestic product (GDP) of the city and its economy index score calculated as a function of the employment levels, productivity, equity, entrepreneurship and gender equality of a city. Table 6 presents the coefficient, i.e., slope of independent

Table 5 Reduction potential of input resources of DMUs

DMU	Efficiency	Supply efficiency			Consumption efficiency			Revenue efficiency		
		Buses held (%)	Staff size (%)	Total cost (%)	Buses held (%)	Staff size (%)	Total cost (%)	Buses held (%)	Staff size (%)	Total cost (%)
BMTC	OTE	14	9	6	42	29	20	4	1	0
	PTE	6	5	3	27	20	12	0	0	0
CSTC	OTE	62	67	62	64	64	72	44	49	42
	PTE	44	63	37	44	63	37	25	41	17
MTC	OTE	0	0	0	0	0	0	2	6	2
	PTE	0	0	0	0	0	0	0	3	1
DTC	OTE	43	44	69	55	56	75	38	42	64
	PTE	40	41	68	55	56	76	33	34	64
AMTS	OTE	44	52	52	56	62	62	9	27	18
	PTE	37	43	29	51	59	42	48	56	40
BEST	OTE	0	0	0	9	0	7	10	4	8
	PTE	40	50	48	54	62	61	1	4	4
CHNTU	OTE	31	31	35	55	50	54	7	4	2
	PTE	0	0	0	0	0	0	0	0	0
PMPML	OTE	31	32	34	50	50	49	3	3	1
	PTE	31	32	34	50	50	49	3	3	1
Average	*OTE*	*28*	*29*	*32*	*41*	*39*	*42*	*14*	*17*	*17*
	PTE	*25*	*29*	*27*	*35*	*39*	*35*	*14*	*18*	*16*

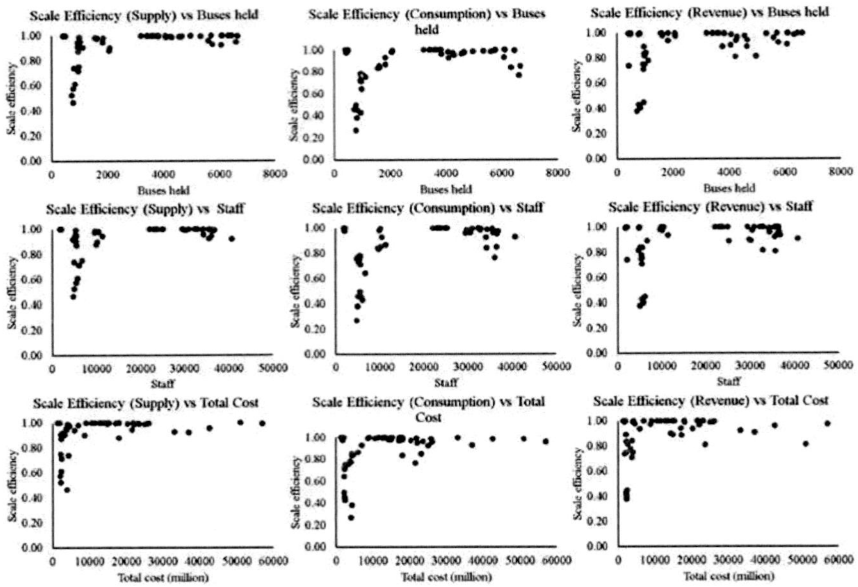

Fig. 2 Scale efficiencies of DMUs versus input resources

Table 6 Results of regression analysis between efficiencies and external variables

External variable	Parameter	Supply efficiency		Consumption efficiency		Revenue efficiency	
		CRS	VRS	CRS	VRS	CRS	VRS
Population	Coefficient	−3E−08	−3E−08	−3E−08	−3E−08	−2E−08	−1E−08
	Correlation	0.51	0.57	0.36	0.48	0.15	0.10
Coverage area of operations	Coefficient	−2E−04	−2E−04	−2E−04	−3E−04	−5E−06	−2E−05
	Correlation	0.17	0.31	0.14	0.44	0.00	0.00
Population density	Coefficient	−3E−06	−1E−06	−2E−06	1E−06	−6E−06	−3E−06
	Correlation	0.08	0.02	0.04	0.01	0.26	0.15
Average trip length	Coefficient	9E−02	7E−02	7E−02	3E−02	8E−02	1E−02
	Correlation	0.14	0.12	0.06	0.02	0.07	0.00
Trips between 5 and 10 km	Coefficient	6E−02	5E−02	8E−02	8E−02	3E−02	2E−02
	Correlation	0.50	0.62	0.60	0.84	0.10	0.13
Buses per million	Coefficient	9E−04	7E−04	1E−03	6E−04	1E−03	6E−04
	Correlation	0.59	0.49	0.50	0.24	0.56	0.44
GDP	Coefficient	7E−06	−4E−05	4E−05	−9E−05	2E−04	8E−05
	Correlation	0.00	0.03	0.01	0.08	0.16	0.10
Economic score	Coefficient	4E−02	3E−02	4E−02	3E−02	6E−02	4E−02
	Correlation	0.41	0.32	0.37	0.18	0.66	0.76

variable and correlation, i.e., R^2 value of the regression analysis for each category of efficiency and the external variables. The following are a few key observations:

Efficiency Versus City Size: Population and city area were taken as the proxies to represent city size. Population exhibited a significant correlation with supply and consumption efficiency and has a negative coefficient, indicating that the service supply efficiency reduces with increasing population. Even the area of the city exhibited a negative relationship, even though the correlation was much lower. Both the findings imply that the larger metropolitan cities have poorer supply efficiencies compared to the small and medium-sized cities.

Efficiency Versus Land-Use Development Characteristics: Population density and trip length characteristics were used as the indicators of land-use development characteristics. Both the variables were found to have a negligible correlation with the consumption efficiencies of bus systems. However, further disaggregation of trip length characteristics revealed that the variable proportion of trips of length between 5 and 10 km was observed to have a significant positive correlation with consumption efficiency. Additionally, the number of buses per million population was tested as a variable for supply density in cities. It exhibited significant positive correlation across efficiency categories. It can be concluded that cities with a combination of a high share of trips in the category of 5–10 km and more supply density of buses are likely to have a higher bus consumption efficiency.

Efficiency Versus Economic Performance of a City: The efficiencies were correlated with the economic performance of the city measured through gross domestic product (GDP) and a composite economic score derived from proxy indicators like employment levels, equity, and growth of new businesses []. The correlation for GDP was not significant while the overall economic score had a significant correlation with revenue efficiency, indicating that cities with good economic performance are also more likely to have more revenue efficient bus systems.

6 Conclusions

We present a comprehensive analysis of the performance of Indian city bus systems and the key internal and external variables that influence their performance. An input-oriented DEA was used to benchmark STUs providing bus services across eight cities as DMUs over a period of seven years between 2009–10 and 2015–16. The service supply, consumption and revenue efficiencies of DMUs under constant and variable return to scale, their scale efficiencies, and the reduction potential for each input resource were derived. Further, the relationship between various categories efficiencies and external variables representing city development and mobility characteristics was tested through correlation analysis.

Buses held, total staff, and total cost incurred on bus service provision were identified as the key input variables for all categories of efficiency measurement. Effective kilometers, passenger kilometers, and total revenue were derived to be the variables to measure service supply, service consumption, and revenue performance of the bus systems. Bangalore, Chandigarh, and Chennai were observed to perform the best across categories, while Kolkata, Ahmedabad, and Delhi had the least efficiencies. Cities performed best on revenue efficiency with a minimum average efficiency of 0.80 followed by supply efficiency with a minimum average efficiency of 0.74 while the consumption efficiency had the least value of 0.62. Low service consumption efficiency even in cities with high supply and revenue efficiency highlights the need for a more detailed analysis of user needs including demand-responsive service planning, accessibility, affordability, passenger comfort, safety, etc. Scale efficiency analysis for the DMUs identified bus fleets between 2000 and 4000 buses, total staff between 10,000 and 30,000, and total annual cost between 10,000 and 30,000 million as the optimal input resources for maximizing efficiency across all three categories.

Correlation analysis with external variables highlighted that as the cities grow in size and economy, they are not scaling up their service supply and revenue outputs adequately, resulting in reducing efficiencies. The consumption efficiency was positively correlated with the percentage of trips between 5 and 10 km in length and the number of buses in the city per million inhabitants. Therefore, STUs can maximize consumption efficiency by deploying maximum buses per capita in areas with trip lengths between 5 and 10 km. Revenue efficiency was positively correlated with the GDP per capita of the city and its economic score, implying that STU revenues perform better as the overall economic strength of the city improves, probably due to higher affordability levels of the population.

The findings from this study provide policy and planning inputs to improve the performance efficiency of bus systems in other developing countries with similar public transport systems, city development, and mobility characteristics. The analysis can be extended further by implementing the proposed methodology in various use cases at the city and regional levels, i.e., to derive a disaggregated assessment of their route and depot-level efficiency, the potential for resource optimization, and impact of external variables impacting bus efficiency. This article has aggregated bus system performance indicators at the city level, thereby comparing cities with different operating environments. A city-level efficiency analysis of disaggregated depot or route level performance is likely to compare services in a similar operating environment, thereby providing a more comprehensive understanding of the factors impacting bus system efficiency.

References

1. UITP (2017) Public transport trends. International Association of Public Transport (UITP), Belgium
2. UITP (2018) Bus benchmarking activity report: Bangalore and Chandigarh. International Association of Public Transport (UITP), India
3. MoRTH (2017) Review of the performance of state road transport undertakings for April 2015–March 2016. Transport Research Wing, Ministry of Road Transport and Highways (MoRTH), pp 65–66

4. Charnes AC, Rhodes WW (1978) Measuring the efficiency of decision-making units. Eur J Oper Res 2(6):429–444
5. Banker RD, Charnes A, Cooper WW (1984) Some models for estimating technical and scale inefficiencies in data envelopment analysis. Manag Sci 30(9):1078–1092
6. Jarboui S, Forget P, Boujelbene Y (2012) Public road transport efficiency: a literature review via the classification scheme. Public Transp 4(2):101–128
7. Emrouznejad A, Parker BR, Tavares G (2008) Evaluation of research in efficiency and productivity: a survey and analysis of the first 30 years of scholarly literature in DEA. Socio-Econ Plann Sci 42(3):151–157
8. Liu JS, Lu LY, Lu W-M, Lin BJ (2013) A survey of DEA applications. Omega 41(5):893–902
9. Agarwal S, Yadav SP, Singh S (2010) DEA based estimation of the technical efficiency of state transport undertakings in India. OPSEARCH 47(3):216–230
10. Daraio C, Diana M, Di Costa F, Leporelli C, Matteucci G, Nastasi A (2016) Efficiency and effectiveness in the urban public transport sector: a critical review with directions for future research. Eur J Oper Res 248(1):1–20
11. Chu X, Fielding GJ, Lamar BW (1992) Measuring transit performance using data envelopment analysis. Transp Res Part A Policy Pract 26(3):223–230
12. Saxena P, Saxena RR (2010) Measuring efficiencies in Indian public road transit: a data envelopment analysis approach. OPSEARCH 47(3):195–204
13. Vaidya OS (2014) Evaluating the performance of public urban transportation systems in India. J Public Transp 17(4):11
14. Cheng G (2014) Data envelopment analysis: methods and MaxDEA software. Intellectual Property Publishing House, Beijing
15. De Borger B, Kerstens K, Costa A (2002) Public transit performance: what does one learn from frontier studies? Transp Rev 22(1):1–38
16. CIRT (2011) State transport undertakings profile and performance 2009–2010. Central Institute of Road Transport (CIRT), Pune
17. CIRT (2012) State transport undertakings profile and performance 2010–11. Central Institute of Road Transport (CIRT), Pune, p 266
18. CIRT (2013) State transport undertakings profile and performance 2011–12. Central Institute of Road Transport (CIRT), Pune, p 252
19. CIRT (2014) State transport undertakings profile and performance 2014. Central Institute of Road Transport (CIRT), Pune, p 254
20. MoRTH (2015) Review of the performance of state road transport undertakings (passenger services) for April 2013–March 2014. Transport Research Wing, Ministry of Road Transport and Highways (MoRTH), p 50
21. MoRTH (2016) Review of the performance of state road transport undertakings (passenger services) for April 2014–March 2015. Transport Research Wing, Ministry of Road Transport and Highways (MoRTH), p 53
22. Pucher J, Korattyswaropam N, Mittal N, Ittyerah N (2005) Urban transport crisis in India. Transp Policy 12(3):185–198
23. Gadepalli R (2016) Role of intermediate public transport in Indian cities. Econ Polit Wkly 51(9):46–49
24. Tiwari G (2002) Urban transport priorities: meeting the challenge of socio-economic diversity in cities, a case study of Delhi, India. Cities 19(2):95–103
25. ISB (2017) Smart cities index: a tool for evaluating cities. Shakti Sustainable Energy Foundation, p 112

An Overview of Approaches and Methods for Evaluating Public Transport Performance

Vishwajeet Verma and Rajat Rastogi

1 Introduction

Public transportation largely fits under sustainable transportation option. Struggling due to its inherent characteristics and defying all odds, the transport system satisfies and balances among all sustainable criteria through economic, social, and environmental means [1, 2]. Nevertheless, in previous decades, a number of factors such as motorization, socioeconomic development, appetite for personalized mobility, increase in private vehicle ownership, urban sprawl, change in city land use, and trip activities have resulted in a switch towards private vehicles and fall in public transportation share in daily commuting [3]. Providing efficient public transport has been a matter of concern for policy-makers, transport engineers, and professionals. In recent years, transit agencies objective has widened from providing service to improvement in service quality, profit earnings, and competitiveness. In such scenario, prior to improving and managing the system service, the system needs to be measured or assessed for its performance to identify gaps and problems in its service levels. Obeng et al. [4] have reported that, "the ability to improve transit performance relies to a great extent on the ability to measure it". In general, transit performance can be assessed either by comparing with benchmark standards or measuring and comparing its relative performance with its peer units. The public transit performance assessment helps in identifying inefficient routes, factors contributing to performance, productivity management, revenue generation, and ridership enhancement while retaining captive riders and attracting choice and potential riders. Moreover, it facilitates monitoring of the service, transit economic evaluation, and development

V. Verma (✉) · R. Rastogi
Transportation Engineering Group, Department of Civil Engineering, Indian Institute of Technology Roorkee, Roorkee, India
e-mail: vverma@ce.iitr.ac.in

R. Rastogi
e-mail: rajat.rastogi@ce.iitr.ac.in

M. Parida et al. (eds.), *Proceedings of the Fifth International Conference of Transportation Research Group of India* , Lecture Notes in Civil Engineering 220,
https://doi.org/10.1007/978-981-16-9925-2_2

of service design standards. This allows addressing the level of achievements of laid goals and challenges which indirectly define community benefits [5]. Nonetheless, it also eases making financial investment decision, allocation of subsidy based on performance, and formulation of policy and public involvement [6]. To do so, the performance evaluation primarily deals with various exogenous, endogenous, managerial, or non-managerial variables which are hierarchically related to operator and user, stakeholders, system environment, and host city characteristics. The use of diverse interrelated criteria, multiple factors and their interrelations, and stakeholders' concerns make the process cumbersome. Primarily, literature suggests three approaches for evaluating public transport performance. First approach considers the variables related to transit operation. Second approach takes into account user perception in the form of importance and satisfaction levels attributable to service quality. Third approach combines the previous two approaches and transit outcome-oriented variables altogether and present them in the form of multiple criteria or perspective.

Few researchers [7–9] had presented a review of public transport performance evaluation methods, approaches, associated variables, and indicators. De Borger et al. [7] focused on parametric and nonparametric frontier methodologies for productivity growth, efficiency estimation, and subsidy allocation techniques. Lee [9] and, Eboli and Mazzulla [8] critically assessed and presented extensive performance indicators. However, review related to combining of three transit performance evaluation approaches and that of performance indicators used for evaluation have not been discussed yet. At the same time, many works in the area have emerged since then which may change the direction of research. The present review study promises to present the extensive literature reviewed since beginning of the practices to the present time. It also presents a generalized methodology of transit performance evaluation, based on the knowledge gained by the authors.

The paper is organized as follows. Section 2 presents brief overview of public transport performance evaluation studies reported over the time. The overview tries to build analogy from earliest approaches of evaluating public transport performance to the present approaches for better understanding of reader. Subsequently, all the three aforementioned approaches are discussed in Sects. 3, 4, and 5, respectively. Section 6 concisely presents performance attributes and indicators considered so far. Section 7 discusses the gaps in the literature and future direction of research based on author's understanding. Section 8 presents a generalized framework for transit performance evaluation. Lastly, in Sect. 9, transit performance evaluation approaches practised in India and future ways ahead are discussed.

2 Evolution of Performance Evaluation Methodology

Prior to 1970s in the USA, profitability was recognized as an overall measure while ridership was considered as system's operating performance measure. Later due to public ownership and government subsidies, profitability was dropped from transit performance measure [6, 10]. In 1974, National Transit Database (NTD) was established in the USA with a motive to monitor transit systems. Towards this, transit was considered as a system, and its various inputs, outputs, and impacts were proposed in a manageable, organized, and interrelated manner [11]. On the another side, many studies assumed transit routes or system as production line or firm where productivity maximization was evaluated in terms of resources such as labour, capital, and fuel utilized (being considered as system inputs), to maximization of revenue in the form of ridership (considered as output) [4, 12–16]. This escalated researchers' interest to explore conceptual framework for measuring transit systems' performance. Subsequently, transit performance evaluation got recognized as a concern by many researchers across the globe [12, 13, 17–19]. Keck et al. [20] summarized number of performance measures-related studies and classified performance measures into two categories, i.e. efficiency and effectiveness. Efficiency was defined as the output produced from the available inputs, whereas effectiveness was defined as the degree at which output is consumed or utilized. Gradually, focus was more confined from efficiency and effectiveness to a set of performance indicators representing efficiency, economy, effectiveness, and user perception to present a complete measure [20, 21]. A number of studies for selection, identification, and application of performance criteria and its indicators have been a topic of debate, and recommendations have been made by some of the researchers [7, 9, 14, 22–25]. Talley and Anderson [23] suggested that efficiency and effectiveness performance criteria should be in accordance with transit firm's objectives. Talley [24] summarized five indicator selection criteria for performance evaluation. He suggested indicators should be goal oriented and concise, economic, timely available and measurable, and controllable and robust. A number of vast and diverse indicators have been observed in those studies which were laborious and confusing. However, no standard sets of indicators were identified or recommended. To overcome this, few researchers had used factor analysis to transform multiple and diverse nature of indicators into few [10, 26, 27]. To list few of these were cost efficiency, cost effectiveness, service effectiveness, overall effectiveness, etc. Due to the use of different empirical models and analysis of diverse transit properties as well as use of data collected during different time periods, inconsistency in the results was observed. The results could not be generalized. Rather, many transit agencies required a simpler way to measure the performance. The new approach required to combine efficiency and effectiveness instead of presenting multiple performance measures. In this attempt, Chu et al. [28] and Boilé [29] used nonparametric data envelopment analysis tool to measure efficiency and effectiveness of the transit system as a single comprehensive measure.

On the other side, contrary to operator's aspects, users are the ultimate consumer of service provided, and based on their experiences, perception, expectations, and satisfaction about the system services, they decide to ride or shift to other travel alternatives. Until 1970, substantive measurement of passenger satisfaction from service quality perception was not available. Around 1970s, British rail suggested that passenger kilometres can be used as proxy indicator for measuring social benefits and accessibility provided by transit to user [30]. Later, stated preferences and univariate measurement of service quality attributes were explored for capturing user perception. Cunningham et al. [31] used satisfaction rating for individual service quality attributes to obtain user satisfaction. In 1999, the Institute of Transport Studies (ITS) explored the ways in which customer satisfaction can indicate service quality of transit industry. In this direction, Prioni and Hensher [32], Hensher and Prioni [30], and Hensher et al. [33] tried to develop an overall service quality index (SQI) as an indicator redefining service effectiveness from passenger perspective. They reported that the service output measured from supply and demand sides (vehicle kilometres and passenger kilometres, respectively) is based on aggregate indicators assuming homogeneity in respect of service quality. Hensher et al. [33] claimed that the revealed preference data reflects experience from existing trip alone. Influences of attributes need to be examined by ascertaining their applicability in the prevailing conditions. Many researchers [30, 32–36] have emphasized on the use of stated preference survey as it allows varying of service quality attributes systematically, as well as, makes it possible to present new opportunities related to the existing service level. This approach leads to the incorporation of passenger perception in the model. However, in real case, the user satisfaction may vary across user's socioeconomic profile, their experience, and expectations. Working in this direction, Eboli and Mazzulla [37] improvised the customer satisfaction index (CSI) reported by Hill et al. [38]. They incorporated user heterogeneity to measure overall service quality through development of a heterogeneous customer satisfaction index (HSCI). User satisfaction and importance rating related to the service quality attribute were used for the same. Weighted scores were obtained after considering rating variance to incorporate user heterogeneity. Dell'Olio et al. [34, 35] used ordered probit modelling approach to incorporate user perception towards service quality. User satisfaction towards service quality was modelled using multinomial logit approach. In one of the studies, the impact of available information on the possible change in the perception towards service quality was examined. Another study examined the service quality that was desired by captive and choice riders. In both of the studies, focus group survey was carried out to identify most relevant service quality variables. Satisfaction scores were taken from users using SP survey method. Eboli and Mazzulla [39] worked in the same direction and used structural equation modelling (SEM) approach to model overall passenger satisfaction related to service quality attributes for evaluating the service quality of bus system operated at the University of Calabria. Based on the discussion made, it can be seen that over the time, estimation of coefficient of factors has been attempted through different modelling approaches like ordered probit model, multinomial logit model, structural equation modelling, etc. These have emerged as analysis approaches to establish relation between overall service quality and passenger rated attributes.

So far, studies pertaining to transit operational efficiency and effectiveness indicators and user satisfaction about transit service quality attributes have been presented separately in preceding paragraphs. The operator and user side have been projected as supply and demand side of transit facility. This creates economic analogy and gives overall transit performance [6, 21, 22, 40]. Further, transit service benefits the community and supports its social and economic exchanges. However, none of the early works [6, 41] had considered ways of involvement of this perspective into transit performance assessment until late 2007. Chapman et al. [42] proposed six categories of neighbouring community criteria and associated factors as another perspective for evaluating performance of transit transfer facilities. Most of these criteria are intangible in nature. In this direction, Sheth et al. [43] and Zhao et al. [44] considered environmental and externalities as indirect measures of community perspective and reiterated that the variables in this perspectives are outcome oriented. Pullen [21] and Takyi [40] suggested involvement of multiple interest groups (operator, user, and service provider) in performance evaluation as these will provide complete and overall relationship of factors in a system. Guidebook for developing a transit performance measurement [5] reported that traditional cost efficiency and cost effectiveness indicators represent performance measures from the transit agency perspective, while these fail to link customer-oriented and community issues, which are fundamental perspectives. Towards this, Eboli and Mazzulla [45] measured performance of Italy suburban bus service both from operator and user perspectives and suggested for further improvisation. Yeh et al. [46] used fuzzy multi-criteria evaluation technique to solve multilevel hierarchies of evaluation criteria which are subjective and imprecise in nature. Sheth et al. [43] proposed a framework which considered multiple goals like maximization of user satisfaction in terms of reasonable access to jobs, societal goals in terms of minimizing noise pollution, emissions, etc., and factors related to transit network infrastructure, transit operational costs, etc. They suggested that these goals should be associated with investment alternatives which can adequately determine transportation service performance. Thus, they presented an integrated framework which considered service provider views, the consumer opinion, and the societal requirements. Few have considered the involvement of various stakeholders' (operators, service providers, community representatives, focused users, transport professionals) perspectives towards public transport. The approach broadens the transit goals and objectives by involving them through surveys and use those responses for calculating various service attributes' weights [47–49].

Based on above detailed overview and discussions, the transit evaluation process can be recognized as ongoing and complex. The complexities evolve due to involvement of multiple factors and goals considered concurrently while capturing interactions of multiple dimensions. In specific words, the evaluation methodology has matured from profitability and ridership to productivity and then to efficiency and effectiveness on one side, while on the other side user's perceptions regarding the quality of service were used as performance measures. Recently, inclusion of both efficiency and effectiveness performance measures and user's perceptions and expectations were combined together to evaluate the performance. Further many researchers have tried evaluating performance of transit as an interrelated system by incorporating multiple stakeholders and related dimensions covering operator, user perception, service infrastructure, societal benefits, and externalities prospective, altogether called as multiple perspective.

The subsequent sections will categorically present the major existing researches towards "Efficiency and Effectiveness", "User Perceptions and Expectations", and "Multiple Perspectives".

3 Efficiency and Effectiveness Measures

The inclusion of efficiency and effectiveness measures for evaluation of performance of transit system has brought in the consumer approach in terms of supply and demand or production and utilization of outputs. The approach outlines the consumption of transit services by users. The research works in this area are further being discussed in following successive paragraphs.

Chu et al. [28] reviewed transit efficiency and effectiveness through mathematical techniques. They used constant return to scale (CRS) DEA method and analysed efficiency and effectiveness as two-dimensional plot (relative effectiveness on x-axis and relative efficiency on y-axis). This helped in evaluating transit performance as a single overall measure. Boilé [29] used both CRS and variable return to scale (VRS) DEA method to calculate relative efficiency and effectiveness scores of both public and private transit firms. The obtained scores were used to calculate scale efficiency which helped in identifying efficient and inefficient systems. On financial side, economics of scale, cost efficiency, and deficit per passenger or passenger per deficit dollar have been used. Few researchers have used stochastic cost frontier model to analyse the performance of public transport [50–55]. Karlaftis [56] combined system efficiency and globally efficient production frontier through DEA while utilizing National Transit Database (NTD) for performance evaluation. Three sets of performance measures, namely efficiency, effectiveness, and combination of both were used. In all of these performance measures, similar input factors and different output factors were considered. Correlation analysis was performed to comprehend the relationship among efficiency, effectiveness, and combined performance measures. Further, to assess economies of scale, efficient production frontier was used. Results suggest that the transit performing well in one dimension generally perform well in other dimension too. Tsamboulas [57] calculated scores through DEA and based on the obtained score transit firms were ranked and grouped among peer units. Karlaftis and Tsamboulas [58] arrived at a conclusion that efficiency and effectiveness performance for a transit varied when different methodologies like neural network, DEA, and stochastic frontier were used.

Most of the researches had estimated technical, scale, and production efficiency as a performance measure considering transit firm as a unit. However, the performance assessment at transit firm level gives little idea about its internal operational behaviour. At the same time, since the transit system environment varies across subunits comparing different routes with different system environment, operational conditions and organizational level become difficult and challenging [59–63]. Some of the researchers have used tobit regression method to identify the environmental factors influencing transit performance [17, 57, 63, 64]. Applicability of DEA remained prevalent while estimating transit performance based on efficiency and effectiveness measures. However, the basic DEA model (including CRS and VRS models) does not have the desirable features of tackling issues like its inability to differentiate among efficient decision-making units (DMUs), use of aggregate data while calculating output scores, and not giving concluding output when identical efficiency scores occur [65, 66]. Recent transit performance studies have observed application of advanced DEA. Barnum et al. [60] used reverse two-stage method coupled with DEA to compare transit subunits (routes) after correcting for system environmental influences. Similarly, Karim and Fouad [63] used two-stage approach to regress the external factors after obtaining efficiency scores using DEA. Chiou et al. [61] proposed two route-based DEA (CRS and VRS) models which can measure transit firm and route-level performance. Zhang et al. [67] used super efficiency-data envelopment analysis (SE-DEA) model to distinguish among DMUs possessing identical efficiency scores.

Yu and Fan [65] applied mixed structure network data envelopment analysis to evaluate Taiwan's bus transit system. The study premises that unlike other production firm where the products can be stored for later usage, service produced by the transit is perishable and can be consumed only at the time it was produced. This perishable nature has impacted on its performance if a proportion of service is only consumed and the unused service is left unaccounted. Thus, a single indicator combining production efficiency, service, and operational effectiveness can give true measure of transit performance.

The next section now presents and discusses the works done in the area of user perceptions and expectations, which is another dimension.

4 Users' Perceptions and Expectations

Since passenger is the supreme evaluator of service quality, their discernment can be measured through satisfaction and importance scores obtained for each service quality attributes as perceived by the user and revealed during interview. Stated importance and derived importance methods are used for identifying key attributes influencing performance. Stated importance method deals with asking user to rate attributes on importance scale, whereas derived importance method deals through statistically verifying relationship of overall service quality and attributes. Due to various limitations and shortfalls associated with stated importance method, derived importance methods are preferred nowadays [68–70]. Ordered probit model [34, 71–73], multinomial logit model [35, 36], structural equation model [39, 67, 74, 75], bivariate correlation analysis [68], factor and regression analysis [68], binary logistic regression model [76, 77] etc., are been used to evaluate influence of service quality attributes on the overall service quality. These statistical methods belong to derived importance method. However, these methods have their own assumptions and relations between dependent and independent variables. If these relations and assumptions are infracted, then this could mislead the result [69, 78]. In such scenarios, data mining techniques are been used nowadays to define relation among variables [69, 79–81]. Stradling et al. [82] described a different approach for variable identification through quadrant analysis. They plotted user dissatisfaction against importance scores obtained from passenger on various performance attributes. Four quadrants were created based on the axis scale. Variables falling under critical quadrant were recommended for improvisation to enhance service quality.

Recent studies have demonstrated ability of nonparametric approaches such as classification and regression tree (CART), neural network (NN) [70], and Bayesian network (BN) [78] to analyse public transport service quality. CART is a nonparametric approach which does not hold predefined relation between the variables. The approach comes with decision rule like "if–then" which helps in categorizing the variables, interpretable through graphs and does not need to specify a functional form of variables [69]. De Oña et al. [69] compared both stated importance and deducted importance method analysis through CART for customer satisfaction survey data collected from metropolitan public transport service of Granada (Spain). They reported that deducted survey method yielded more precise results. Garrido et al. [70] applied NN to determine the influence of attributes on overall service quality of a bus system operating in Granada (Spain). The study revealed that similar results were obtained when compared with regression techniques, decision tree, and relative importance methods. Another researcher has applied BN to assess passenger satisfaction about bus system of Nanjing, China [78].

Zhang et al. [77] developed a combined comfort model to evaluate overall bus comfort of existing bus system of Nanjing, China. Passenger's experience about noise, vibration, thermal comfort, and acceleration attributes was modelled with overall bus comfort through regression modelling.

Next, the works which combine perspectives of all stakeholders are discussed.

5 Multiple Perspectives

Researchers have also considered other attributes related to transit system infrastructure, system operation, system usage, etc., for evaluating the performance of transit systems. The works related to these aspects are discussed in the successive following paragraphs.

Yeh et al. [46] considered five different criteria, namely safety, comfort, convenience, operation, and social duty for performance assessment of ten bus companies in Taipei through fuzzy multi-criteria techniques. Sheth et al. [43] referred literature for various measures and dimensions related to evaluating public transport performance and concluded that these measures and dimensions reflect multiple perspectives such as that of passengers', the service provider's, and the society. According to the service provider's viewpoint, adequate service along a route at least cost is efficient for transit agency, whereas, for the customer, service along a route which consists of the most quality attributes like shortest travel time or highest seating comfort is the efficient one. Finally, the system environment conditions within which the bus route provides service vary significantly. Based on these, a framework was prepared which consisted of multiple goals like maximization of user satisfaction in terms of reasonable access to jobs, societal goals in terms of minimizing noise pollution, emissions, etc., and factors related to transit network infrastructure, transit operational costs, etc. The approach proposed an integrated framework considering views of the service provider, the consumer, and the society. To achieve this, network and goal programming DEA formulations were combined with due consideration given to environmental effects.

Lao and Liu [83] reported that the performance evaluation of transit agencies has majorly been done from management perspective, and little consideration is given to spatial aspects. Productivity is measured by transport system's operating costs and revenue and then compared across agencies so as to determine the relative efficiency of the agency. It was highlighted that spatial aspects, such as local population, transportation network, and commuting patterns within which a transit system operates, influence the passenger demand and extent of transit operation [84]. In this work, the performance of individual bus lines was measured and compared relatively with highest performing bus line. Both operational efficiency and spatial effectiveness were measured in terms of productivity and number of riders, respectively. These were calculated through VRS-DEA model. Lastly, due to no clear difference in relative efficiency, K-means clustering is used to understand the categorized performance among routes.

Eboli and Mazzulla [45] proposed methodology based on derivation of an indicator for each service quality aspect which combines both user perception scores as subjective and transit agency performance measures as objective indicators. A total of 26 service attributes were considered to evaluate the suburban bus service in towns of Cosenda and Rende, Italy. Users expressed level of satisfaction on a rating scale of 0–10 while objective indicators were calculated from data provided by transit agency. The methodology assumed an intermediate value between the subjective and objective measures of service quality at first. Then the indicator was obtained through optimization, minimizing the distance from assumed intermediate value to the subjective and objective attributes.

Li et al. [85] recommended six aspects, namely safety, speediness, punctuality, comfort, economy, and convenience, from the operators' and passengers' perspectives. After a comprehensive consideration of data acquisition and index calculation, four indexes were selected to evaluate the operation of Beijing bus routes. Few of the above indexes can be directly quantified, while others cannot. For bus operational evaluation, DEA model was used to calculate the efficiency score for each DMU route. Further, to understand the influence of each index, sensitivity analysis of indexes was conducted. Results showed that the operational performance for all routes in both upward and downward direction was better during off-peak hours than during peak hour.

Hassan et al. [86] developed public transit system evaluation framework, wherein passengers' perceptions and operator-related variables were incorporated in the form of evaluation criteria. Further, performance indicators for each criteria were identified. The analyses were performed at two levels such as route level and system level (which consists of various routes). Data were collected for each level and were normalized to bring these to a similar scale. After normalizing these variables, weightage was assigned, and further the values were obtained for each individual route. These values were used through TOPSIS evaluation techniques, and scores were created. Based on these scores, K-means clustering was performed to group the routes based on their performance ranges. Routes performing low on the clustering range need more measures to perform better, whereas routes performing well were observed as best routes among all. Further, user opinion in terms of suggestions and feedbacks was also used as a subjective measure of improving performances of routes.

Boujelbene and Derbel [87] used analytical hierarchical process (AHP) to calculate scores based on weighted criteria to compare four transit operators in Tunisia. Economic, efficiency, effectiveness, and quality of service were four criteria considered for performance evaluation. The relative priorities for an alternative are multiplied by the importance of the corresponding criteria and summed over all criteria.

Zhang et al. [67] integrated the information entropy theory and super efficiency-data envelopment analysis (SE-DEA) model to reduce the risk of the SE-DEA model being affected by the dimensions of indicators and to improve the method's discrimination capability and obtain more objective evaluation results. SE-DEA showed improved results than basic DEA. The public transit industry regulator, transit operators, and passenger travel requirements were used to construct an evaluation indicator system based on satisfaction and efficiency to measure public transit service performance of 13 transit operators in Yangtze Delta Region (YDR) of China.

Verbich et al. [88] assessed performance of 14 transit agencies, including two from Canada and 12 from the USA. The agencies operate at least two modes, namely bus and rail, including light rail, heavy rail (metro or subway), and/or street rail (cable car, streetcar, etc.). The evaluation was based on transit users, societal prospects, and transit agencies viewpoint. By using multi-criteria evaluation techniques, composite service quality index, and rider satisfaction index from users' prospects, society index (extent by which the agencies can carry people replacing car trips) and agency index (fare box recovery ratio) were calculated. Eventually, sum of all the indices was used as an overall performance index. Trade-off among riders' satisfaction and society and agency requirements with respect to affordability were presented. It was concluded that agency performing well in one performance parameter also performs well in other parameter.

Apart from the above studies, TCRP 165 [89] grouped various transit service quality performance measures into two areas, namely availability, and comfort and convenience. Service frequency, service operation hours, and access to the transit station facilities are considered as the subperformance measures under the first group. The second group's subperformance measures are passenger load for estimation of level of crowding, service reliability, and travel time. Each subperformance measures are further described corresponding to their level of service viewed both from transit agency and passenger perspectives.

The discussion so far has focused on the use of analysis approaches, tools, and their combination to arrive at methodological frameworks for the evaluation of performance of transit systems. One point that can be noticed is that final outcome has been in terms of indicators and indices. These have been discussed specifically in the following section.

6 Performance Attributes and Indicators

The researchers over the years have considered different attributes related to transit systems. Table 1 represents those factors considered by the researchers under different category relevant to transit performance assessment. Operational parameters, service provider, quality of service, user, system infrastructure, and externalities are the transit relevant category used by the researchers.

Further Table 2 represents several indices developed by researchers for evaluating transit performance. Majority has considered user perception and satisfaction about transit quality of service.

7 Discussion

The literature indicates that user, operator, and service provider are the most referred perspectives for transit performance evaluation [45, 67, 86]. Societal viewpoints along with that of users' and operators' were incorporated by Sheth et al. [43] and Verbich et al. [88]. Lao and Liu [83] examined the performance from both the managerial and geographic perspectives and stated that the local population characteristics, transportation network, and commuting patterns largely determine passenger demand as well as transit operational scale. Hassan et al. [86] based their premise for performance evaluation on performance-related criteria and factors recommended by service providers, transit user representatives, and operator and transit specialists. Zhang et al. [67] paid attention to government, transit operator, and passenger perspectives as a part of multiple perspectives for performance evaluation. These multiple perspectives possessing different goals were essence of performance evaluation of a transit system. Measurement of all these perspectives altogether constitutes overall transit performance. However, considering number of perspectives and their associated criteria and factors is challenging and difficult. This is due to the diverse, multiple, and varying nature of goals and factors which define different perspectives. The complexities arise due to data unavailability, limitations of data collection approaches, and appropriateness of methodological tools to analyse them. Thus, a transit model which can correlate diverse goals of multiple perspectives and transform those into a single component can be explored. In that aspect, hierarchical indices for each perspectives aiming to create performance index on top level can be proposed. At second level, various quality attributes-related indices similar to their characteristics such as societal benefits and externalities index, operational and service quality index can be created. Affordability, reliability, access environment,

Table 1 Factors studied to evaluate public transport performance

Category	Factor	Description	References
Operational parameter	Operation time	Total hours of service being provided out of 24 h a day	[43, 45, 46, 83, 86]
	Service frequency	No. of vehicles run in an hour	[43, 45]
Societal benefits	Transit ridership per capita	Extent to which transit agencies are able to carry people as a proportion of the total population	[88]
	Service usage	No. of passengers/population under coverage area	[6]
Service provider	Passenger travelled	Total kilometre of passenger travelled	[43, 86]
	Revenue	Revenue generated from transit service	[67, 86]
	Fleet size	No. of standard vehicles used	[67]
	Employees	No. of employee involved full time	[67]
	Financial viability	Farebox recovery ratio to assess financial viability	[88]
	System cost	Comprises operation, maintenance, and fuel and investment cost	[49]
Quality of service	Accessibility	Potential ability to reach desired destinations	[43, 49, 88, 89]
	Travel time	To measure speediness	[43, 85, 49, 86, 90]
	Scheduled reliability	Measures the reliability of the transit schedule	[43, 45, 85, 49, 90]
User perception	Affordability	Transit fare	[45, 85, 83, 88]
	Comfort	In terms of load factor which is the ratio of passenger kilometres to carrying capacity kilometres	[45, 85, 46, 88, 89]
	Cleanliness	Transit interior and exterior cleanliness	[45]
	Information	Availability of transit-related information to rider	[45, 49]

(continued)

Table 1 (continued)

Category	Factor	Description	References
	Safety and security	Safety and security against crimes at stops and inside vehicle	[45, 46]
	Personnel	Personal appearance and helpfulness of driver	[45]
	Customer services	Helps in tackling complaints and ease in using service	[45]
	Convenience	Hour of service and transfer time	[85, 46, 89]
System infrastructures	Priority lanes	No. of lanes having priority facility	[43]
	Connectivity	No. of stops, its location, and station spacing defining coverage area	[43, 45, 83]
	Parking space availability	Facility for providing parking facility to the user	[43]
Externalities	No. of crashes	Vehicle accident rate; no. of passenger accident rate	[43, 85, 49]
	Noise pollution	Noise produced due to vehicle operation	[43, 49]
	Air pollution	GHGs emitted in environment due to operation of vehicle	[43, 49]
	Resources degraded	Monetary loss to the facility provided, i.e. forest, land area consumed, etc.	[43]

and passenger satisfaction from quality are the few indices which can be develop at the third level of evaluation hierarchy.

Similarly, estimation of public transport sustainability is demand of present time. Transit sustainability can be measured through its contribution and impact towards social, economic, environmental, and system effectiveness domains [1, 2, 94–96]. Many researchers have recommended different indicators in these domains [97–103], and many researchers have developed frameworks [104, 105] for calculating transit sustainability. Noticeably, transit performance criteria and its constituent factors related to multiple perspectives such as societal, externalities, and operational are much similar to sustainable performance criteria such as social, environmental, and system effectiveness domain [95]. Thus, estimation of transit performance incorporating sustainability can be a matter of interest.

Table 2 Performance index used for evaluation of public transport

Performance index	Description	Factors considered	References
Service quality index	Empirical approach used to combine revealed preferences and stated preferences, i.e. selection of stated alternative preferences based on user perception and experience	Reliability, travel fare, access distance, safety at stops, travel time, facilities at stops, air conditioning, information at bus stop, frequency, safety inside bus, cleanliness, driver attitude, and bus entry	[30, 32, 33]
Transit service indicator	Ratio of total travel time perceived by user through transit as compared to auto travel time	Auto travel time Transit travel time	[91]
Heterogeneous customer satisfaction index	Extension of customer satisfaction index (Hill et al. 2003), to consider user heterogeneity (As cited in [38])	Route characteristics, service characteristics, service reliability, comfort; cleanliness, fare, information, safety and security, personal, customer services, and environment	[37, 45]
Service quality index	For monitoring transit service quality considering time trends	Frequency, punctuality, travel speed, stops accessibility, ticket fare, cleanliness, space inside vehicle, air conditioning, information, safety, personnel, ease of entering bus, timetable, and overall service	[92]
Composite service index	Combined measure of service quality from service operator and the user point of view, taken as intermediate value between those	Availability, accessibility, time, customer care, comfort and amenities, and safety and security	[93]
Overall performance index	Combination of four indexes incorporated from quality of service, user satisfaction, societal benefits, and operators revenue	Quality of service, rider satisfaction, and society	[88]

In India, Ministry of Urban Development (MoUD) had revised "Service-Level Benchmark" toolkit in 2013 [106]. This is used for benchmarking service quality of city-level transport while addressing public transport, intermediate transport, pedestrian facilities, non-motorized facilities, etc., as separate components rather than considering them all as one transit system. In contrast, aggregated level of performance estimation would not reflect the individual transit performance. Moreover, the toolkit calculates level of service by comparing with its benchmark or standard value obtained which may be difficult. Another method presented by Ministry of Road Transport and Highways, Government of India [107], states the process to calculate operational efficiency for performance assessment of state road transport undertakings. The method only considers transit operational relevance factors sidelining users' satisfaction about quality of service. Thus, standardization of factors which can be applicable in Indian context can be the future research area. However, from author's understanding gained from literature, most of the variables are applicable in Indian context. Nonetheless, societal benefits, noise pollution, number of crashes, and resources degraded due to transit facility factors are difficult to quantify in Indian context. In addition, there are issues with sharing of data or information of operators' side attributes, which makes operator's side assessment difficult. In the light of these points, the revision of the existing evaluation method and tools seems required.

Regarding tools and techniques used to study transit performance, nonparametric and econometric approaches mainly DEA and MCET have been used which involve diverse stakeholders' goals and objectives. A compromise among these diverse goals and objectives will yield appropriate solution satisfying all criteria. Talley and Anderson [23] derived performance standards to compromise among these through mathematical equations. Data envelopment analysis (DEA) is applicable for estimating efficiency of system through input–output ratio. Multi-criteria evaluation techniques are applied with weights ascertained based on priority and preferences of individuals. In these cases, the transit performance problem solution might deviate from exact one. Recently, nature-based optimization tools seem at best in dealing with multiple goals and objective-oriented optimization problem [108–112]. Application of these to solve transit performance evaluation problem can be a matter of interest.

Based on the discussion done so far, a generalized framework for evaluation of transit system performance is developed. This is discussed in the next section.

8 Generalized Framework for Performance Evaluation

Problem of public transport performance evaluation has been addressed widely and diversely in terms of factors, criteria, tools, and methodologies. However, a simpler and generalized way of evaluating transit performance has not been addressed. Figure 1 presents a flowchart of public transport performance evaluation problem in a generalized, applicable, top-down sequential approach. It majorly consists of five stages elaborated one after another subsequently.

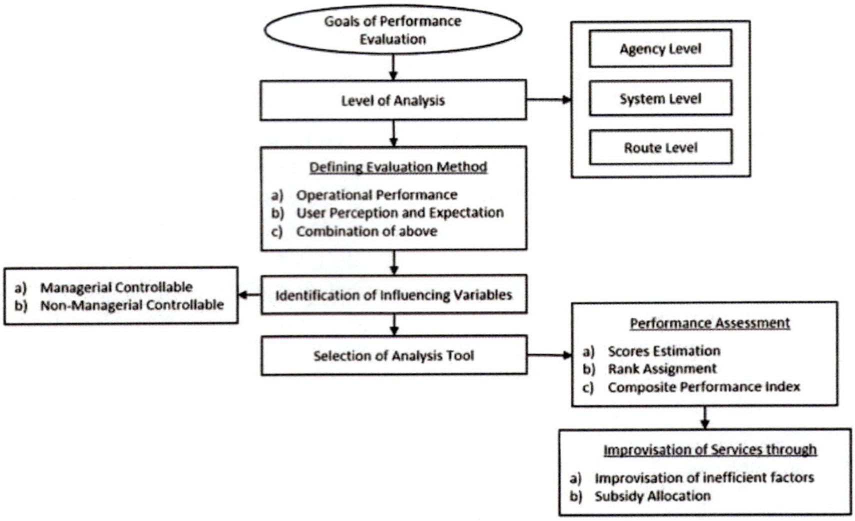

Fig. 1 Generalized approach for public transport performance evaluation

8.1 *Goals of Performance Evaluation*

It is suggested to have purpose of the performance evaluation beforehand since the purpose shapes further evaluation strategies. Dajani and Gilbert [11] have mentioned general strategies for performance evaluation which can be allocating funds to one or many transit system, assessing effectiveness of transit improvement programmes, diagnosing and alleviating causes of transit problems, or identifying the attractive transit features among passenger. The perspectives depend on the goals and objectives of an authority which undertakes the evaluation. Based on these, succeeding stages and their sub-elements are decided.

8.2 *Level of Analysis*

Earlier studies show public transport performance evaluation was conducted at national, regional, and state levels involving public and private ownership. Moreover, recent studies demonstrate that performance evaluation has been done on system, as well as, on routes within the system for its service improvement [40, 43, 60]. In particular, the level of evaluation depends on the goals and objectives for which performance evaluation is to be done.

For instance, government interest might be to compare state transit service agencies for subsidy allocation, while private agencies interest may be to improve or monitor system by better utilizing funds and taking investment-related wise decisions so that it can compete among alternatives. Thus, transit agency is considered as a unit, and performance evaluation is performed at agency level. However, considering transit agency as single unit do little about its subunit performance [60]. Evaluating performance for transit subunit is done by evaluating its routes for their internal operational characteristics. This requires performance evaluation on system as well as on routes level. On contrary, there can be instances when service provider interest is towards only internal activities of transit system. It can be to improve inefficient routes performance such as increasing frequency, providing comfort, installing passenger information system, etc. In this case, routes should be compared with peer routes based on its performance so that the inefficient one can be identified. In sum, three categories of evaluation level have been observed, first among agencies, second on system, as well as, on routes within the system, and last among routes alone.

8.3 Defining Evaluation Method

From previous studies, the public transport evaluation method has emerged as comprehensive multidimensional nature problem involving multiple stakeholders' inputs, multidimensional criteria, and factors related to efficiency and effectiveness performance. Three kinds of evaluation methods have been observed from literature. First considers users' perception towards transit quality of service. Second method performs performance estimation using transit operational parameters related to transit efficiency and effectiveness, thus presenting operators' side. Third approach is the combination of both users' perspectives and operational performance.

8.4 Identification of Performance Measures and Indicators

A number of performance measures and indicators have been used in various studies. These can be broadly classified as controllable and manageable by service provider on one hand and non-manageable and non-controllable on the other. For instance, frequency, operation hour, etc., can be controlled and managed by service provider. Talley [24] presented an improvised selection criteria for transit performance measures and indicators. According to him, the selected performance indicators should reflect transit operator's objectives with respect to efficiency and effectiveness measure. As well as, the chosen indicators should be concise, available, economic, timely available, measurable, controllable, and robust. Concise in sense that it should be small and non-overlapping and reflect different aspects. Data collection should be easy and possible with little investment of resources. The impact should be measurable and should minimize the influence of factors which are not in control of transit management. Robust means that the developed approach can be used for evaluating various transit service scenarios.

8.5 Choice of Data Analytic Technique

Usually three kinds of evaluation methods have been found in literature: analytical, statistical, or estimation of coefficients by modelling and multi-criteria based. Analytical methods utilize mathematical optimization techniques to estimate performance level. These are based on minimizing inputs with constant output or maximizing output with constant input. Data envelopment analysis (DEA) method is used where efficiency frontier is created based on best performing decision-making unit among all. Statistical analysis includes quadrant and gap analysis, factor analysis, scatter graphs, bivariate correlation, cluster analysis, conjoint analysis, the estimation of coefficients by modelling say regression model, structural equation model, and logit models. The third and the simplest way of estimating performance when the number of criteria and factors involved is high is through multi-criteria assessment techniques.

9 Approaches in India and Way Forward

In India, efforts to improve public transport services have been emphasized in recent years [113, 114]. There are altogether 54 State Road Transport Undertakings (SRTUs) comprising 24 State Road Transport Corporations (SRTCs), 12 Companies, eight Government Departmental Undertakings, and 10 Municipal Undertakings, providing passenger transport services for rural, intercity, and urban areas. The revenue, cost structures, and their net profits/losses are determined by the operational efficiency broadly covered through physical and operational parameters [107]. Fleet vintage, fleet utilization, capacity utilization profile, occupancy ratio, staff productivity, and fuel efficiency are the indicators which are used to estimate operational efficiencies of SRTCs. Besides, Ministry of Urban Development, Government of India, in 2013, published guidebook for benchmarking urban transport level of service as performance measure to measure quality of service. Benchmarking of urban transport service levels is done in terms of LOS 1 to LOS 4 [106]. The guidebook benchmarks service coverage, efficiency, accessibility, affordability, financial sustainability, safety, and environment as individual LOS. These individual LOS are summed together to derive urban transport service level. In short, the SRTC's operational efficiency is estimated through physical and operational parameters, whereas urban transport's service quality as a whole (including pedestrian, paratransit, public transports) is estimated through service-level benchmarking. In both of the methods, the attention to evaluate public transport at system or route level is neglected. Also the considered parameters or factors were not emulated and do not provide flexibilities to incorporate variables from stakeholders' perspectives (user, service operator, service provider, etc.). Recently, IRC: 124-2017 standard for Bus Rapid Transit (BRT) Design Guidelines for Indian cities is published [115] which mainly suggests design and evaluation-related criteria for BRT system evaluation. The design criteria list design-related parameter considering overall BRT system infrastructure. The performance evaluation criteria list operational-related variables. However, applicability and inclusion of these variables are questionable since the lists are limited only to design and operational aspects. City's local hosting characteristics, user perceptions and expectations, and service providers' objectives also play a major role in assessing service performance. For example, Delhi public transport ridership can be enhanced through comfort, punctuality, and safety which are user variables [116]. Similarly, safety, comfort, and timely performance need to be improved as compared to accessibility factors for Agartala (India) city bus service [75]. Thus, with these studies [75, 116] and as discussed previously, it becomes clear that performance factors and variables considering multidimensional criteria have significant role while evaluating the public transport performance. Inclusion of these in the guidelines is one aspect that can be looked at.

The transit performance evaluation approach is generally observed as five sequential stages, i.e. identification of study goals, analysis level, method identification, factors selection, and tool selection. It involves diverse range of goals, criteria, factors, and techniques. Literature suggests three different approaches to evaluate transit service performance broadly classified through transit users' perception or satisfaction-based approach, where the different transit service aspects are rated by the users through conduct of a satisfaction survey. The most common transit service aspects are reliability, capacity, frequency, fare, cleanliness, comfort, security, staff, information, and the ticketing system. Second approach considers different variables relevant to the transit system demand and operation such as loading/ridership, travel time, travel distance, frequency, service duration, revenue, manpower, cost, accident data, fuel consumption, and emission to calculate the efficiency and effectiveness indicators. Third approach is the combination of the first two approaches. This gives both user-desired quality of service (QOS) and operation relevant variables influencing performance from operator's perspective.

The user-desired quality of service can be measured by their satisfaction about service which can be finally accepted or rejected. On the other hand, operational variables reflect the magnitude of service that is available for users. Both users' and operators' perspectives can be formulated as service perceived to be acceptable and service provided as available, respectively. Apart from these two aspects, the societal benefits, service infrastructure, and externalities are the other factors which also influence the transit performance. Societal benefits in terms of number of car trips replaced by transit, user attraction by examining station coverage area, etc., can be considered. System infrastructure includes attributes such as availability of parking facility, type of carriageway, i.e. exclusive or restricted, connectivity, i.e. number of stops, fare collection method, i.e. automatic or manual, etc. Externalities are products in the system environment like emissions. Overall, multidimensional performance evaluation approach prevails at present. Further, the identification of significant attributes influencing context-specific transit service quality and its associated factors for various stakeholders' perspectives need to be determined. This can be achieved by first identifying specific variables from literature. Sequentially, important attributes need to shortlist through focused survey group, pilot survey, importance rating survey, or a combination of all these as ad hoc method. For that, standardized survey form, types of questions, and attributes rating scales, keeping stakeholders and users viewpoints in mind, need to be designed. Regarding the estimation of attributes' weight, apart from data mining and statistical models, the applicability of Monte Carlo analysis [95], gravity, urgency, and trend matrix [117] can also be studied. Examination of conditions that lead to shift in travel mode choice is also required to be addressed.

Moreover, to solve such performance evaluation problem, recent development of index incorporating user and operator-related factors can pave way for further researches in developing index which can confine diverse influencing criteria and factors to a single dimension representing transit performance as a whole. The authors intend to work on developing a universal indicator that can not only provide information regarding the state of a transit system but also allows comparison across transit system.

10 Conclusion

The present study reviews literature on public transport performance evaluation. The literature on the topic is widely scattered and sometimes overlaps among level of service, service quality, efficiency and effectiveness, and performance evaluation. To simplify this, the evolution of the transit performance evaluation approaches and methods is represented broadly from the beginning of the topic to the present time. Based on the approaches and methods being reported in the literature, three approaches are observed. First, efficiency and effectiveness are based on transit operational-related variables. Second, service quality approach accounts user perception and satisfaction levels. Third, multiple perspectives approach, which prevails at present, combines the previous two along with system outcome-oriented variables such as societal and externalities altogether considering various stakeholders viewpoints. In retrospect, a number of wide, conflicting and diverse service quality criteria, and factors have been observed. In this aspect, a more general and applicable category of specific attributes and its associated factors are described. Nonetheless, data availability for societal and environmental attributes remains as challengeable as was before. Instead, indirect, proxy variables can help to overcome this problem. For user-related factors, standard survey methods according to given transit system characteristics should be devised. Thus, standardization of approaches, methods and service quality attributes, and factors are required. Future research works can be the development of multilevel hierarchical indices aiming to create a universal indicator that can provide information regarding the state of the transit systems. Standardization can be looked at under the influence of Internet and Communication Technology (ICT), both in response collection and in enhancing transit service quality standards or for the identification of conditions that lead to modal shift say change in level of attractiveness due to technological advancements.

References

1. Gudmundsson H, Höjer M (1996) Sustainable development principles and their implications for transport. Ecol Econ 19:269
2. Kennedy C, Miller E, Shalaby A, MacLean H, Coleman J (2005) The four pillars of sustainable urban transportation. Transp Rev 25:393
3. Badami MG, Haider M (2007) An analysis of public bus transit performance in Indian cities. Transp Res Part A Policy Pract 41:961
4. Obeng K, Benjamin J, Addus A (1986) Initial analysis of total factor productivity for public transit. Transp Res Rec 48
5. Transportation Research Board (2003) A guidebook for developing a transit performance-measurement system, p 368
6. Allen WG, DiCesare F (1976) Transit service evaluation: preliminary identification variables characterizing level of service. Transp Res Rec 606:47
7. De Borger B, Kerstens K, Costa Á (2002) Public transit performance: what does one learn from frontier studies? Transp Rev 22:1
8. Eboli L, Mazzulla G (2012) Performance indicators for an objective measure of public transport service quality. Eur Transp Trasp Eur 1
9. Lee D (1989) Transit cost and performance measurement. Transp Rev 9:147
10. Karlaftis MG, Mccarthy PS (1997) Subsidy and public transit performance: a factor analytic approach. Transportation (Amst) 24:253
11. Dajani JS, Gilbert G (1978) Measuring the performance of transit systems. Transp Plan Technol 4:97
12. Gathon H-J (1989) Indicators of partial productivity and technical efficiency in the European urban transit sector. Ann Public Coop Econ 60:43
13. Wunsch P (1996) Cost and productivity of major urban transit systems in Europe. J Transp Econ Policy 30:171
14. Hatry HP (1980) Performance measurement principles and techniques: an overview for local government. Public Prod Rev 4:29
15. Viton PA (1998) Changes in multi-mode bus transit efficiency, 1988–1992. Transportation (Amst) 25:1
16. Farrell MJ (1957) The measurement of productive efficiency. J R Stat Soc Ser A 120:253
17. Boame AK (2004) The technical efficiency of Canadian urban transit systems. Transp Res Part E Logist Transp Rev 40:401
18. Kerstens K (1996) Technical efficiency measurement and explanation of French urban transit companies. Transp Res Part A Policy Pract 30:431
19. Fielding GJ (1992) Transit performance evaluation in the U.S.A. Transp Res Part A 26:483
20. Keck CA, Zerrillo RJ, Schneider NR (1980) Development of multimodal performance measures for transit systems in New York state. Transportation (Amst) 9:355
21. Pullen WT (1993) Definition and measurement of quality of service for local public transport management: foreign summaries. Transp Rev 13:247
22. Tomazinis AR (1977) A study of efficiency indicators of urban public transportation systems, p 318
23. Talley WK, Anderson PP (1981) Effectiveness and efficiency in transit performance: a theoretical perspective. Transp Res Part A Gen 15:431
24. Talley WK (1986) A comparison of two methodologies for selecting transit performance indicators. Transportation (Amst) 13:201
25. Fielding GJ, Glauthier RE, Lave CA (1978) Performance indicators for transit management. Transportation (Amst) 7:365
26. Fielding GJ, Brenner ME, Faust K (1985) Typology for bus transit. Transp Res Part A Gen 19:269
27. Fielding GJ, Babitsky TT, Brenner ME (1985) Performance evaluation for bus transit. Transp Res Part A Gen 19:73

28. Chu X, Fielding GJ, Lamar BW (1992) Measuring transit performance using data envelopment analysis. Transp Res Part A 26:223
29. Boilé MP (2001) Estimating technical and scale inefficiencies of public transit systems. J Transp Eng 127:187
30. Hensher DA, Prioni P (2002) A service quality index for area-wide contract performance assessment. J Transp Econ Policy 36:93
31. Cunningham L, Young C, Lee M (1997) Developing customer-based measures of overall transportation service quality in Colorado: quantitative and qualitative approaches. J Public Transp 1:1
32. Prioni P, Hensher DA (2000) Measuring service quality in scheduled bus services. J Public Transp 3:51
33. Hensher DA, Stopher P, Bullock P (2003) Service quality—developing a service quality index in the provision of commercial bus contracts. Transp Res Part A Policy Pract 37:499
34. Dell'Olio L, Ibeas A, Cecín P (2010) Modelling user perception of bus transit quality. Transp Policy 17:388
35. Dell'Olio L, Ibeas A, Cecin P (2011) The quality of service desired by public transport users. Transp Policy 18:217
36. Eboli L, Mazzulla G (2008) A stated preference experiment for measuring service quality in public transport. Transp Plan Technol 31:509
37. Eboli L, Mazzulla G (2009) A new customer satisfaction index for evaluating transit service quality. J Public Transp 12:21
38. Hill N, Brierley J, MacDougall R (2017) How to measure customer satisfaction. Routledge, p 160
39. Eboli L, Mazzulla G (2007) Service quality attributes affecting customer satisfaction for bus transit. J Public Transp 10:21
40. Takyi IK (1993) A multidimensional methodology for evaluating public transportation services. Transp Res Part A Policy Pract 27:395
41. Vuchic VR, Kikuchi S (1974) Design of outlying rapid transit station areas. Transp Res Rec 1
42. Chapman B, Iseki H, Miller M, Ringler A, Smart M, Taylor BD (2007) Evaluating connectivity performance at transit transfer facilities. Los Angeles
43. Sheth C, Triantis K, Teodorović D (2007) Performance evaluation of bus routes: a provider and passenger perspective. Transp Res Part E Logist Transp Rev 43:453
44. Zhao Y, Triantis K, Murray-Tuite P, Edara P (2011) Performance measurement of a transportation network with a downtown space reservation system: a network-DEA approach. Transp Res Part E Logist Transp Rev 47:1140
45. Eboli L, Mazzulla G (2011) A methodology for evaluating transit service quality based on subjective and objective measures from the passenger's point of view. Transp Policy 18:172
46. Yeh CH, Deng H, Chang YH (2000) Fuzzy multicriteria analysis for performance evaluation of bus companies. Eur J Oper Res 126:459
47. Li X, Liu Y, Liu D, Gao Z (2016) Multidimensional assessment of developing an urban public transit metropolis in China. J Urban Plan Dev 142:04015021
48. Falcocchio JC (2004) Performance measures for evaluating transportation systems: stakeholder perspective. Transp Res Rec 220
49. Pedroso G, Bermann C, Sanches-Pereira A (2018) Combining the functional unit concept and the analytic hierarchy process method for performance assessment of public transport options. Case Stud Transp Policy 6:722
50. Filippini M, Kuenzle M (2005) Cost efficiency in regional bus companies: an application of alternative stochastic frontier models. J Transp Econ Policy 40:95
51. Jørgensen F, Pedersen PA, Volden R (1997) Estimating the inefficiency in the Norwegian bus industry from stochastic cost frontier models. Transportation (Amst) 24:421
52. Holmgren J (2013) The efficiency of public transport operations—an evaluation using stochastic frontier analysis. Res Transp Econ 39:50

53. Karlaftis MG (2010) Ownership and competition in European transit: assessing efficiency. Transportmetrica 6:143
54. Holmgren J (2018) The effects of using different output measures in efficiency analysis of public transport operations. Res Transp Bus Manag
55. Ayadi A, Hammami S (2015) An analysis of the performance of public bus transport in Tunisian cities. Transp Res Part A Policy Pract 75:51
56. Karlaftis MG (2004) A DEA approach for evaluating the efficiency and effectiveness of urban transit systems. Eur J Oper Res 152:354
57. Tsamboulas DA (2006) Assessing performance under regulatory evolution: a European transit system perspective. J Urban Plan Dev 132:226
58. Karlaftis MG, Tsamboulas D (2012) Efficiency measurement in public transport: are findings specification sensitive? Transp Res Part A Policy Pract 46:392
59. Yu M-M, Chen L-H (2016) Assessment of transportation performance: a network structure. In: International series in operations research and management science
60. Barnum DT, Tandon S, McNeil S (2008) Comparing the performance of bus routes after adjusting for the environment using data envelopment analysis. J Transp Eng 134:77
61. Chiou YC, Lan LW, Yen BTH (2012) Route-based data envelopment analysis models. Transp Res Part E Logist Transp Rev 48:415
62. Georgiadis G, Politis I, Papaioannou P (2014) Measuring and improving the efficiency and effectiveness of bus public transport systems. Res Transp Econ 48:84
63. Karim Z, Fouad J (2018) An analysis of public bus transport performance and its determinants factors: the case of major Morocco's cities. In: 2018 international colloquium on logistics and supply chain management (LOGISTIQUA). IEEE
64. García Sánchez IM (2009) Technical and scale efficiency in Spanish urban transport: estimating with data envelopment analysis. Adv Oper Res 2009
65. Yu MM, Fan CK (2009) Measuring the performance of multimode bus transit: a mixed structure network DEA model. Transp Res Part E Logist Transp Rev 45:501
66. Seiford LM, Zhu J (1998) Sensitivity analysis of DEA models for simultaneous changes in all the data. J Oper Res Soc 49:1060
67. Zhang C, Juan Z, Luo Q, Xiao G (2016) Performance evaluation of public transit systems using a combined evaluation method. Transp Policy 45:156
68. Weinstein A (2000) Customer satisfaction among transit riders: how customers rank the relative importance of various service attributes. Transp Res Rec J Transp Res Board 1735:123
69. De Oña J, De Oña R, Calvo FJ (2012) A classification tree approach to identify key factors of transit service quality. Expert Syst Appl 39:11164
70. Garrido C, De Oña R, De Oña J (2014) Neural networks for analyzing service quality in public transportation. Expert Syst Appl 41:6830
71. Tyrinopoulos Y, Antoniou C (2008) Public transit user satisfaction: variability and policy implications. Transp Policy 15:260
72. Bordagaray M, Dell'Olio L, Ibeas A, Cecín P (2014) Modelling user perception of bus transit quality considering user and service heterogeneity. Transp A Transp Sci 10:705
73. Abenoza RF, Cats O, Susilo YO (2017) Travel satisfaction with public transport: determinants, user classes, regional disparities and their evolution. Transp Res Part A Policy Pract 95:64
74. De Oña J, De Oña R, Eboli L, Mazzulla G (2013) Perceived service quality in bus transit service: a structural equation approach. Transp Policy 29:219
75. Deb S, Ali Ahmed M (2018) Determining the service quality of the city bus service based on users' perceptions and expectations. Travel Behav Soc 12:1
76. Mahmoud M, Hine J (2016) Measuring the influence of bus service quality on the perception of users. Transp Plan Technol 39:284
77. Zhang K, Zhou K, Zhang F (2014) Evaluating bus transit performance of Chinese cities: developing an overall bus comfort model. Transp Res Part A Policy Pract 69:105
78. Xu C, Wu J, Rasouli S, Yang M (2016) Exploring passenger assessments of bus service quality using Bayesian networks. J Public Transp 19:36

79. De Oña R, Eboli L, Mazzulla G (2014) Key factors affecting rail service quality in the Northern Italy: a decision tree approach. Transport 29:75
80. d'Ovidio FD, Leogrande D, Mancarella R, Schinzano A, Viola D (2014) A multivariate analysis of the quality of public transport services. Procedia Econ Financ 17:238
81. de Oña J, de Oña R, Eboli L, Mazzulla G (2015) Heterogeneity in perceptions of service quality among groups of railway passengers. Int J Sustain Transp 9:612
82. Stradling SG, Anable J, Carreno M (2007) Performance, importance and user disgruntlement: a six-step method for measuring satisfaction with travel modes. Transp Res Part A Policy Pract 41:98
83. Lao Y, Liu L (2009) Performance evaluation of bus lines with data envelopment analysis and geographic information systems. Comput Environ Urban Syst 33:247
84. O'Neill WA, Ramsey RD, Chou J (1992) Analysis of transit service areas using geographic information systems. http://onlinepubs.trb.org/Onlinepubs/trr/1992/1364/1364-015.pdf
85. Li J, Chen X, Li X, Guo X (2013) Evaluation of public transportation operation based on data envelopment analysis. Procedia Soc Behav Sci 96:148
86. Hassan MN, Hawas YE, Ahmed K (2013) A multi-dimensional framework for evaluating the transit service performance. Transp Res Part A Policy Pract 50:47
87. Boujelbene Y, Derbel A (2015) The performance analysis of public transport operators in Tunisia using AHP method. Procedia Comput Sci 73:498
88. Verbich D, Badami MG, El-Geneidy AM (2017) Bang for the buck: toward a rapid assessment of urban public transit from multiple perspectives in North America. Transp Policy 55:51
89. Brinckerhoff P (2013) Transit capacity and quality of service manual, 3rd edn. Transportation Research Board, Washington, DC
90. Zak J (2011) The methodology of multiple criteria decision making/aiding in public transportation. J Adv Transp 45:1
91. Fu L, Xin Y (2007) A new performance index for evaluating transit quality of service. J Public Transp 10:47
92. De Oña J, De Oña R, Eboli L, Mazzulla G (2016) Index numbers for monitoring transit service quality. Transp Res Part A Policy Pract 84:18
93. De Oña J, De Oña R, Diez-Mesa F, Eboli L, Mazzulla G (2016) A composite index for evaluating transit service quality across different user profiles. J Public Transp 19:128
94. Miller P, de Barros AG, Kattan L, Wirasinghe SC (2016) Public transportation and sustainability: a review. KSCE J Civ Eng 20:1076
95. Miller P, de Barros AG, Kattan L, Wirasinghe SC (2016) Analyzing the sustainability performance of public transit. Transp Res Part D Transp Environ 44:177
96. Gudmundsson H, Hall RP, Marsden G, Zietsman J (2016) Sustainable transportation. Springer Berlin Heidelberg, Berlin, Heidelberg, p 304
97. Gilbert R, Irwin N, Hollingworth B, Blais P (2002) Sustainable transportation performance indicators (STPI), p 1
98. Gudmundsson H (2002) The role of performance indicators in sustainable transport policy. In: Traffic and transportation studies. American Society of Civil Engineers, Reston, VA
99. Jain D, Tiwari G (2017) Sustainable mobility indicators for Indian cities: selection methodology and application. Ecol Indic 79:310
100. Johnston R (2008) Indicators for sustainable transportation planning. Transp Res Rec J Transp Res Board 2067:146
101. Litman T (2018) Well measured: developing indicators for sustainable and livable transport planning, p 108
102. Buzási A, Csete M (2015) Sustainability indicators in assessing urban transport systems. Period Polytech Transp Eng 43:138
103. Castillo H, Pitfield DE (2010) ELASTIC—a methodological framework for identifying and selecting sustainable transport indicators. Transp Res Part D Transp Environ 15:179
104. Mahdinia I, Habibian M, Hatamzadeh Y, Gudmundsson H (2018) An indicator-based algorithm to measure transportation sustainability: a case study of the U.S. states. Ecol Indic 89:738

105. Ramani T, Zietsman J, Gudmundsson H, Hall R, Marsden G (2011) Framework for sustainability assessment by transportation agencies. Transp Res Rec J Transp Res Board 2242:9
106. MoUD (2013) Service level benchmark: urban transport, p 82
107. MoRTH (2017) Review of the performance of state road transport undertakings for Apr 2015–Mar 2016. New Delhi, p 65
108. Canca D, De-Los-Santos A, Laporte G, Mesa JA (2017) An adaptive neighborhood search metaheuristic for the integrated railway rapid transit network design and line planning problem. Comput Oper Res 78:1
109. Fan W, Machemehl RB (2006) Optimal transit route network design problem with variable transit demand: genetic algorithm approach. J Transp Eng 132:40
110. Nikolić M, Teodorović D (2013) Transit network design by bee colony optimization. Expert Syst Appl 40:5945
111. Pattnaik SB, Mohan S, Tom VM (1998) Urban bus transit route network design using genetic algorithm. J Transp Eng 124:368
112. Zhao H, Xu W, Jiang R (2015) The Memetic algorithm for the optimization of urban transit network. Expert Syst Appl 42:3760
113. MoUD (2014) National urban transport policy, p 34
114. MoHUA (2018) National urban policy framework, p 98
115. IRC: 124-2017 (2017) Bus rapid transit (BRT) design guidelines for Indian cities. Indian Road Congress, New Delhi
116. Suman HK, Bolia NB, Tiwari G (2017) Comparing public bus transport service attributes in Delhi and Mumbai: policy implications for improving bus services in Delhi. Transp Policy 56:63
117. Venanzi D, Gali JVA, Hasegawa HL, Faustino DL, Da Silva OR (2018) Application of quality tools in solving problems in the production process of an ABC—case study company. Prod Foco 8:520

Identification and Prioritization of Crash-Prone Locations for Selected Road Corridor in Kerala

Ebin Sam, S. Shaheem, Arun Chandran, and Deepa Radhakrishnan

1 Introduction

State Highway 1 (SH 1), popularly known as Main Central Road (MC Road), is an inter-district highway connecting Ernakulam and Thiruvananthapuram districts providing regional connectivity. The study stretch, Adoor–Kazhakkuttom passes through three districts, is in rolling terrain characterized with sharp bends/curves and inhabitation along either side. The length of the corridor is about 80 km, which has been developed under the Kerala State Transport Project. It carries a heavy volume of traffic during the day and night, increasing tendency of over speeding and high road fatalities during the last few years. The average annual crash rate of the study road of 80 km covering three districts is 6.4 crashes per km, which is nearly four times higher than the average annual crash rate for state highways in Kerala. Hence, it is imperative to identify crash-prone locations on the study corridor and to prioritize them so that scarce resources can be utilized more judiciously for implementing road safety measures.

2 Scope and Objectives

The scope and objectives of the study are limited to the identification and prioritization of crash-prone locations (CPL) on the Adoor–Kazhakkuttom study corridor of length 80 km.

E. Sam (✉) · S. Shaheem · A. Chandran · D. Radhakrishnan
KSCSTE-National Transportation Planning & Research Center (NATPAC), K. Karunakaran Transpark, Ulloor - Akkulam Road, Thuruvikkal P.O., Thiruvananthapuram, Kerala 695011, India

M. Parida et al. (eds.), *Proceedings of the Fifth International Conference of Transportation Research Group of India* , Lecture Notes in Civil Engineering 220,
https://doi.org/10.1007/978-981-16-9925-2_3

3 Literature Review

Srinivasan et al. evaluated three scientific methods for the identification of accident blackspots on National Highways in Kerala, namely quantum of accident method, accident-prone index method, and weighted severity index method [1].

Geurts et al. compared the characteristics of the crashes occurring in the "black" zones to those crashes scattered all over the road in Belgian peri-urban region. In the study, the association algorithm was used to profile "black" zones in terms of crash data and location characteristics. The frequent item sets were generated to identify crash concentrations and to find the factors that explain the occurrence of the crashes in the "black" zones [2].

Chand et al. computed accident risk index and accident severity index for different states in India, based on the accident indicators. Indices computed were compared across different states in India [3].

Bobade et al. identified crash-prone locations by ranking the parameters and computing the severity index for selected stretches of Mumbai–Pune Expressway and Pune–Solapur Highway. The parameters which have the potential to cause maximum number of crashes were assigned maximum weightage and top rank [4].

Ghadi et al. selected two types of blackspot identification methods for identifying dangerous locations: spatial autocorrelation method and sliding window screening method. It was concluded that SPA method is suited in low-speed urban roads, where crashes tend to be more aggregated around real hazardous points or conflict areas. In the case of a high-speed road types, distribution pattern of the crashes tends to be more scattered from the exact hazardous points, and SLW method identifies blackspots more efficiently [5].

4 Study Stretch

Road stretch from Adoor High School Junction to Vettu Road Signal Junction, part of State Highway 1 of length 80 km, passes through semi-urban areas and rural section, connects major centers like Pothencode, Venjaramoodu, Vamanapuram, Kilimanoor, Nilamel, Chadayamangalam, Ayoor, Kottarakkara, Enath, and Adoor. Entire study corridor has a dual-lane carriageway width with paved shoulders on both sides. The study stretch comes under the jurisdiction of nine police stations, and the sub-sections based on police station limits are given in Table 1.

Map showing the sub-sections according to police station jurisdiction is shown in Fig. 1.

Table 1 Sub-sections according to police station jurisdiction

S. No.	Name of police station	Name of the road section	Road length (km)
1	Kazhakkuttom	Vettu Road Jn–Chanthavila	1.75
2	Pothencode	Chanthavila–Koliyakode	7.91
3	Venjaramood	Koliyakode–Karette	10.6
4	Kilimanoor	Karette–Vazhodu	10.3
5	Chadayamangalam	Vazhodu–Agamon Bridge	12.8
6	Kottarakkara	Agamon Bridge–Puthoormukku	23.7
7	Puthoor	Puthoormukku–Enath Bridge	3.07
8	Enath	Enathu Bridge–Kilivayal	3.8
9	Adoor	Kilivayal–Adoor High School Jn	5.7

5 Crash Scenario and Trend

Data pertaining to crashes for the previous three calendar years (2015–2017) has been collected from nine local police stations. The crash data collected were sorted and refined for in-depth analysis.

In the study stretch (Adoor High School Junction to Vettu Road Signal Junction), 1521 crashes occurred during three calendar years (2015–2017) resulting in 235 fatal crashes, 1179 grievous injury crashes, and 84 minor injury crashes. During the last three calendar years, 252 people died; 1528 people were grievously injured, and 383 people suffered minor injuries in road crashes in the study stretch.

86% of road crashes on the study stretch occurs between 6 am and 9 pm. Around 60% of crashes is of "Vehicle to Vehicle" crashes on the study stretch, followed with vehicle-to-pedestrian crashes (28%) and single-vehicle crashes (10%). Hit pedestrians, head-on, and hit from back collisions contributed to around 71% of road crashes. 38% of road users involved was two-wheelers followed by cars (28%) and pedestrians (15%).

The number of crashes and the severity of victims on the study road are showing an increasing trend, and the trend chart is shown in Figs. 2 and 3, respectively.

6 Mapping of Crash Attributes

Crash data were compiled and geocoded for creating geographic information system-based (GIS) crash database for the study stretch and for identification of crash-prone locations. Crash-spot GPS survey was conducted to establish geographic locations of crashes with the help of compiled FIR data and police personnel. A total of 1521 crash locations were attributed with individual crash characteristics. All the crashes along with the attributes were digitized and mapped on GIS software for performing further geospatial analysis and identification of crash-prone locations.

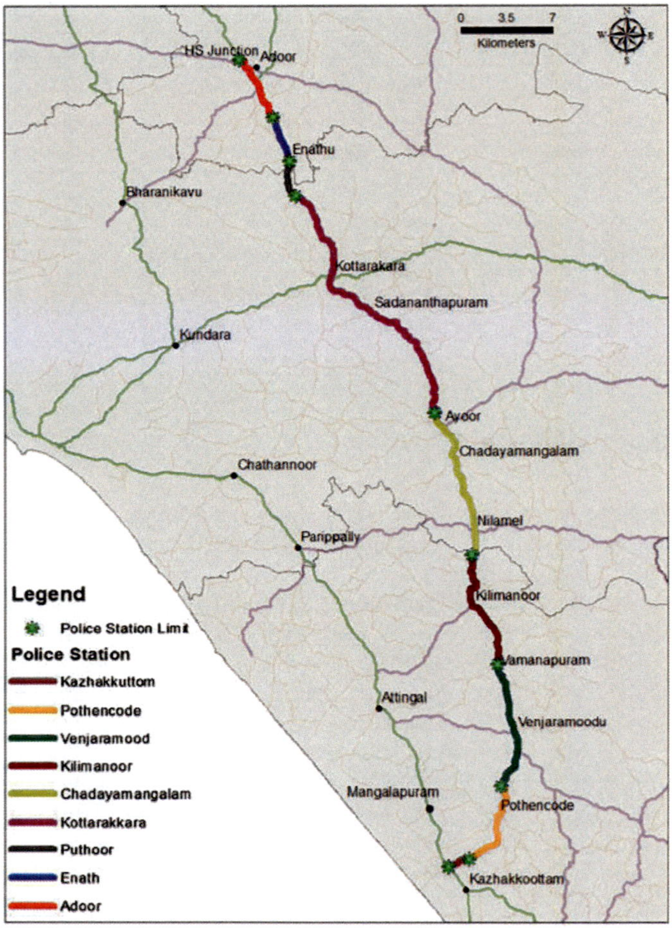

Fig. 1 Map showing sub-sections according to police station jurisdiction

Map showing the fatal crash spots and grievous injury crash spots on the study stretch is shown in Figs. 4 and 5, respectively.

7 Crash-Prone Locations

Crash-prone locations (CPLs) may be nodes (junctions/intersections) or links (mid-block between adjacent nodes) or cells (areas). Black zones are identified from the

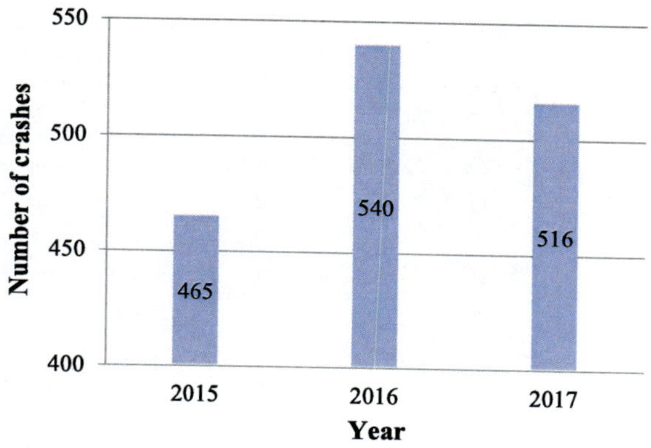

Fig. 2 Trend of road crashes on the study road (2015–2017)

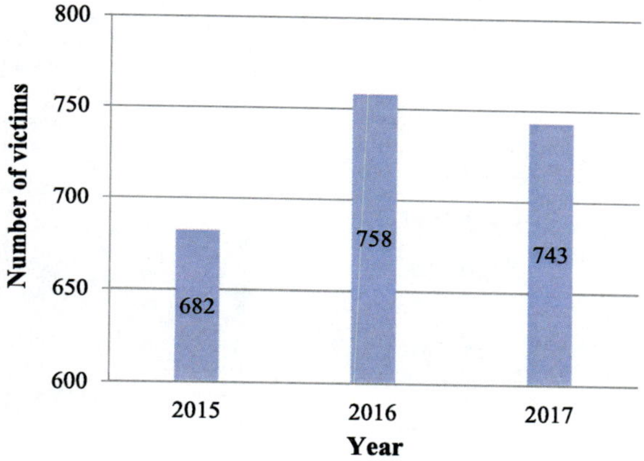

Fig. 3 Trend of the number of victims on the study road (2015–2017)

spatial interaction existing between proximate crash locations. The use of geographical information systems and point pattern techniques, which can identify such zones, is deployed in this study.

GIS-based crash information systems provide a platform to conduct spatial analysis of the crash data, which are almost impossible by using a non-spatial database [6]. GIS-aided spatial analysis provides insights about crash-prone locations, hotspots, etc. It is possible to merge crash and highway data, geocode the crash data and its attributes, calculate number, frequency, crash rates, and threshold values.

The road network and crash data were loaded in the ArcGIS software for identifying crash-prone locations. Each data points represent individual crashes of those locations, which had recorded road crashes on the study road. Mapped and attributed crash data were analyzed using ArcGIS software to identify the crash accumulation of particular type and to analyze similarities if any exists. Locations with a high number of crashes were identified as hazardous location or crash-prone location.

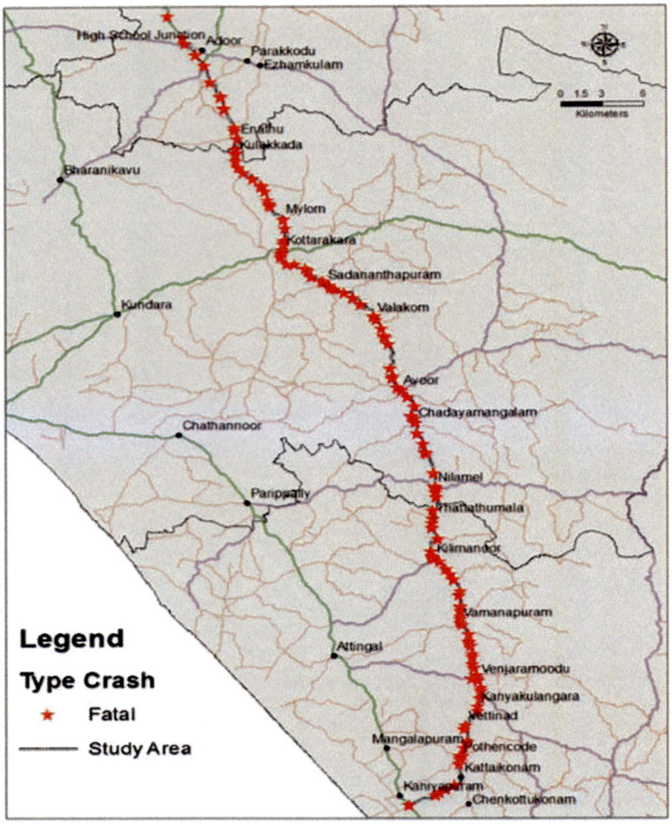

Fig. 4 Map showing fatal crash spots on the study corridor (2015–2017)

7.1 Hazardous Stretches

MoRTH (2015) definition of black spots was used to identify hazardous locations in the study corridor. A total of 85 locations were identified as hazardous locations on the study corridor based on MoRTH definition.

Definition of road crash black spot (MoRTH): Road crash black spot on National Highways is a road stretch of about 500 m in length in which either five road crashes or 10 fatalities took place during last three calendar years [7].

Map showing the hazardous stretches on the study stretch is shown in Fig. 6.

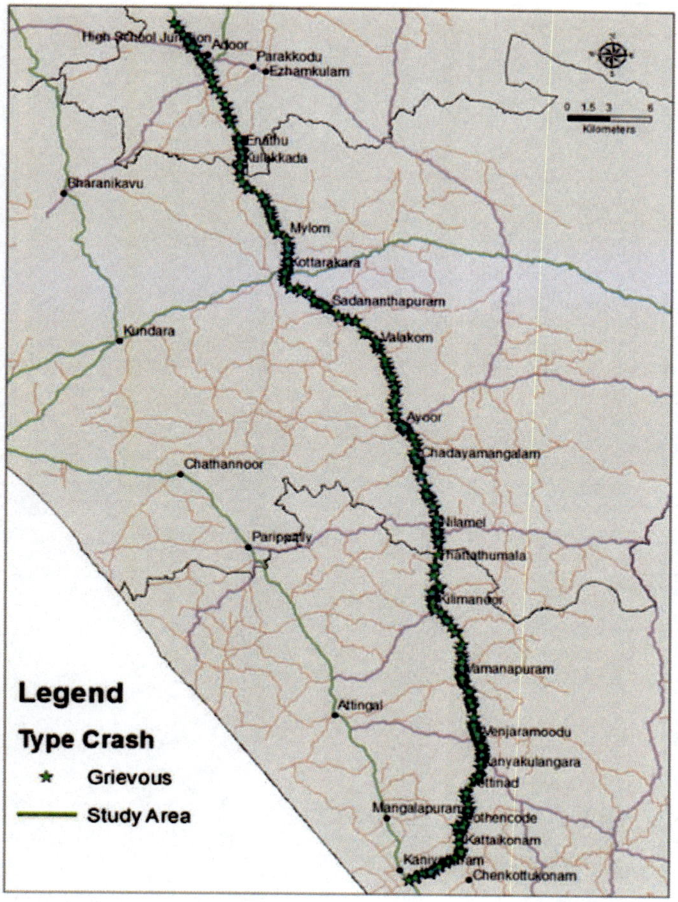

Fig. 5 Map showing the grievous crash spots on the study corridor (2015–2017)

7.2 *Priority Crash Prone Locations*

Hazardous stretches were evaluated based on crash severity index (CSI) for identification of priority crash-prone locations. For estimation of CSI, the weight–age to fatal crashes will be assigned as 7 and to grievous injury crashes as 3. Minor injury and non-injury crashes were not considered in the analysis as they are either not reported or under-reported [8].

The hazardous stretches on the road stretch were prioritized for implementing remedial measures, based on crash severity index (CSI) values. All hazardous stretches, whose CSI value above the threshold value of 98, were considered as first-order crash-prone locations. The stretches that are classified under first-order crash-prone locations are highly crash prone, which should be given highest priority

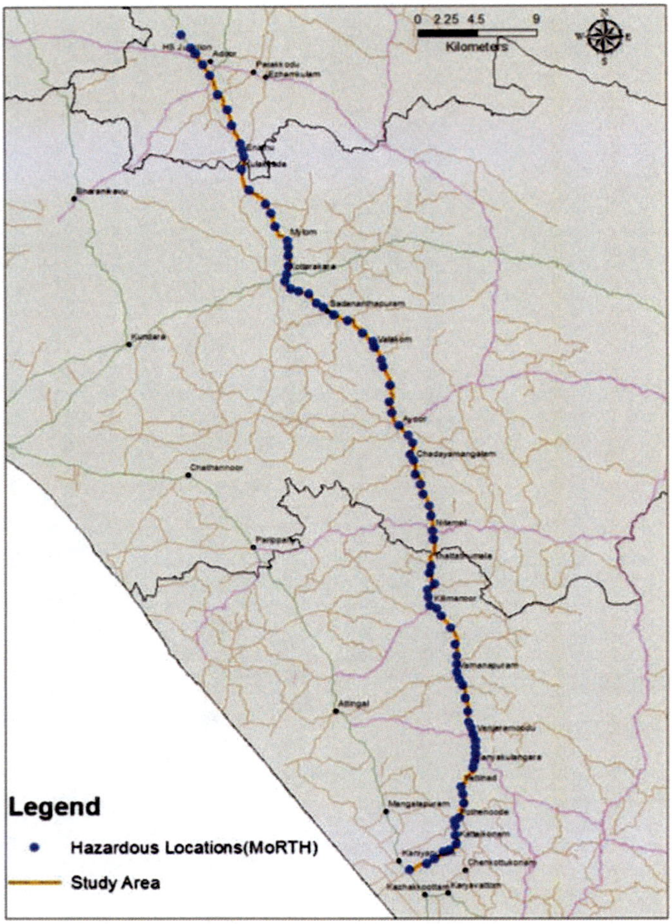

Fig. 6 Map showing the hazardous stretches on the study stretch

for implementing suitable countermeasures. CSI values of hazardous spots, which lie between 84 and 97, were considered as second-order crash-prone locations. CSI values in between 56 and 83 were considered as third-order crash-prone locations.

Table 2 shows the threshold value estimation used for prioritizing crash-prone locations.

A total of 34 priority crash-prone locations were identified and were classified into three orders based on average severity. List of priority crash-prone locations is given in Table 3. Map showing the list of priority crash-prone locations according to the order of priority on the study stretch is given in Fig. 7.

Table 2 Threshold value estimation for prioritizing crash-prone locations

S. No.	Priority order	Threshold formula	Threshold CSI value
1	I	AverageSeverity + 1.5 × StandardDeviation	98
2	II	AverageSeverity + StandardDeviation	84
3	III	AverageSeverity	56

7.3 Characteristics of Priority Crash-Prone Locations

The priority crash-prone locations have some common deficiency with respect to other location, which can be identified from the type of collision, road, and traffic characteristics. Characteristics of road crashes in crash-prone location will give an insight into the type of treatment required at the particular stretch. Characteristics of crash-prone locations were extracted from GIS software after performing geospatial analysis.

The total length of the priority crash-prone locations is 17 km, which is around one-fifth of the total length of the study corridor. It was found that 847 crashes occurred on the identified crash-prone locations, i.e., around 56% of total crashes occurred at the identified crash-prone locations. In case of the number of victims involved in road crashes, 1175 people were either killed or injured at CPLs among the total of 2183 victims on the study stretch, i.e., around 54% of total victims occurred at the identified crash-prone locations. In other words, more than half of the road crashes on the study stretch can be addressed by focusing on the priority crash-prone locations, which forms only one-fifth of the total length of study stretch.

Some of the characteristics of priority crash-prone locations are as follows:

- More than 40% of crashes at Thandrampoika, Kulakkada, Valakom, Kilimanoor Junction, and Vayakkal occur at nighttime, which necessitates suitable nighttime safety enhancements/interventions.
- More than 50% of crashes in Vamanapuram, Velavoor, Poruthamon, Kilimanoor, Ambalamukku, Kattaikonam, Kilimanoor Shilpa Jn, and Vettu Road Signal Jn occur near to the intersection.
- The major share of collisions is vehicle–vehicle collision pattern of around 59%, followed by vehicle–pedestrian collision pattern of 29%.
- 29% of crashes occurred is of pedestrian hit type, followed by head-on collisions (27%), and rear-end collisions (15%).
- 54% falls under vulnerable road user group comprising of pedestrians (16%) and motorized two-wheelers (38%). 27% of road users involved in crashes at CPLs is of car/jeep category.

Crash characteristics of individual crash-prone locations were computed from the output of GIS Software. A sample of the output is given in Table 4 and shown in Figs. 8, 9, and 10.

Table 3 List of priority crash-prone locations according to the priority order

S. No.	Crash prone locations	Priority order	Crash severity index (CSI)
1	Thandrampoika	I	159
2	Venjaramoodu	I	134
3	Karette	I	121
4	Keezhayikonam	I	119
5	Ayoor	I	115
6	Vayyette	I	111
7	Pothencode	I	98
8	Kulakkada	II	95
9	Kalayapuram	II	93
10	Vamanapuram	II	89
11	Nilamel	II	88
12	Poolanthara	II	85
13	Nellimoottilpadi	II	85
14	Velavoor	III	83
15	Inchacadu	III	83
16	Chadayamangalam	III	82
17	Valakom	III	82
18	Karikkom	III	79
19	Porunthaman Jn	III	78
20	Pulamon	III	77
21	Kuriyode	III	76
22	Kilimanoor Jn	III	75
23	Puthussery Bhagam	III	75
24	Thycadu	III	72
25	Ambalamukku	III	72
26	Policode	III	71
27	Kattaikonam	III	69
28	Shilpa Jn, Kilimanoor	III	69
29	Kannankode	III	66
30	Panaveli	III	66
31	Kilivayal	III	65
32	Vetturoad Signal Jn	III	59
33	Ayoor bridge	III	57
34	Vayakkal	III	57

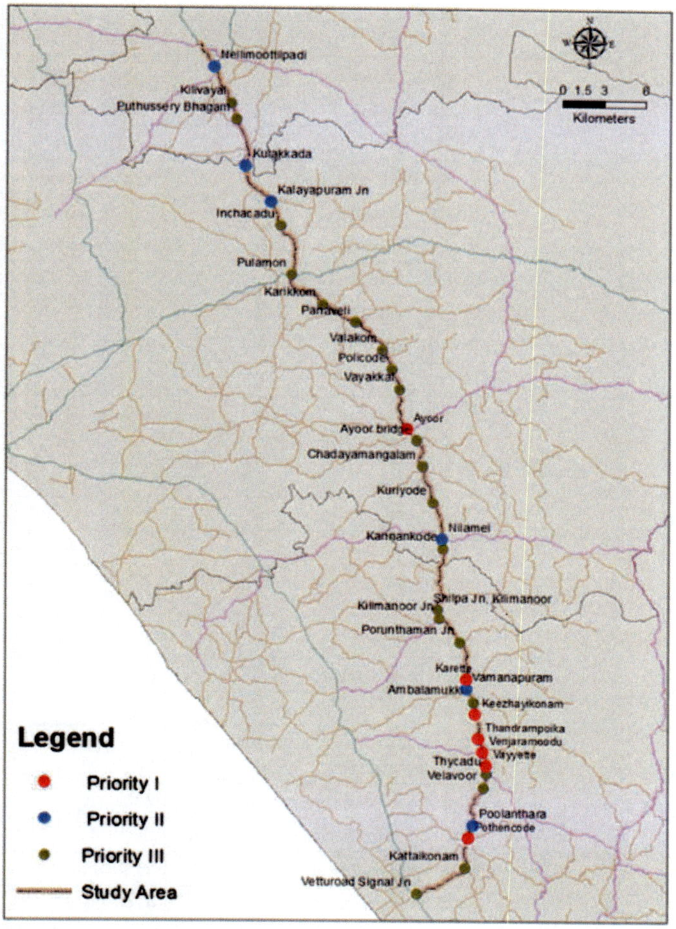

Fig. 7 Map showing crash-prone spots according to order of priority

8 Conclusion

Priority crash-prone locations and their characteristics were identified with the help of crash data and spatial analysis in GIS software. A total of 34 priority crash-prone locations were identified and were classified into three orders of priority based on average severity. It was found that more than half of the crashes (56%) occurred at the identified CPLs, which is only around one-fifth of the total length of the study corridor. Characteristics of individual CPLs have been established to identify common deficiency with respect to other locations. Suitable short-term crash reduction treatments have been recommended based on the identified potential road safety issues faced by road users.

Table 4 Crash characteristics of crash-prone location—Thandrampoika

Serial number	1
Place of occurrence	Thandrampoika
Name of road	Main Central Road (SH 1)
Type of road	SH (KSTP)
Police station limit	Venjaramoodu
Starting landmark	Indian Coffee House
Ending landmark	Alanthara Petrol Pump
Starting coordinates	8.685914°, 76.906983°
Ending coordinates	8.690412°, 76.905768°
Starting chainage	27.730
Ending chainage	28.230
Road length	500 m
Crash data–year	2015–2017
Priority order	1
Crash severity index	159
Number of fatal crashes	6
Number of grievous injury crashes	39
Total number of fatal and grievous injury crashes	45
Number of fatalities	6
Number of grievous injuries	58
Total number of fatalities and grievous injuries	64
Crashes at road intersections	24%
Crashes at road sections	76%
Crashes at daytime	53%
Crashes at nighttime	47%

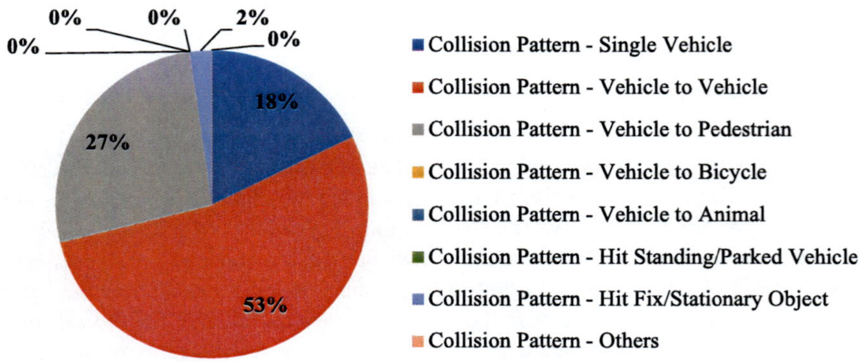

Fig. 8 Pattern of collision at Thandrampoika

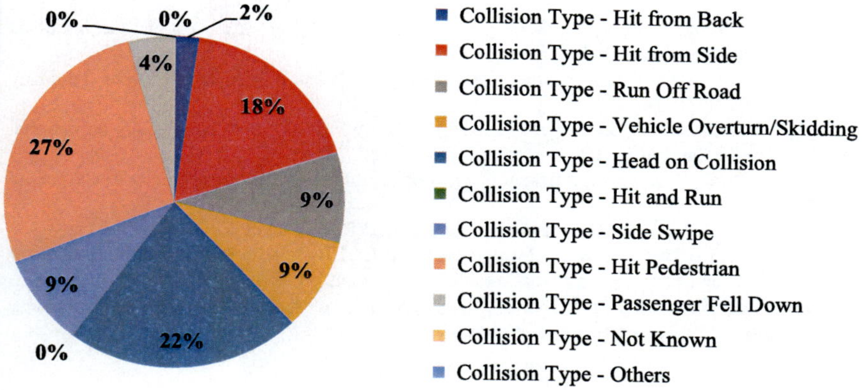

Fig. 9 Pattern of collision

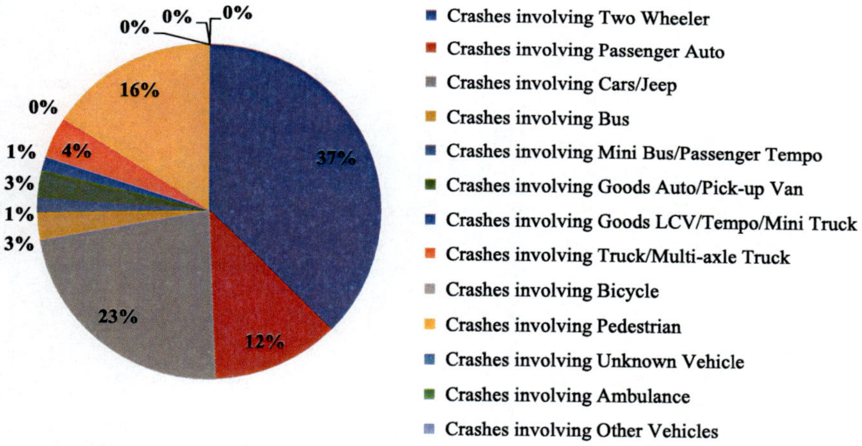

Fig. 10 Type of collision at Thandrampoika

Acknowledgments The authors acknowledge the support rendered by Kerala Police in providing the crash information. The authors also acknowledge Kerala State Transport Project for the support rendered for the smooth conduct of the study.

References

1. Srinivasan NS, Iyer VS, Chand M, Srinath K (1987) Scientific identification and improvement of accident prone locations on national highways in Kerala. J Ind Road Congr 48(3):1–10

2. Geurts K, Thomas I, Wets G (2005) Understanding spatial concentrations of road accidents using frequent item sets. Accid Anal Prev 37(4):787–799
3. Chand M, Alex AP (2007) A comparative analysis of accident risk of states in India. Highway Res Bull 105–116 (2007)
4. Snehal UB, Jalindar RP, Raviraj RS (2015) Identification of accident black spots on national highway and expressways. IOSR J Mechan Civil Eng 12(3):61–67
5. Ghadi M (2017) Comparison different black spot identification methods. Transp Res Procedia 27:1105–1112
6. Mitchell A (2015) The ESRI guide to GIS analysis. ESRI Press, vol 2
7. Ministry of Road Transport & Highways (MoRTH) (2015) Government of India: protocol for identification and rectification of road accident black spots on National Highways
8. National Highway Authority of India (NHAI): Terms of reference for "Appointment of safety consultants for PPP projects on DBFO basis"

Impact of Accessibility on Mobility and Socioeconomic Levels of Slum Dwellers of Kolkata

Suchismita Nayak and Sanjay Gupta

1 Introduction

Transport impact significantly on the well-being of individuals and communities as lack of suitable and affordable transportation can marginalize different groups within society by creating barriers to access to various urban facilities [1, 2]. While Indian cities are major contributors to country's (nearly 70% share in the Gross National Product (GNP) of the country), urban poor (nearly 50% of the city's population) can gain access to urban facilities (i.e., source of income, health, and educational institutions) by investing immense physical effort and long travel time [3, 4]. Urban poor encounters a complex trade-off among residential location, travel distance, and travel mode, intending to minimize the social exclusion resulting from low earning potential [5]. Besides, the growth of motorization impacts urban poor well-being directly (e.g., increase of travel time and travel cost to workplaces travel time) and indirectly (e.g., inadequate access to public transport, low priority to infrastructure planning for non-motorized transportation, biases in investment, and regulatory policies) [6, 7]. Thus, general urban mobility policies without inclusion of urban poor targeted mobility strategies aiming to address transport inequity would result in no benefits to the poor [8–14]. Due to lack of narratives regarding urban poor, the master plans or city development plans have very little emphasis to address the mobility issues of urban poor specifically through more rationale location of job centers in proximity to

S. Nayak (✉)
Indian Institute of Technology Kharagpur, Kharagpur 721302, India
e-mail: suchismita@iitkgp.ac.in; suchsimitanayak@gmail.com

S. Gupta
School of Planning and Architecture, Delhi 110002, India

M. Parida et al. (eds.), *Proceedings of the Fifth International Conference of Transportation Research Group of India* , Lecture Notes in Civil Engineering 220,
https://doi.org/10.1007/978-981-16-9925-2_4

their residential areas. In addition, narrow range of researches on how limited accessibility and mobility can result in decreased quality of life and well-being of urban poor especially in developing country like India fails to give insight into application of appropriate inclusive transportation planning interventions. In this context, this research would be helpful for evolving scientific slum specific planning strategies and special schemes focusing on improvement of mobility of urban poor using the concept of accessibility.

2 Methodology

A four-phase methodology has been adopted in this research to assess the impact of accessibility on the socioeconomic condition of urban poor. This paper is based on an empirical study conducted on slum dwellers Kolkata. The study based on surveys of 200 slum dwellers spread over eight slums has analyzed socioeconomic characteristics and mobility pattern of urban poor in case of slums in Kolkata. In the second phase, descriptive statistical analysis has been conducted to give an overview regarding socioeconomic condition of slums. In the third phase, accessibility of those slums have been assessed based on three types of accessibility indexes such as transport accessibility index, public transport accessibility index, and Hansen's access to employment opportunity index.

Transport accessibility index reflects the walking time from the point-of-interest to the transport access point, the reliability of the service modes available, the number of services available within the catchment, and the level of service at the public transport access points, i.e., average waiting time. It has been calculated for all mode separately which includes access time and waiting time for mode [15].

$$\text{Total Access Time} = \text{Walk Time} + \text{Average Waiting Time}. \tag{1}$$

For each selected route, the scheduled waiting time (SWT) is calculated. This is estimated as half the headway (i.e., the interval between services).

$$\text{SWT} = 0.5 * (60/\text{Frequency}). \tag{2}$$

The access time is converted to an equivalent doorstep frequency (EDF).

$$\text{EDF} = 30/\text{Total Access Time (minutes)}. \tag{3}$$

Thus, for a single transport mode, the AIs can be calculated using the following formula:

$$\text{AImode} = \text{EDFmax} + (0.5 * \text{All other EDFs}). \tag{4}$$

Calculating the overall accessibility index is a sum of the individual AIs over all modes:

$$\text{AIpoi} = \Sigma(\text{AImode } 1 + \text{AImode } 2 + \text{AImode } 3 \cdots \text{ AImode } n). \tag{5}$$

The measure does not include the speed or utility of accessible services, crowding, the ability to board services, and ease of interchange.

Public transport accessibility index X_j of zone j is defined as: $X_j = \surd(N_{ij}/A_j)$, where N_{ij} is off-peak frequency of route I passing through zone j and A_j is an area of zone j. Hansen's access to employment opportunity index provides accessibility to various employment opportunities in an urban area [16]. It is calculated on the basis of a number of formal job opportunities and distance from slum location to the job locations. It is calculated as follows:

$$\epsilon E_j / D_{ij}. \tag{6}$$

where E_j is a number of jobs and D_{ij} is the distance from zone i (slum location) to zone j (job location) and only formal job opportunities are included for index calculation.

After calculating three types of accessibility index value for eight slums, the impact of accessibility on socioeconomic condition has been assessed. This analysis has aimed to given an insight to conclude among three types of accessibility which one is the more impactful for determination of quality of life. Finally, the paper has been concluded with the highlights on the assumptions and limitations of this research.

3 Case Study Profile

The study area is the Kolkata Metropolitan Corporation Area (KMC) which has an area of 1875 km^2 with a population of 4.58 million people at a density of 7978 persons per km^2 and comprising 141 wards. One third of Kolkata's population lives in approximately 7000 notified and un-notified slums in Kolkata Metropolitan Areas and 1236 notified slums within KMC. A wide range of modal alternatives starting from old transportation systems (e.g., trams, ferries, hand-driven rickshaws, circular rail, especially found in inner city) to advanced transportation systems (e.g., e-rickshaw, public bicycle sharing, OLA/UBER, and taxi) is available in this city along with other common public transportation system and paratransits (e.g., bus, train, metro, shared auto, and rickshaw).

4 Database

For the study, a sample of 200 slum dwellers from eight slums in the city was chosen considering various factors like accessibility to transit services, geographically location, and access to work centers and local shopping areas besides other social infrastructure facilities such as health centers, schools, and community halls (Fig. 1). The city of Kolkata has an estimated 33% of its population inhabited by slum dwellers. The case slums in the city were selected considering factors like accessibility to transit services, geographically location, and access to work centers and local shopping areas besides other social infrastructure facilities such as health centers, schools, and community halls. Households were selected randomly to perform household survey in each selected slum to collect household and individual level socioeconomic characteristics and trip details for a representative day of the week. In addition, bus stop boarding alighting survey was performed at the nearest bus stop of household survey locations which helped in computing public transport accessibility index (readers are requested to refer Nayak and Gupta [17] for further insights).

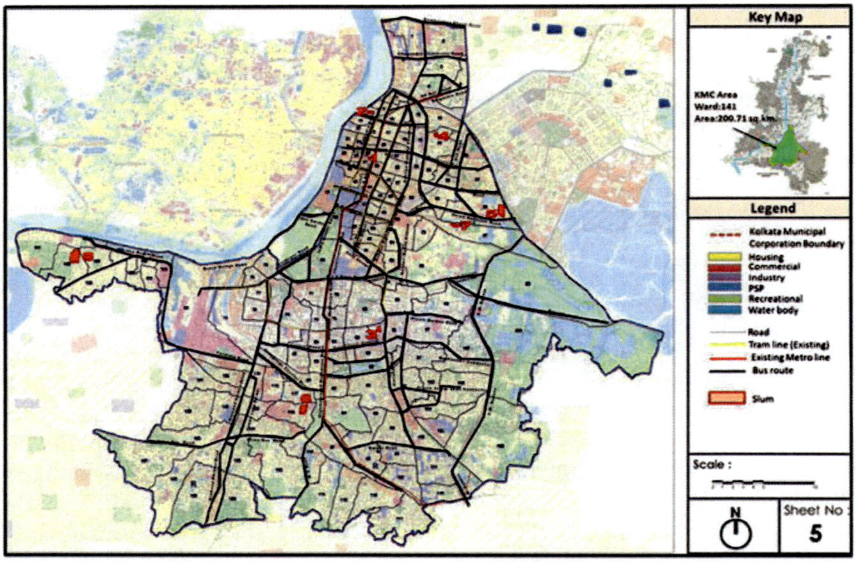

Fig. 1 Location of selected slums for survey

5 Socioeconomic Characteristics and Mobility Pattern of Case Slum Dweller

Eight case slums from KMC area have been selected for this research which are located in three different areas with the different socioeconomic pattern. While three slums are located in the inner area (Zajaria road slum, Garcha road slum, and Gossain para basti) which is old CBD area of Kolkata, two are located in the middle area (Suren Sarkar street and Jojbagan basti) and three are in outer area (Ghol para basti, Tangra, and Ayub nagar basti) which is relatively undeveloped low-density outskirts area, respectively.

From the descriptive statistical analysis (Fig. 2), it is observed that the ratio of male–female is almost same in all the three areas slums, but literacy rate decreases in slums which are located at outer areas in comparison with inner area slums. Further, it has been observed that per capita income and the number of earners in family decreased in slum dwellers residing away from the city center in comparison with inner area slum dwellers, while the expenditure on transport is higher in outer area slums compared to inner area slums owing to inaccessibility factors. In terms of

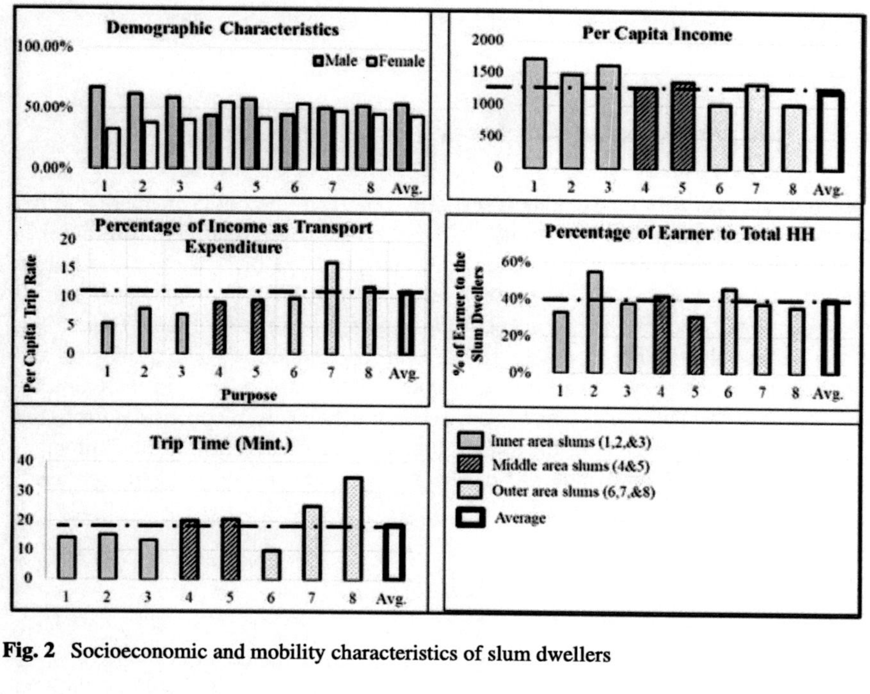

Fig. 2 Socioeconomic and mobility characteristics of slum dwellers

mobility pattern, it has been found that per capita trip rate decreases in slums located away from the city center, while the trip length and travel time increase for outer area slums. For three areas, poor people mainly commute by walk followed by bus. Among three areas in the inner area, the percentage of slum dwellers commute on foot is higher since facilities and job centers are located within walking distance, and in most of part of the inner area, cycle is banned, so walk is an only suitable option for poor. In middle and outer areas where the cycles are allowed, nearly 10 and 6% of slums dwellers use cycle, respectively.

6 Assessment of Accessibility Levels

In order to assess the accessibility levels of different slums and its impact on socioeconomic condition and empowerment of slum dwellers, three types of accessibility indices are chosen such as overall transport accessibility index, public transport accessibility index, and Hansen's access to employment opportunity index, respectively. Based on the cited accessibility indices, the impacts of accessibility indices on socioeconomic and mobility pattern of slum dwellers have been analyzed in this section.

6.1 Transport Accessibility Index

Transport accessibility index reflects walking time from the point-of-interest to the transport access point, the reliability of the service modes available, the number of services available within the catchment, and the level of service at the public transport access points—i.e., average waiting time. It has been calculated for all modes separately which include access time and waiting time for mode. Depending on this index value (Table 1), case slums are categorized into a good, moderate, and poor accessible category, and accordingly slum no 1to 3 are under good accessibility, 4 and 5 moderate, while 6 to 8 have poor accessibility levels, respectively. From the spatial distribution (Fig. 3) of the slums, it can be concluded that as the slums are located away from the core of the city transportation accessibility has been decreased.

6.2 Public Transport Accessibility

As poor people mostly use bus to commuter longer distances, the assessment of public transportation accessibility separately is important. Zajaria Slum (Slum 1) which is located in the inner zone has the highest accessibility, while Ayub Nagar Basti, located at outskirts, has the lowest accessibility (Table 1). It is observed that with the increase in transport accessibility and public transport accessibility, the

Table 1 Accessibility levels of case slum dwellers

Range of Index	Map colour	Description	Slums	Off-peak frequency of routes passing through	Area (sq.km.) (A)	PT Accessibility Index	Slum No	Employment Index
Transport accessibility index				**Public transport accessibility index**				**Hansen's access to employment opportunity index**
0.01 –2.5		Very poor	6,7 and 8	19	0.28	8.23	1	1,38,292
2.51–5.00		Very poor		3	0.5	2.45	2	1,08,979
5.01–10.00		Poor		5	0.28	4.23	3	89,228
10.01–15.00		Moderate	4 and 5	19	8.04	1.54	4	83,972
15.01–20.00		Good	1, 2, 3	10	0.5	4.47	5	56,051
20.01–25.00		Very good		1	3.14	0.56	6	52,239
25.01–40.00		Excellent		2	3.14	0.79	7	31,504
40.01+		Excellent		2	18.1	0.33	8	12,106

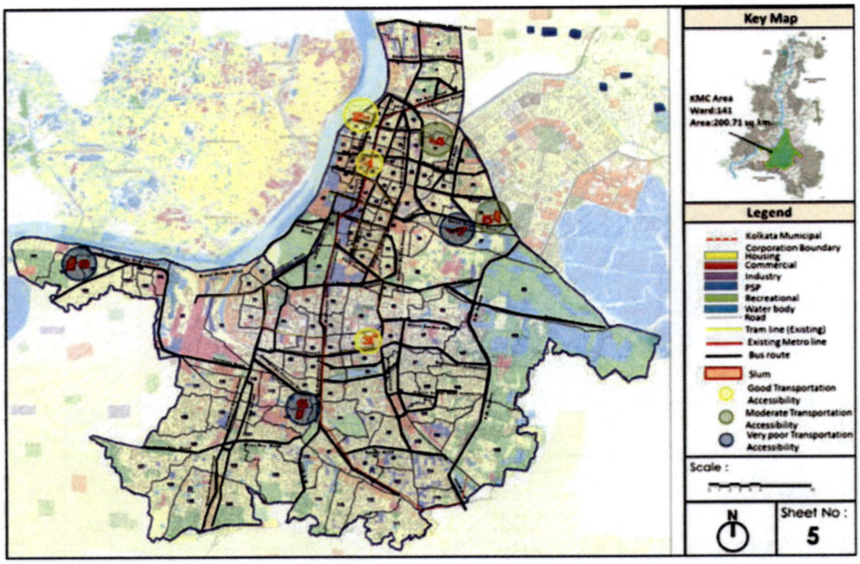

Fig. 3 Spatial distribution of transportation accessibility

average travel time and waiting time decrease, while per capita trip rate increases (Fig. 4). Also, people from good accessible areas can travel a longer distance and access to an area with more job opportunities which results in an increase of workers in the formal sector, higher per capita monthly income with reduced transportation expenditure, thus resulting in increased savings.

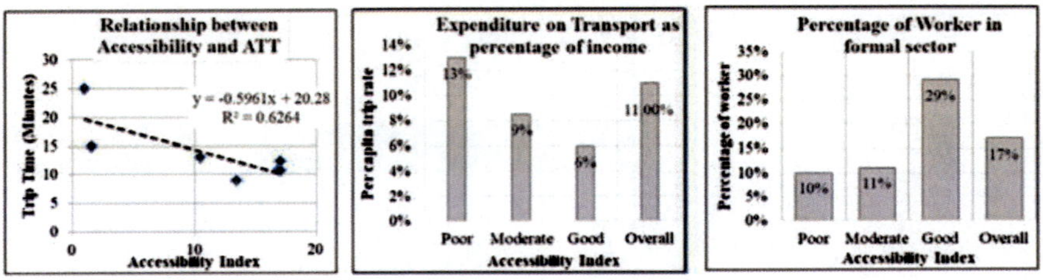

Fig. 4 Impact of transportation accessibility and public transport accessibility on socioeconomic and mobility of slum dwellers

6.3 *Hansen's Access to Employment Opportunity Index*

Hansen's access to employment opportunity index provides the accessibility to various employment opportunities. It is calculated on the basis of a number of formal job opportunities and distance from slum location to the job location. The access to employment opportunity calculation procedure for slum 1 is given in Table 2.

It is observed that the average monthly per capita income increases with the increase of access to employment opportunity using Hansen's accessibility index formula. As a result, with improved income and reduced trip expenditure, the household savings increases which help to improve the quality of life of slum dweller reflected by an increase of child literacy rates, a decrease of child labor rate, and increase of a number of people involved in formal secure jobs. In all the regression equations, the high R-square value (> 0.6) and robust sign and small constant value have established good statistical fit (Fig. 5).

It is further observed that (Table 3) when both transport accessibility index and employment opportunity indexes both are high, then the income of slum dwellers is highest (slum number 1–3). The average monthly savings of slum dwellers residing

Table 2 Hansen's accessibility index calculation for Slum 1

Job location	No. of jobs	Distance (Km)	Hansen's employment index
45 (B.B.D. Bag)	26,562	2.2	12,074.07
46 (Esplanade)	22,500	2	11,250
47 (chadni)	35,000	1	35,000
150 (Saltlake)	24,300	17.4	1396.55
Within Ward	55,000	0.7	78,571.42

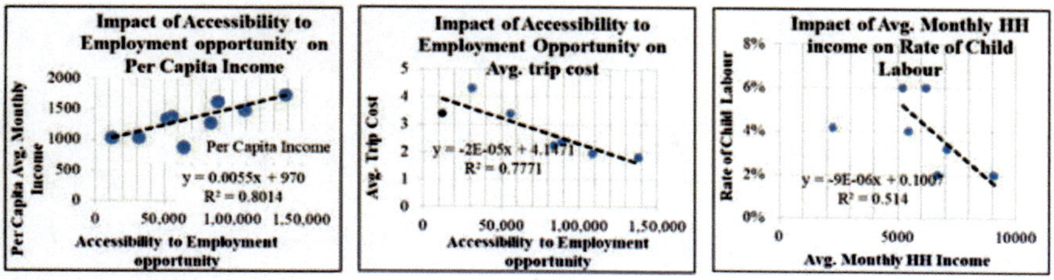

Fig. 5 Impact of Hansen's accessibility to employment's opportunity index on socioeconomic and mobility of slum dwellers

Table 3 Impact of transport accessibility and access to the job opportunity on slum dwellers

Slum no.	No. of inhabitants	Transport accessibility index	Accessibility to employment opportunity index	ATL (Km.)	ATT (Mint.)	Per capita income	Avg. monthly savings /head
1	505	17.06	138292	6.03	29.53	1731	401
2	330	17.04	108979	4.70	42.52	1479	183
3	3137	16.96	89228	2.85	25.06	1621	226
4	1264	10.2	83972	5.42	49.32	1343	82.5
5	1340	13.5	56051	7.18	42.78	1375	89
6	2100	1.6	52239	4.83	32.23	1274	66.4
7	7298	1.1	31504	8.09	48.55	1027	10.27
8	3054	1	12106	14.68	84.66	1030	12.3

in slum number 5 which has good transport accessibility but moderate accessibility to employment opportunity is higher than slum dwellers residing in slum number 4 which has moderate transport accessibility but good accessibility to employment opportunity. The per capita average monthly savings is more sensitive with transport accessibility than accessibility to employment opportunity highlighting the importance of transport accessibility.

7 Conclusion

This paper has attempted to analyze the accessibility of slum dwellers of Kolkata and its impact on their socioeconomic condition and mobility pattern. The paper using the indices of transport accessibility, public transport accessibility, and Hansen's access to employment opportunities index, respectively, brings out the impact of varying accessibility levels on the economic well-being of case slum dwellers. It is observed from the findings that an improvement of accessibility to transport, particularly especially public transport, followed by accessibility to the employment opportunity is indispensable for holistic improvement of case slum dwellers. It is concluded that transport accessibility is more sensitive with regard to economic upliftment of slum dwellers as a consequent to enhanced mobility levels in comparison with accessibility to employment opportunity. This reveals that policy planners must lay more emphasis on the enhancement of transportation accessibility as part of pro-poor mobility policies and strategies. The impact of different policies on transportation accessibility as well as quality of life of urban poor may be analyzed in future which will assist policy makers for selection of the right mix of policies. This research has been limited to the slum dwellers only. The other section of urban poor such as homeless, people lived in squatter settlements could be included. In future, this study can be extended for the other socioeconomic strata also and interface among them can be established. In addition, this study can be conducted in other cities in India for preparation of national level policy to create an inclusive travel environment.

Acknowledgements This study is part of a Master's thesis funded by the Ministry of Education (MoE), Government of India. We are thankful to the editor for providing us the opportunity to submit for further consideration and review. We are also thankful for the constructive comments of the anonymous reviewers. The authors also report no conflict of interest.

Appendix: Survey Questionnaire

SCHOOL OF PLANNING AND ARCHITECTURE, NEW DELHI
MASTER OF URBAN PLANNING
RESEARCH STUDY ON PLANNING FOR INCLUSIVE MOBILITY OF URBAN POOR

Household Survey

Name of the surveyor: .... Date/day: ....

Name of respondent: .... Address ....

A. Socio-economic profile (Household Level)

1. **No. of persons (HH Size):** a. 1......b.2......c.3..........d.4.........e. 5..........f. 6.........g. More than 6

2. **Duration of stay in this area**
   (a) Since Birth (b) If not since birth, then the reason for migration
   (b.1) Employment (b.2) Education (b.3) Search of employment (b.4) Others, specify

3. **When did you move to this area?** From where did you migrate (District name ....)
   (a) 0 - 2 Years (b) 2 - 5 Years (c) 5 -10 Years (d) More than 10 Years4

4. **Tenure of Housing** a. Owned b. Rent ....(Rent /Month)

5. **Monthly Household Income** .... (Rs./Month)

6. **Number of student in the household** a. 1 b.2 c.3 d.4 e. >4

7. **Number of earner in the household** a. 1 b.2 c.3 d.4 e. >4

8. **Vehicle Owned and No.** : a. Cycle .... b.2 Wheeler......... c. cycle rickshaw.......... d. Auto rickshaw........
   e. Others (Specify)..

9. **Monthly Household Expenditure** ....(Rs./Month)

Sl No.	Heads	Monthly Expenditure (Rs.)
1	**Food**	
2	**Cloth**	
3	**Home**	
4	**Transport**	
5	**Health**	
6	**Others**	

10. **How much do you save?** ....(Rs./Month)
    If any savings, how do you invest your savings?
    i. bank (with an interest rate) ii. buy immovable properties iii. Invest in business
    iv.Other (please specify) ....

B. Socio- Economic Data (Personal Level)

Sl. No	Relation with Head of the family	M/F	Age	Education	Occupation	Income (Rs. /Month)
1.						
2.						
3.						
4.						
5.						
6.						
7.						

C. Trip Information:

1. **O-D Survey**

Person No.	Trip No.	Origin	Destination	Purpose	Mode 1	Length (Km)	Waiting time	Time of travel	Cost (Rs.)	Interchange Point	Mode 2	Length (Km)	Waiting time	Time of Travel	Cost (Rs.)
1															
2															
3															
4															
5															
6															

2. **Opinion on Service on different mode used:**

Sl. No.	Facility	Good	Satisfactory	Poor	Remarks
1.	Public Transport				
2.	Accessibility				
3.	Availability				
4.	Affordability				
5.	Safety				
6.	Comfort				
7.	Condition of Road				Kutcha:……….. Pucca:……… Water loggiong:…………………
8.	Condition of Footpath				
9.	Connection to the city				
10.	Quality of Infrastructure				i. Shelter at bus stop a. Y…..b. N….. ii.Cycle Track a. Y……. b. N…….
11.	Availability of alternate modes				Specify:

D. Attitudinal Information:

i. **Issues: a. ………………………… b. …………………….. c.…………………… d.……………………**

ii. **Suggestions:**

………

………

E. Slum Catchment Area Study: (Community Level):

i. **Slum Size: …………………… sq. m.**

ii. **Inhabitants: ……………………………**

iii. **Year of Establishment: ………………………………**

iv. **Road Network: a. Density……………(Km/Sq.km.) b.Pucca……… . Kutcha ………………**

v. **Facilities within slum pockets:**

Sl. No.	Facility	Distance (Km.)
1.	Daily market	
2.	Shopping area	
3.	Primary School	
4.	Secondary School	
5.	Private doctor	
6.	Govt. Hospital	
7.	Private Hospital	
8.	Bank	
9.	Recreational/ Social visit	

vi. **Availability of Modes:**

Modes	Availability (Y/N)	Distance from stop (Km.)
Buses		
Cycle Rickshaw		
Auto Rickshaw		
Metro		
Rail		
Circular rail		
Tram		
Ferry		

Code:

Source of Income	Code	Mode of Travel	Code	Purpose	Code
Industrial Labour	1	Walk	W	Work	W
Contraction Labour	2	Bicycle	C	Education	E
Rickshaw puller	3	2 Wheler	2W	Shopping	S
Auto driver	4	Auto rickshaw	A	Recreational	R
Street vendors	5	Bus	B	Health	H
Shop owner	6	Train	T	Social	S
Business man with no employee	7	Metro	M	Others	specify
Business man with employee	8	Ferry	F		
Self employed professional (Specify)	9	Tram	Tr	**Origin/ Destination**	**Code**
Maid/Servant	10	Taxi	Ta	Home	H
House wife	11	Own Car/Jeep/Van	J	Workplace	W
Student	12	Others car/Jeep/Van	J1	Institution	I
		Others (Specify)	O	Market	M
				Hospital	H
				Others	specify

Bus-stop Survey at.............................. **Date:..................................**

Sl. No.	Bus No.	Origin	Destination	Time of Arrival	Time of Departure	No. of board	No. Of Alighting	Avg. Occupancy/ Bus
1								
2								
3								
4								
5								
6								
7								
8								
9								
10								
11								
12								
13								
14								
15								
16								
17								
18								
19								
20								
21								
22								
23								
24								
25								
26								
27								
28								
29								
30								
31								
32								
33								
34								
35								
36								
37								
38								
39								
40								
41								
42								
43								
44								
45								
46								
47								
48								
49								
50								
51								
52								
53								
54								
55								

References

1. Social Exclusion Unit (2003) Making the connections: final report on transport and social exclusion. Science 294(5544):1–139. https://doi.org/10.1126/science.294.5544.953d
2. Pyrialakou VD, Gkritza K, Fricker JD (2016) Accessibility, mobility, and realized travel behavior: assessing transport disadvantage from a policy perspective. J Transp Geogr. https://doi.org/10.1016/j.jtrangeo.2016.02.001
3. Pucher J, Korattyswaropam N, Mittal N, Ittyerah N (2005) Urban transport crisis in India. Transp Policy 12(3):185–198. https://doi.org/10.1016/j.tranpol.2005.02.008
4. The World Bank (2011) Urbanization in India: integral part of economic growth. World Bank, pp 5–9
5. Starkey P, Hine J (2014) Poverty and sustainable transport: how transport affects poor people with policy implications for poverty reduction. A literature review. UN-Habitat, p 72
6. Cheng L, Bi X, Chen X, Li L (2013) Travel behavior of the urban low-income in China: case study of Huzhou City. Procedia Soc Behav Sci 96:231–242. https://doi.org/10.1016/j.sbspro.2013.08.030
7. Salon D, Gulyani S (2010) Mobility, poverty, and gender: travel 'choices' of slum residents in Nairobi, Kenya. Transp Rev. https://doi.org/10.1080/01441640903298998
8. Preston J, Rajé F (2007) Accessibility, mobility and transport-related social exclusion. J Transp Geogr. https://doi.org/10.1016/j.jtrangeo.2006.05.002
9. Delbosc A, Currie G (2011) Exploring the relative influences of transport disadvantage and social exclusion on well-being. Transp Policy 18(4):555–562. https://doi.org/10.1016/j.tranpol.2011.01.011
10. Lucas K (2012) Transport and social exclusion: where are we now? Transp Policy. https://doi.org/10.1016/j.tranpol.2012.01.013
11. Palmer CJ, Astrop A, Maunder DAC (1997) Constraints, attitudes and travel behaviour of low income households in two developing cities. TRL Rep 263
12. Rahman A, Barter P (1998) UNCHS (Habitat) regional symposium on urban poverty in Asia Fukuoka. In: Transport and urban poverty in Asia: a brief introduction to the key issues, 27–29 October 1998, pp 27–29
13. Stanley JK, Hensher DA, Stanley JR, Vella-Brodrick D (2011) Mobility, social exclusion and well-being: exploring the links. Transp Res Part A Policy Pract 45(8):789–801. https://doi.org/10.1016/j.tra.2011.06.007
14. The World Bank (2002) Urban transport and poverty reduction. Urban Transp Strateg Rev 25–38
15. Wu BM, Hine JP (2003) A PTAL approach to measuring changes in bus service accessibility. Transp Policy. https://doi.org/10.1016/S0967-070X(03)00053-2
16. Dalvi MQ, Martin KM (1976) The measurement of accessibility: some preliminary results. Transportation (Amst) 5(1):17–42. https://doi.org/10.1007/BF00165245
17. Nayak S, Gupta S (2020) Travel probability fields—an approach to understand travel behavior: case study of slum dwellers in Kolkata, India. Transp Res Procedia 48:1546–1554. https://doi.org/10.1016/j.trpro.2020.08.197

Small River Bridges to Serve as Cost-Effective Water Storage Structures

P. L. Bongirwar

1 Introduction

Bridges are an essential requirement for highway as they enable us to cross the natural river course. By changing the design with marginal additional cost of 8–10%, they can also serve as water storage structure. Maharashtra state in India has a tradition of constructing *Bridge cum Bandhara* (BCB). These structures are popular as they serve dual purposes of providing a bridge for crossing the river as well as for storing water. The tradition started in 1960 in Kolhapur city where sugar cane is grown in large quantity. It was observed that the irrigated water would percolate into the riverbed and was thus not utilized. Hence, they started to recycle water, and structure to collect the water came to be known as 'Kolhapur Bandhara' which is shown in Fig. 1.

It consists of closely spaced piers having a clear distance of 2 m. Gates are fixed in slots to create the water storage. The gates are fixed at the end of the monsoon and removed before onset of the next monsoon. Such structures have been used to tap post-monsoon flow to create storage not exceeding 3.5 m in depth. They are ideal structure for situations such as:

(a) To tap post-monsoon flow to create storage of up to 3.5 m depth. The storage is created by fixing needles/gates between Bandhara piers to tap the last flow.
(b) Stored water is used for drinking and irrigation.
(c) It recharges the ground water which is then available for irrigation as well as domestic use.
(d) Surplus water that percolates into the ground on account of irrigation activities again goes into the river and is get stored and reused for irrigation.
(e) Such stored water can be used for artificially recharging the nearby bore wells as well as open wells to augment the ground water.

P. L. Bongirwar (✉)
PWD, B 1102 Pataliputra CHS Four Bungalow Signal, Andhari West, Mumbai, Maharashtra 400053, India

M. Parida et al. (eds.), *Proceedings of the Fifth International Conference of Transportation Research Group of India* , Lecture Notes in Civil Engineering 220,
https://doi.org/10.1007/978-981-16-9925-2_5

Fig. 1 Typical Kolahpur type BCB

Fig. 2 Bridge cum Bandhara on Purna River, T. Daryapur

The water is stored within the riverbanks; hence, it does not require additional land acquisition, and therefore, the scheme is popular. These structures were not necessarily built along the regular roads but at suitable convenient locations. Later on, designs were developed to construct new bridges on roads or to convert existing bridges on planned roads also to be used as bridge cum Bandhara (BCB). Several BCBs have been constructed on bridges having length of more than 150 m. A few photographs of Bandhara constructed are shown in Fig. 2.

2 Limitations of Traditional BCBs

The traditional BCB design has certain advantages and disadvantages. The major limitation of the traditional structure is that the gates are required to be fixed when the monsoon subsides and removed before the onset of the next monsoon. The gates need to be stored at a suitable place for around 8 months. The government therefore encouraged formation of societies of beneficiaries who would undertake the responsibility of the operations and authorized them to recover the charges from the beneficiaries. However, the potential benefits of the system were not fully used

Fig. 3 Typical cement plugs

due to lack of interest by farmers. Damages are reported due to non-removal of gates or storage not created due to delay in fixing. Theft of the gates is also common. Government has also prescribed financial norms related to storage for such structure. Besides, about 15–20% water is lost due to evaporation.

3 Cement Plug/Cement Weir

It is practice of irrigation department to construct permanent small weir known as cement plug to store water within riverbank. Generally, storage planned is up to 2.0 m. These structures also create a pool of water which helps in increasing ground water. The water is used as source for drinking water and limited irrigation. Such typical structures are shown in Fig. 3.

4 Problems of Small Rivers/Nallas

In rivers having a width of up to 30–35 m, maximum depth of water during flood does not generally exceed 4.5 m, and their behavior is also similar. Once the rain stops, the water drains off in a few hours, and river becomes virtually dry. For nallas (small streams) up to 25 m width, it may require only a few hours to be completely empty. Even in monsoon, these minor rivers can be crossed after a few hours of stoppage

Fig. 4 Typical cement plugs

of rain. Each village has such a natural small river course as water is a prime need for survival. Villages settled near the natural course of these rivers do not play an active role in conserving and recharging the ground water. Some of the photographs as shown in Fig. 4 show the dry riverbeds. Even the well in the riverbed is dry.

5 Innovative, Affordable, and Cost-Effective Elliptical-Shaped Curved Weir as BCB

A new design of an elliptical-curved concrete weir which was deemed to be ideally suitable for small rivers as a BCB structure was evolved. It has removed the limitation of traditional BCBs which require fixing and removal of gates. It can be considered as substitute to cement plug. The design involves constructing elliptical shape arches made of reinforced cement concrete touching the bridge to serve as weirs to store water up to 2.4 m. This structure then ensures water for the entire monsoon period and also helps to pond the water on the upstream side. In this innovative design, all forces, i.e., overturning or sliding due to standing water, are not transferred to main bridge. In cement plug, the design is based on assuming that it is a gravity structure. Elliptical arch is structural-efficient system. In cement plug, the length of weir is decided to allow full flood discharge over weir so that afflux, highest flood level (HFL), is within the bank. This requires increasing length beyond the normal width of river, and on both ends, guide walls are required to be made. Besides, extensive protection measures are required to be done. Extensive study has been done by Indian Institute of Technology (IIT), Roorkee for evolving formulae for discharge for a curved weir and has published paper 'Discharge Characteristics of Sharp Crested Weir of Curved Plan' form. It is seen that the discharge capacity of curved weir is at least 25–30% more compared to straight weir. We do not require extensive separate guide bunds. Protection measures are normally provided for bridges. As such, all this cost is saved, and cost of this structure is 40% of that of cement plug and generally lies between Rs. 60–75,000 per running meter. We are thus making use of additional potential at marginally increased cost.

Fig. 5 Construction sequence

5.1 *NeriPeth Project*

The culvert crossing is constructed using multicell Hume pipes at NeriPeth. Thereafter, the elliptical arch weirs were constructed in RCC which act as the weirs. Once the water reaches the top of the arch, it spills over the crest and passes through the pipe opening. The construction sequence is shown in Fig. 5.

The NeriPeth structure was completed in one month at a cost of ₹ 0.2 million, and there was standing water for the entire monsoon period thus helping the recharging of the ground water in a big way.

5.2 *Elliptical BCB on Mum Marodi Road*

Recently, a structure based on this design is completed in 20 days on a bridge having three spans of 5 m using the technology of precasting (Fig. 6). Minor bridge of three spans of 5 m has been converted in to Bridge cum Bandhara in 20 days using precast arches. Each arch consists of six pieces not weighing more than 1 ton which are placed with crane and in situ joined.

Fig. 6 Elliptical arch wit standing water and back water also shows construction technology

6 BCB and Minor Bridges

The typical structures for which the arch type BCB can be constructed at affordable cost are as follows:

1. Multi-cell Hume pipe culverts
2. Culverts/minor bridges having multiple spans with each span of less than 6 m.
3. Single/multicell box culvert of size 2 × 2 m, 3 × 3 m, 2 × 3 m, etc.
4. 2 m semicircular arch culverts
5. Kolhapur type weirs

These are shown in Fig. 7.

Fig. 7 Arch type BCB constructed at affordable cost

7 Functioning of the Elliptical Arch

By constructing elliptical arches, the linear waterway of the bridge is not reduced. The arches function as curved weirs and only change the direction of flow. Extensive model studies have been done for determining the formulae for discharge over such curved weirs. The model study has proved that the normal formula for discharge over a straight weir is applicable except that equivalent length could be 1.20–1.30 times that of the straight length depending on the curved shape. The height of such curved weir can be decided to create standing storage and find the afflux for the designed highest flood. The afflux level for maximum designed flow must be within the banks of the river/nalla for the proposed storage level. Standard formulae to calculate afflux for weir can be used.

8 Model Study

Extensive model studies have been done at (1) Pune Engineering College (Fig. 8), (2) Amravati Engineering College, and (3) Chandra–Bhaga canal (Fig. 9) to get a better understanding of the behavior of the flow. All these model studies have revealed that there are no adverse flow conditions such as heavy turbulence, hydraulic jump,

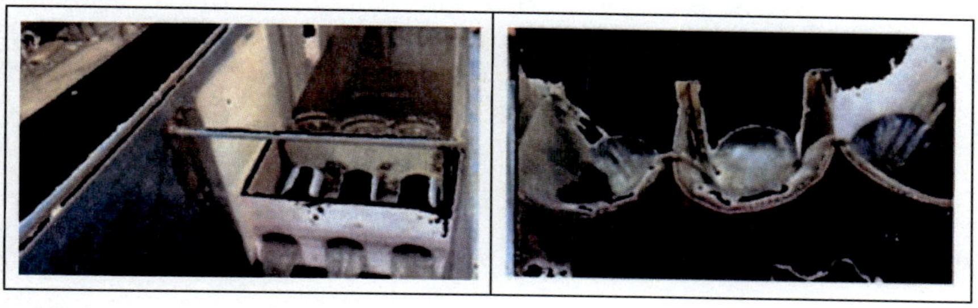

Fig. 8 Model study at Pune Engineering College

Fig. 9 Study of flow over the elliptical weirs at Chandra–Bhaga canal

and vortex formation. On the contrary, the flow is smooth. The model study also confirmed that afflux as calculated by the curved weir formula is reasonably correct.

9 Efficient Structure System Leading to Cost-Effective and Affordable Solution

The curved weir is an alternate to cement concrete plug or low height straight weirs which are all gravity structures and constructed to ensure standing water to help ground water recharging. The length of the straight weirs must be kept slightly more than the natural width of the river to keep the afflux within the banks. Wing walls are must to contain flood waters with in bank. Besides, as there is free fall of the water over the straight weirs, elaborate protection measures are required to be provided on the downstream side.

However, the elliptical arch weir system is structurally more efficient and hence cost-effective. All the forces acting on the arch weir are not transferred to the main bridge; hence, the weir structure is light. Elliptical close shape is structurally efficient; hence, only, 100 mm thick arch with nominal steel is enough. There is also no need for separate wing walls. As the water overflowing the curved weir falls on impounded water, hence elaborate protection measures are not required. Extensive study has been done by IIT, Roorkee to evolve discharge formulae for curved weir as mentioned above in Para 5. The curved length of weir increases the flow at least by 30% compared to straight length; hence, afflux is also less. Due to all these advantages, the cost of the curved weir as a storage structure is 30–40% that of a similar straight concrete weir serving the same purpose. The average cost of the elliptically curved weir comes to Rs. 60,000–70,000 per running meter.

10 Shape of Arch

The arch shape has been selected based on findings of the model studies and analyses for hydraulic behavior of bridges. With the objective to have a safe structure, the arch shape was selected so that the cumulative plan area of the arches is at least 1.25 times and preferably 1.5 times that of the waterway of the straight span to cater for the designed flood lineal waterway. The elliptical shape was found to be fulfilling the condition. Various shapes, as parabola, circle, hyperbola, etc., were studied, and it was found that in an elliptical shape the curve length is least and hence would prove cost-effective. As individual spans may vary the shape of the arch was, so selected that a common formwork with minimum changes would suffice to keep cost of form works low. An efficient design of ellipse to suit all spans with minimum changes would reduce cost of construction is shown in Table 1.

Table 1 Optimize form work

Shape	Water way	Required area	Base	Height	Area	Chord
2-m-semicircle arch, 2 spans	9.4	11.75	4.5	4.2	13.8	10.36
KT weir, 2 spans	14	17.5	5.5	4.2	19	11.36
5-m span, 3-m height	15	18.75	5.5	4.2	19	11.36
6-m span, 3-m height	18	22.5	6.5	4.2	23.2	12.36
4-m span, 3-m height	12	15	4.5	4.2	14.8	10.36
We may prepare formwork as per fist case and add straight plate to get base						
For smaller spans						
2-m-semicircle arch, 1 spans	4.7	5.88	2.2	4.5	7.8	9.6
KT weir, 1 span	7	8.76	2.4	4.5	8.7	9.8
2 * 2 Box	4	5	2.4	4.5	8.7	9.8
2 * 3 Box	6	7.5	3	4.5	11.4	10.4
3 * 3 Box formwork for first case to be decided and straight plate added to get required base	9	11.25	3	4.5	1.4	10.4

11 Construction Methodology

The cost of each of the elliptical arch structure does not exceed Rs. 4–6 lakhs. Since the weir works required to be constructed are scattered over a large area, hence to keep the cost low, the works need to be got executed through local contractors. Good quality formwork is essential along with M25-grade self-compacting concrete. The cost of formwork and concrete is virtually the same. To achieve economy through repetitive use, it is intended to encourage an agency to own the formwork and rent it to local contractors. To ensure better quality and accelerate construction, it is proposed to precast the arches in sections at a centralized place/factory or at site. The cast sections would be erected at site and joined by in situ concreting. This technique is now fully developed. The three arches for bridge on Mul Marodi are completed in three days. Self-compact concrete only is used. Surface vibrators are used. Mold is kept in sloping position to ensure full flow of concrete (Fig. 10).

Shape of arch is elliptical plan area shall be at least 1.25-times waterway of bridge.

12 Potential Use of Elliptical BCB

As the cost is low and affordable with many benefits, the technology is expected to be used by many. The potential usages are as follows:

Groundwater Recharging: In the areas under consideration, rainfall occurs on an average for 20–25 days in a year. During each of these rainfalls, 7000–8000 cubic

Fig. 10 Formwork and mold

meter of water is collected which translates to about 1.5 lakh cubic meter per annum. The water needs to be used for recharging the ground water through simple techniques which are readily available. Even if 30% of the water is saved, a village can be made tanker free.

Potable Water at Affordable Price in Saline Tracks of Land: In saline areas, an ordinary sand filter costing around Rs. 1.5 lakh can supply potable water at. Rs. 1.0 for 20 L. Since rainwater is being purified, costly reverse osmosis (RO) systems are not required. At village 'Apoti' in Akola District, one scheme has been sanctioned by the Government of Maharashtra and is under implementation.

Artificial Ponds/Malgujari tanks: The Government of Maharashtra is providing subsidy and encouraging farmers to construct farm ponds (Shettale: small water bodies) which can be filled with water using solar pumps (Fig. 11) to create irrigation facilities. In Chandrapur district, there are several natural small lakes called 'Malgujari' tanks. The cost of small tanks, solar systems, pipelines, etc., normally does not exceed Rs. 30,000 per acre which is much less compared to the average cost of Rs. 0.1.5–4.5 lakhs per acre of a major irrigation project.

13 Bridge cum Bandhara Across Andhari River

A bridge cum Bandhara on Andhari River (Fig. 12) is almost completed bridge which comprises of nine spans of 8 m. 3.5 m deep storage is planned. There are four Malgujari tanks having area of 40 ha having capacity of 10 lakh cubic meter. These will be filled using solar pumps which will be operate from June 15th after December. The water available in Bandhara will be pumped. Thus, 16 lakh-cubic meter water storage will be created which is enough to irrigate 1500-acre land. Further irrigation will be by gravity. Thus, virtually, there is no reusing cost. The total cost of scheme is Rs. 350 lakhs which works out to Rs. 24,000 per acre which is less than 15%

Fig. 11 Irrigation facility created for 61 acres with zero recurring cost

Fig. 12 Bridge cum Bandhara across Andhari River

compared to cost of irrigation in major dam which works out to 1.5–4.5 lakh per acre The scheme thus would fulfill government's policy to double the income of farmers.

14 Funding for Implementation

As water is a priority sector, the government is providing funds for water conservation projects. The solution suggested to augment ground water is affordable and cost-effective. They can be implemented in many areas. Few pilot projects demonstrate the economic advantage of the scheme enumerated above. Funds, therefore, would be available through budget or through the Member of Assembly (MLA) or Member of Parliament's discretionary funds on account of low cost of individual scheme. Corporate social responsibility (CSR) funds could also be available. The government has announced that they want to double the income of farmers. Thus, this scheme will enable the government to fulfill this important objective. The Government of India has decided that for all the works of National Highways and 'Pradhan Mantri Gram Sadak Yojana' (PMGSY), the funding for BCB will be done by the Government of India. Large-scale National Highways improvement program is in progress; hence, the BCB structures can be constructed from the contingencies of the ongoing works.

Acknowledgements The authors are thankful to Shri Debadwar, Chief Engineer; Shri Jaiswal, Executive Engineer and other staff of Maharashtra Public Works Department (PWD) who were involved in the implementation of projects. Shri Jangade, Dr. R. K. Ingale, Head VNIT for doing the design and preparing standard plans. Former Secretary, Maharashtra PWD and former Member MPSC took lot of efforts in scrutinizing the design and gave critical comments. Shri Garibdas of Innovative Construction has played a major role in developing construction technology. Shri Dhawale, Superintending Engineer, Minor Irrigation Project has extended all help in conducting model study at a large scale similarly. Prof. Mrs. Thube of Pune Engineering College and Dr. Tatewar of Amravati Engineering College took lead in laboratory studies which helped to draw useful conclusions. The authors are also thankful to all the experts involved in making the scheme successful.

References

1. Kumar S, Ahmad Z, Mansoor T, Himanshu SK, Discharge characteristics of sharp crested weir of curved plan-form
2. Nilgunde P, A mini project report on "study of arch type bridge cum Bandhara. Submitted By (121796011) under the guidance of Dr. A. D. Thube, College of Engineering, Pune
3. Modi PN, Seth SM, Hydraulics and fluid mechanics including hydraulic machines
4. Bongirwar PL, Namjoshi AG, Jaiswal AM, Dual purpose bridge structure-crossing and storage-need of the day, Paper number 496 of 2004 of IRC
5. Provision of Bandhara on the bridges under construction under PMGSY roads-amendment to para 8.5 (V) of PMGSY guidelines P-1701/S/2011-RC(part)(FMS-337815) dated 16-10-04
6. MORTH RW-NH-34066/59/2015-S&R(B) dt 18-4-17: Construction of Bridge cum Barrage structure on national highways serving dual purpose, i.e., to cross water body as well as to store water

Bus Bunching Control Strategies: A Case Study of Delhi

K. Ramachandra Rao, Raghu Kumar Yadav, and Saurabh Agarwal

1 Introduction

In the environment with high pollution levels and traffic congestion in many cities, one of the contributors is the private vehicles. Thus, an efficient bus system would encourage people to reduce the usage of personal travel modes. However, the current bus system is not in good shape, i.e., most of the buses do not seem to follow the prescribed schedule. Hence, two or more buses which should be evenly spaced according to the schedule have a varied distribution. In an extreme case, two buses may arrive at the same spot together. This phenomenon is known as bus bunching, and it has various negative effects. It leads to a longer waiting time for commuters and also increases the cost of bus service providers since the frequency with which they are sending the buses is not being experienced by commuters. Thus, there is a need to reduce bus bunching in order to improve commuter experience and increase revenues.

As stated above, there is a strong need to address the bus bunching problem to reduce the commuter wait time while reducing the cost of bus operators. Further, there is also a need to test the strategies of controlling bus bunching. Rest of the paper is presented in six sections. Section 2 presents the literature review and study objectives and scope. Section 3 presents the time–space diagram. Section 4 presents the details of the PrefixSpan algorithm. Section 5 presents the simulation results. Section 6 presents the control strategies and conclusions of the paper.

K. Ramachandra Rao (✉)
Department of Civil Engineering and Transportation Research and Injury Prevention Program (TRIPP), IIT Delhi, Hauz Khas, New Delhi 110016, India
e-mail: rrkalaga@civil.iitd.ac.in

R. K. Yadav · S. Agarwal
Department of Mechanical Engineering, IIT Delhi, Hauz Khas, New Delhi 110016, India

M. Parida et al. (eds.), *Proceedings of the Fifth International Conference of Transportation Research Group of India* , Lecture Notes in Civil Engineering 220,
https://doi.org/10.1007/978-981-16-9925-2_6

2 Literature Review

2.1 Bus Bunching

The problem of bus bunching is well defined in the literature. Andres and Nair [1] looked at the problem of bus bunching on a popular route in the city of Dublin, Ireland. They explain the problem of bus bunching as being cyclic. Once a bus enters into the cycle, the problem magnifies until the buses bunch together.

The field of bus bunching has been studied over the past few decades but has got new attention due to the data provided by GPS technology. Through this data, we are now able to make better predictions and come up with better solutions [2]. This problem is also discussed elsewhere [3].

2.2 Time–Space Diagram

In order to analyze bunching, we need to visualize the model, and it can be done using time–space diagrams. We transform the position of stops to one-dimensional representation, i.e., offset distance from the depot. We can create such a diagram which shows all the trips occurring on a single day and use it to visualize bunching.

2.3 PrefixSpan Algorithm

Bus bunching causes various problems like delays, decrease in comfort for passengers, and higher waiting times. Nowadays, it is important for public transportation companies to improve their service reliability because of the higher competition with private companies [4]. This is because reliability is an important metric, and if people know that a certain bus is going to come at a certain time, then they are more likely to use them. It is known that bus bunching begins with a headway deviation from a stop [5]. Therefore, we need to identify stations where instances of bunching occur in a frequent manner and from where all the irregularities begin. Such bus stops are known as bunching black spots and are highly problematic regions [6]. We can do this by using the PrefixSpan algorithm. It is a sequential pattern mining method, and it mines the complete set of patterns [7]. It is faster than other methods which are used in the problem of sequential pattern mining.

2.4 Control Strategies

The purpose of control strategies is that of making a schedule which if followed would result in better regularity of frequency of buses on a given route using historical and real-time data. If all the buses are sent from the depot in a proper manner, then we will get much better results. Another way to reduce bunching is by identifying particular stations on a route, which on a regular basis cause headway deviation from the target headway. To address this problem of identifying problem stations, a data mining approach is used by authors in [6].

They apply a PrefixSpan algorithm to identify patterns in headway deviation across all stations. These stations are termed as bunching black spots (BBS) which are then assigned certain slack times in the schedule to overcome the problem. Slack times are a holding method which state that the bus should not leave all stations before the scheduled departure time. One of the problems with this approach is that service quality of buses goes down even though reliability may go up. Also, it may happen that the delays occurring in the route are more than the slack times introduced [1]. The performance measures like mean absolute error (MAE) and root mean square error (RMSE) to compare our simulation model to actual data were used as mentioned in [8].

2.5 Problem Motivation

Most of the works have been done in the area of bus bunching has focused on developing static control strategies for reducing bus bunching. We try to implement some of these strategies and verify the results using a simulation tool.

2.6 Objectives of the Work

Based on the review of literature, we outline the objectives of this study:

1. Identifying bunching black spots using the past data of bus.

This involves using PrefixSpan algorithm to mine patterns in data. Using this, we compute several bunching black spots (BBS) in the route which frequently give rise to bunching further in the route.

2. Preparing a computer simulation of the model.

This involves simulating the problem on AnyLogic software. The model should be as close as possible to the reality for which we use historical data.

3. Use different methods to reduce bunching and observe the same through simulation. The methods to reduce bunching are implemented on the simulation, and results are compared to verify the effectiveness of the methods.

3 Time–Space Diagram

3.1 Theory

Time Headway and Space Headway

Time headway is the difference between the arrival times of two buses at a particular station. Space headway is the distance between two consecutive buses at a given time. Thus, the distribution of time and space headway is a good indicator of the performance of a given route. Further, mention of headway in the paper would imply time headway unless otherwise stated.

Time–Space diagram

A space–time diagram is a good tool to visualize the bus bunching phenomenon. It is the plot of '*offset distance*' of the bus vs the '*time*'. Thus, steeper the curve, faster is the bus. If any two lines intersect, it means the buses have met or the space headway between them is zero. The vertical distance on time–space diagram between any two curves gives the space headway between the buses, and the horizontal distance gives the Time headway.

3.2 Data Acquisition

The data on the position of different buses were provided by Delhi integrated multimodal transit system (DIMTS) in the form of time-stamped Global Positioning System (GPS) data. The position of the bus in terms of latitude and longitude was given along with the time. Data for three different bus routes over a period of month of January, 2013 were provided by DIMTS.

We got the information of routes, stations, trips, and scheduled arrival times of buses at different stations from open transit data (OTD) created by IIIT Delhi (Indraprastha Institute of Information Technology) in association with the Delhi Government, Delhi Transport Corporation (DTC) and DIMTS.

3.3 Cleaning the Data

Then, we started with cleaning the data and extracting meaningful information out of it. We worked out the arrival time of all buses on different stations using MATLAB which was just sufficient for all further work to be done. This was done using the data of exact locations of bus stations from OTD and matching it with the actual positions of buses given by DIMTS with a tolerance of $\pm$ 10 m around the bus stop. Figure 1 shows the space–time plot for January 20th.

4 PrefixSpan Algoirthm

As discussed earlier in the paper, we intend to use PrefixSpan Aagorithm [6] to identify bunching black spots. PrefixSpan is a sequence mining algorithm used to mine sequential patterns in data. Our aim is to find those bus stations from where frequent headway deviations start and result in bus bunching further in the route. Step-by-step procedure of this implementation is discussed below.

Let, there be vectors **B** and **n** representing all buses on a given day and stations, respectively.

1. Calculate the headways of all buses on all stations throughout the span of 29 days.

   Headway sequence ->

$$\{X_1\}, \{X_2\}, \{X_3\}, \ldots, \{X_n\}$$

   where each $\{X_i\}$ is a $[(B - 1) * 1]$ vector.

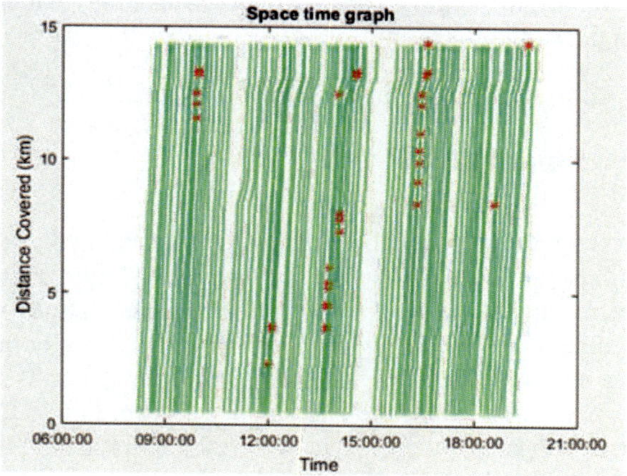

Fig. 1 Time–space diagram of buses on January 20th, 2013 with red * showing the bunching instances where buses have a headway less than 1 min

Table 1 Sample data of sequential patterns fed into the PrefxSpan algorithm

Trip	Sequence
1	0 0 0 1 1 1 1 1 0 0 0 0....
2	0 0 0 0 0 −1 −1 −1 1 0 1 1....
3	0 0 0 0 0 0 0 0 1 1 1 1 1....

2. From these headways, calculate the headway deviation (change in headway in going from station *i* to station *i + 1*) for all bus trips.
   Headway deviation sequence ->

$$\{H_1\}, \{H_2\}, \{H_3\}, \ldots, \{H_n\}$$

where $\{H_1\} = 0$ and $\{H_i\} = \{X_i\} - \{X_{i-1}\} \forall i \in [2, n]$

3. All those headways lying within an acceptable threshold are assigned 0; those with increasing headways more than threshold are assigned 1, and those with decreasing headways less than threshold are assigned −1. This forms sequences of integers 1, 0, and −1 which are analyzed by the algorithm to mine repeated patterns. A sample table is shown in Table 1.

$$h_i = \begin{array}{l} 0 \text{ if } |H_i| \leq -\text{threshold} \\ 1 \text{ if } H_i \geq -\text{threshold} \\ -1 \text{ if } H_i \leq -\text{threshold} \end{array}$$

4. Input this table into the algorithm to obtain all the patterns. The result obtained from the algorithm is shown in Table 2 taking a minimum support of 100 (i.e., those sequences which occurred in at least 100 trips)

As we can see from Table 2, stations 5, 6, and 20 produce the longest bunching, and thus, we would try and introduce slack times to them in our simulation presented in the coming sections.

5 Simulation

After getting done with identifying the bunching black spots using PrefixSpan algorithm, we proceeded to setting up an AnyLogic simulation which would help us verify that introducing slack times to these stations would improve the bunching problem. We did an agent-based simulation with both flowchart and statechart for the purpose. The main flowchart for the simulation is shown in Fig. 2.

Source generates the buses at depot according to the given schedule. The buses currently in operation, i.e., between the depot and final station, are queued in *inService* block. *Hold* block holds them till the time they reach their destination where they are then passed into *sink* block. During the service, buses are controlled by the vehicle

Table 2 Mined sequential patterns with minimum support = 100

Sequence number	Sequences
1	4, 5, 6
2	5, 6, 7
3	5, 6, 7, 8
4	6, 7, 8
5	6, 7, 8, 9
6	7, 8, 9
7	8, 9, 10
8	9, 10, 11
9	12, 13, 14
10	13, 14, 15
11	14, 15, 16
12	15, 16, 17
13	18, 19, 20
14	19, 20, 21
15	20, 21, 22
16	20, 21, 22, 23
17	21, 22, 23
18	22, 23, 24
19	23, 24, 25
20	24, 25, 26
21	25, 26, 27
22	26, 27, 28
23	27, 28, 29

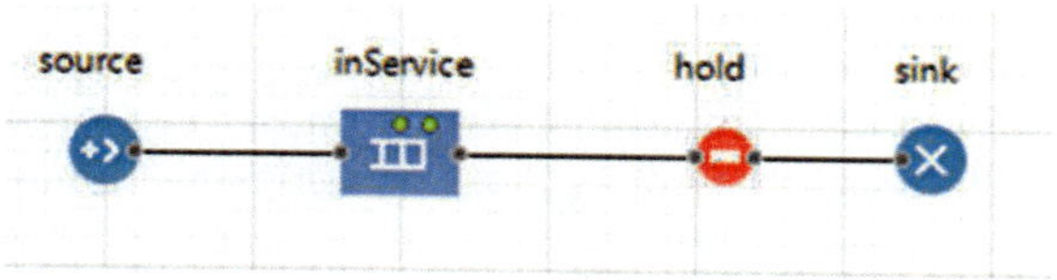

Fig. 2 Main (agent) flowchart of the vehicle

statechart which moves them from the starting station to the final station incorporating traffic conditions and waiting time at the stations. The vehicle statechart is shown in Fig. 3.

To simulate the real-life traffic conditions, we divided the entire day into 3-time windows. First one from 7 to 11 am, second from 11 am to 4 pm, and the third one from 4 to 9 pm. We then calculated, from the real data provided by DIMTS, the average speed and standard deviation of the buses between each pair of station. The result is shown in Table 3.

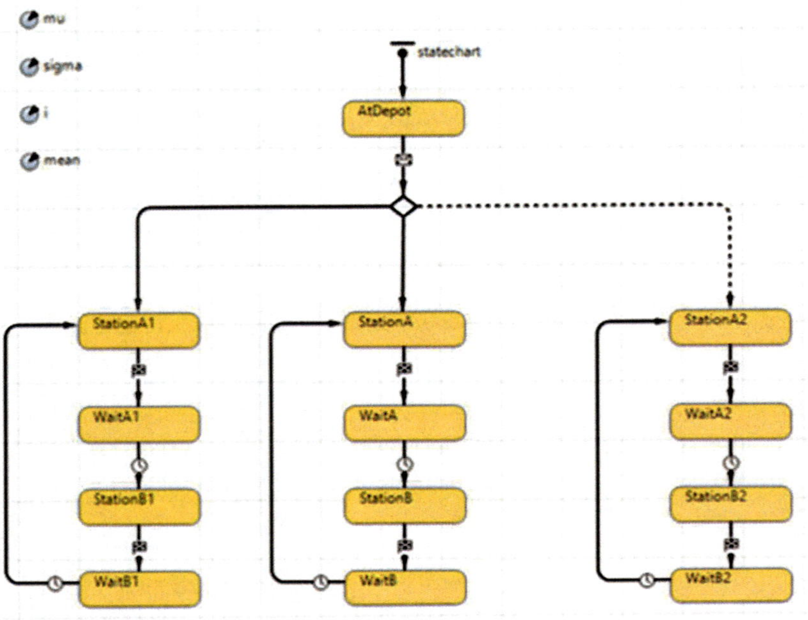

Fig. 3 Vehicle statechart with three different time windows

As we see from Table 3, a good variation of speeds between different stations is obtained. The average speed of buses is almost same throughout the day although their variation is on the higher side in the evening.

The mean and standard deviation are used to approximate the speed of the buses as a normal distribution with a lower and an upper limit. As shown in Fig. 3, the buses are dispatched from the depot and sent into one of the three time windows. The waiting time at stations is approximated by exponential distribution. We iterated for different values of lambda (1/mean), the deviation of the simulation from real data. This was achieved by comparing the actual arrival time of the buses with those predicted by the simulation at each station. The result is summarized in Table 4.

From Table 4, we see that the predicted arrival time varies on average 33 s from the actual arrival time.

Space–time diagram of actual vs the predicted bus trajectories of some random buses is also plotted to visualize the prediction better and is shown in Fig. 4.

Table 3 Average speed and standard deviation of speed between bus stations

Station		Morning		Afternoon		Evening	
From	To	Average (km/h)	Deviation (km/h)	Average (km/h)	Deviation (km/h)	Average (km/h)	Deviation (km/h)
2	3	25.5	5.6	22.2	5	18.8	9
3	4	38.1	7.3	40	11.9	40.2	20.8
4	5	26.3	8.9	27.4	10	27.2	13.7
5	6	24	7	24.6	8.2	23.6	12.1
6	7	7.9	2.9	7.7	3.1	6.4	3.5
7	8	27.9	5.5	26.5	4.3	24.4	11.5
8	9	11.4	4	9.9	3.5	9.2	4.6
9	10	33.5	14.2	26.7	13.6	26.1	14.3
10	11	23.7	5.1	21.9	4.2	20.7	9.9
11	12	26.3	6.5	27.1	5.7	25.2	12
12	13	31.7	5.7	29.6	7.2	27.9	13.5
13	14	32.1	14.5	34.9	12.8	33.6	17.4
14	15	28.8	5.4	28.8	5.5	27.3	13.1
15	16	23.5	3.9	23.6	4.1	21.5	10.6
16	17	45.4	12.5	41.3	12	40.5	20.2
17	18	43.2	16.9	40.8	16.1	40.7	21.5
18	19	28.2	15.9	30	14.3	26.8	14.9
19	20	18	3.7	15.9	3.3	14.7	7.7
20	21	29.6	4.5	29.1	6.9	27.1	13.2
21	22	39	11.9	38.1	11.5	37.5	19.2
22	23	27.1	7.6	25.7	7.5	24.4	12.2
23	24	22.3	5.2	20.9	5.8	21.3	10.6
24	25	41.4	6.1	34.8	13.5	39.7	19.7
25	26	33.8	9.6	26.4	17.7	33.2	17.7
26	27	13.5	9.7	8.2	4.3	11	6
27	28	19.4	6.5	18.4	7.3	16.6	8.8
28	29	43.9	16.1	39.2	14.8	39.8	21.3
Mean		28.3	8.2	26.6	8.6	26.1	13.2

6 Simulation

Having obtained a reasonably accurate model, we then implemented the iterations of various control strategies discussed above.

Table 4 Different standard errors at all stations between actual and simulated data

Station	Mean absolute error (min)	Root mean square error (min)
1	7	7.7
2	0.5	0.8
3	0.3	0.5
4	0.4	0.5
5	0.4	0.6
6	0.7	1
7	0.7	1
8	1.1	1.3
9	0.2	0.4
10	0.7	1.2
11	0.5	0.5
12	0.4	0.6
13	0.3	0.3
14	0.4	0.5
15	1.1	1.9
16	0.3	0.4
17	0.2	0.4
18	0.5	0.6
19	0.9	1.3
20	0.4	0.5
21	0.3	0.5
22	0.6	0.9
23	0.6	0.8
24	0.4	0.8
25	0.4	0.7
26	1.4	1.7
27	0.5	0.7
28	1	1.5
Mean	**0.56**	**0.81**

6.1 PrefixSpan

As discussed earlier on PrefixSpan implementation, stations 5 and 20 were identified as bunching black spots. We introduced a slack time of 1 min on each of these stations. The number of bunching instances reduced but not significantly. This is because much of the bunching takes due to the irregular dispatching inter-arrival time from the depot. Also, the holidays (Republic Day) are not taken into account

Table 5 Bunching instances observed versus predicted by AnyLogic

Date (Jan 2013)	Day	Observed	Predicted
1	Tuesday	126	143
2	Wednesday	138	185
3	Thursday	75	161
4	Friday	130	137
5	Saturday	67	122
6	Sunday	72	135
7	Monday	157	164
8	Tuesday	151	151
9	Wednesday	107	148
10	Thursday	60	175
11	Friday	119	228
12	Saturday	178	196
13	Sunday	77	174
14	Monday	91	137
15	Tuesday	115	169
16	Wednesday	183	196
17	Thursday	180	174
18	Friday	155	180
19	Saturday	172	184
20	Sunday	47	129
21	Monday	125	244
22	Tuesday	128	144
23	Wednesday	68	181
24	Thursday	79	160
25	Friday	24	65
26	Saturday	26	16
27	Sunday	83	137
28	Monday	136	149
29	Tuesday	76	111

during which the number of buses run actually is significantly less resulting in less bunching instances. The result is tabulated in Table 6.

6.2 *Dispatching Schedule*

Since most of the bunching problem is caused due to irregular dispatching from depot, we next tried to improve this. Since DIMTS aims at a target headway of

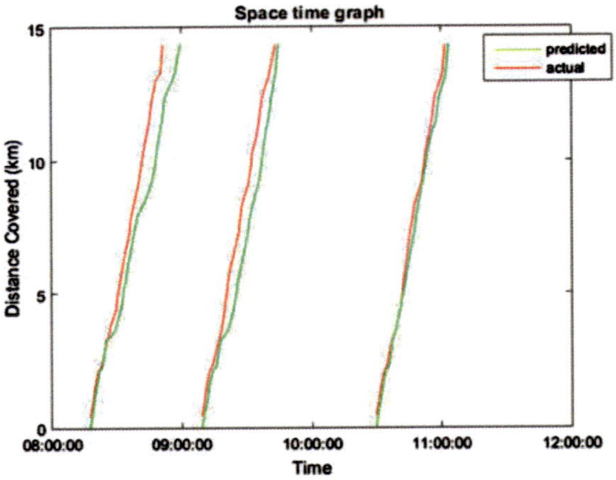

Fig. 4 Actual vs predicted bus trajectories of some random buses

6 min between its buses, we set the inter-arrival time of the buses at depot equal to 6 min keeping the slack times on stations 5 and 20 equal to 1 min each. The result significantly improved time headways between the buses. The result is shown in Table 6.

From Table 6, it is clear that by introducing slack times does not reduce bus bunching by a significant amount. It only decreases bunching by around 13%. Regulating the dispatching schedule while keeping the slack times as it is, it reduces bunching by around 80% as is expected. Further, Figs. 5, 6, and 7 highlight the difference in bunching using red an asterisk.

6.3 Conclusion and Further Scope

The key problems addressed were as follows:

- Found out those points (stops) where bunching is taking place through a threshold of one minute.
- Using the PrefixSpan algorithm to determine the bunching black spots.
- Making a simulation of the model in AnyLogic using bus speeds from real data and comparing both the models to determine the level of accuracy.
- Using different strategies like introducing slack times at bunching black spots and rescheduling buses to reduce bus bunching.

Comparing the results obtained from these strategies to actual data to determine the best strategy to reduce bus bunching,

the results obtained are as follows:

Table 6 Bunching instances using different control strategies

Date	Day	AnyLogic predicted	Using PrefixSpan	PrefixSpan and dispatching schedule
1	Tuesday	143	126	13
2	Wednesday	185	124	46
3	Thursday	161	168	14
4	Friday	137	109	4
5	Saturday	122	107	31
6	Sunday	135	184	28
7	Monday	164	135	35
8	Tuesday	151	156	20
9	Wednesday	148	142	35
10	Thursday	175	174	42
11	Friday	228	157	37
12	Saturday	196	137	61
13	Sunday	174	143	18
14	Monday	137	149	42
15	Tuesday	169	129	53
16	Wednesday	196	133	18
17	Thursday	174	112	24
18	Friday	180	138	38
19	Saturday	184	115	30
20	Sunday	129	131	26
21	Monday	244	145	34
22	Tuesday	144	129	13
23	Wednesday	181	113	31
24	Thursday	160	115	38
25	Friday	65	158	32
26	Saturday	16	132	24
27	Sunday	137	190	16
28	Monday	149	114	32
29	Tuesday	111	119	54
	Average	**155**	**137**	**30**

- Stops 5, 6, and 20 on route 78 DOWN frequently caused headway deviations which resulted in bus bunching further ahead in the route. These stations are thus the bunching black spots of the route.
- Adding slack times of one minute each to these stations reduced the bunching occurrence by approximately 13%.

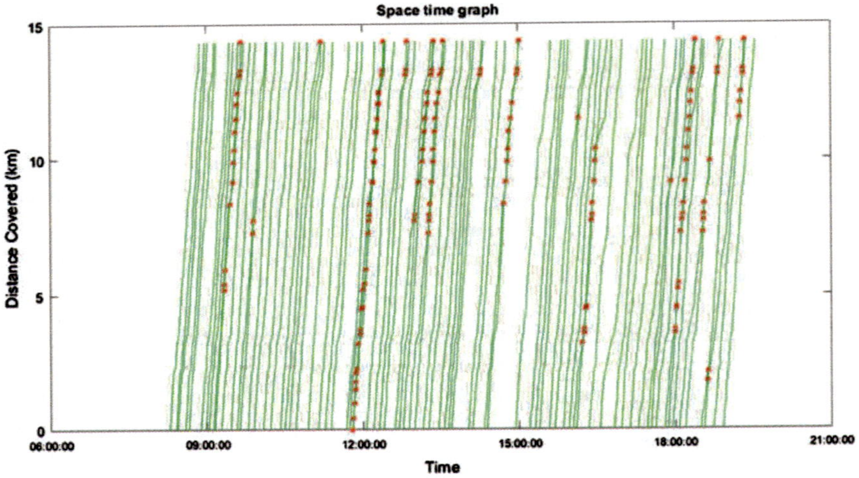

Fig. 5 Bunching instances without any control strategies on January 01, 2013

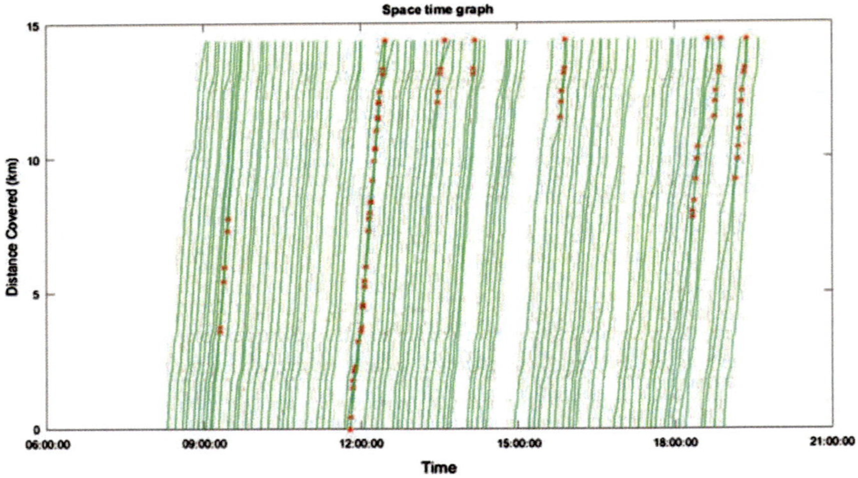

Fig. 6 Bunching instances using slack time control

- Regulating the departure of buses from depot further reduced the bunching occurrences by 80% highlighting this as the major cause of bunching.

  The conclusions drawn are as follows:

- Lesser bunching instances were found by rescheduling the buses compared to introduction of slack times.

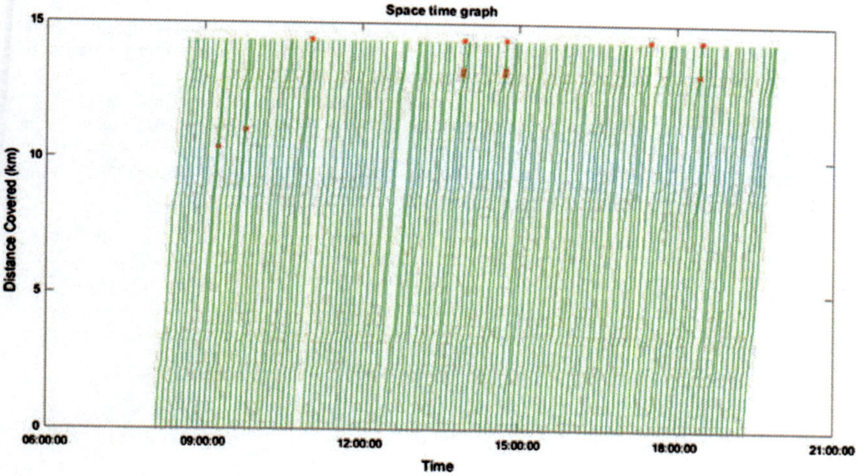

Fig. 7 Bunching instances using both slack time and dispatching control

- It is important to control the departure of buses from the depot to minimize the problem of bus bunching.

  Future work.

- Validation of the results obtained by testing them on the bus route.
- Create a dynamic model which uses real-time data to provide solutions for bus bunching.
- Formulating various combinations of dynamic and static control strategies to reduce bunching.
- Verifying the above obtained solutions in real buses and improving them further for better service reliability.

References

1. Andres M, Nair R (2017) A predictive-control framework to eliminate bus bunching. Transp Res Part B 104:123–148
2. Nair R, Eric B, Yiannis G, Oliver V (2014) Data as a resource: real time predictive analysis for bus bunching. In: Transportation research record, presented at 2015 TRB 94th annual meeting
3. Daganzo CF (2009) A headway based approach to eliminate bus bunching: systematic analysis and comparisons. Transp Res Part B 43:913–921
4. Strathman J, Kimpel T, Dueker K (1999) Automated bus dispatching, operations control and service reliability. Transp Res Rec 1666:28–36
5. Newell G, Potts R (1964) Maintaining a bus schedule. In: 2nd Australian road research board, pp 388–393

6. Luis MM, Daganzo CF, Joao G, Joao MM, Jorge FS (2012) Bus bunching detection: a sequence mining approach. In: Conference paper, ubiquitous data mining (UDM) workshop in conjunction with the 20th European conference on artificial intelligence—ECAI 2012
7. Pei J, Han J, Mortazavi-Asl et al (2001) PrefixSpan: mining sequential patterns efficiently by prefix-projected pattern growth. In: Proceedings of the 17th international conference on data engineering. IEEE Press, New York, pp 215–224
8. Balasubramanian P, Rao KR (2015) An adaptive long-term bus arrival time prediction model with cyclic variations. J Public Transp 18:1–18

Effect of Two-Lane Rural Highway Horizontal Curves on Driver's Galvanic Skin Resistance

Anitha Jacob, Jisha Akkara, K. J. Jinesh, and Jose P. Therattil

1 Introduction

Driving is a complex task which needs to handle inputs from road, vehicle, and environment. Along with the above factors, the condition of vehicle and driver himself will be controlling the driver expectations. All these are input which acts as stimuli to a driver. Workload is a measure of the effort required by a driver to complete the task. It is generally observed that very low or very high workload are undesirable. On one hand, it will cause monotonous driving, and on the other hand, it causes fatigue. Eventually, it may result in erroneous driving action that may lead to an accident. The objective of the study is to identify the geometric features that influence workload and to determine the desirable specification of these features for an optimum workload condition.

One of the most sensitive measures for emotional arousal is galvanic skin response or galvanic skin resistance (GSR), also referred to as electrodermal activity (EDA) or skin conductance (SC). Whenever a person is emotionally aroused, his body exhibits physiological changes with increase in sweating. As a result, the electrical conductivity of the skin subtly changes. Due to stimuli, the sweat glands automatically get activated more prominently on forehead, cheek, palm, fingers, and sole of foot. Whenever sweat glands are triggered and become more active, they secrete moisture through pores toward the skin surface. By changing the balance of positive and negative ions in the secreted fluid, electrical current flows more readily, resulting in

A. Jacob (✉)
Government Polytechnic College, Chelakkara, Thrissur, Kerala, India

J. Akkara · K. J. Jinesh · J. P. Therattil
Jyothi Engineering College, Cheruthuruthy, Thrissur, Kerala, India

M. Parida et al. (eds.), *Proceedings of the Fifth International Conference of Transportation Research Group of India* , Lecture Notes in Civil Engineering 220,
https://doi.org/10.1007/978-981-16-9925-2_7

measurable changes in skin conductance. With increase in conductivity, skin resistance decreases. This change in skin conductance is generally termed as galvanic skin response (GSR) and is measured in terms of micro Siemens. GSR offers distinctive insights into physiological and psychological processes of a person [1]. That is why this parameter is considered as one of the measure for describing the workload of a driving person.

Driving is a task demanding activity. Depending on the complexity of the road or experience of the driver, this task will arouse emotions in a driver [2]. The goal of the research work is to provide a highway horizontal curve design guideline which is safe and comfortable from the driver's point of view. To achieve this goal, driver's real-time workload in terms of galvanic skin resistance and their interaction with the geometry is analyzed. The study analyzes workload of car and bus drivers during their drive along two-lane rural highways. The study limits to the workload measurement of car and bus drivers on isolated horizontal curve sections.

2 Literature Review on Driver Workload Based Geometric Design Consistency

Today, research on rural highway safety is focused on providing the users with a set of geometrically consistent roads that meet their expectations. When the geometry satisfies the desires of users, the chances for driving errors are limited. This will promote safety on highways. Assessment of how far a given geometry conforms to user expectations is a measure for evaluating consistency of a highway. Measures like operating speed, vehicle stability, alignment index and driver workload are different tools used for the evaluation [3–8]. Among the tools, driver workload-based evaluation can be considered as the direct assessment of a highway from its very users. A driver driving through a highway is more or less continuously processing visual and kinesthetic information, making decisions, and carrying out control movements. This forms the basis for driver workload method of evaluating highway geometry. Workload has been defined as a measure of the 'effort' expended by a human operator while performing a task, independently of the performance of the task itself [9]. It is the time rate at which drivers must perform a given amount of driving tasks that increase with the increase of the complexity in highway geometric features [10]. Locations with high workload or large positive change in workload were found to be associated with high accident rates [11].

A subjective rating methodology for evaluating and improving design consistency of two-lane rural highways, was established by Messer [10, 12]. Wooldridge et al. [13] applied visual occlusion technique for measuring driver workload. He measured a driver's visual demand on horizontal curves as a percentage of time which a driver observes the roadway. But it is found difficult to capture or observe the change in driver mental workload. Moreover, compared to other consistency measures, this measure is much more complex. Hence, research based on direct mental workload measurement is very few. Some studies are available on galvanic skin response (GSR), a psycho-physiological measure for mental workload [14, 15]. This study is an attempt to assess the mental stress (through galvanic skin resistance) that the geometry imparts on a driver while driving through various curved sections.

3 Data Collection

3.1 Site Selection

Data were collected from 114 horizontal curve sections of various State Highways of Kerala that satisfies the following criteria:

- No intersections near the vicinity of the study curves.
- No physical features such as narrow bridges, schools, factories, or recreational parks that may interfere the smooth flow of the vehicle.
- Not located close to towns or built-up areas that may significantly affect the driver travel patterns on the curves.
- Adequate tangent length with flat gradient is present preceding the study curve.
- Condition of pavement has to be good. As the driver data collection could be completed after the repair work on road sections were over, this criterion could be ensured.

The selected stretches are

(1) Kulappully–Pattambi–Koottupatha–Perumbilavu (SH 39, 23–31 km).
(2) Vazhakode–Pazhayannur (SH 74–20 km).
(3) Ottupara–Pannithadam (SH 50–30 km).

3.2 Geometric Data Collection

Major cross-sectional details of curves were collected using total station survey. Width of shoulder and preceding tangent length were measured using rodometer of 0.01 m accuracy. Availability of sight distance (SD) at each site was determined by conducting field surveys. Geometric survey yielded data regarding radius of curve (R), length of curve (CL), deflection angle (DA), rate of superelevation (SE), preceding tangent length (PTL), width of carriageway (W), and width of shoulder (SW). The summary statistics of geometric data is given in Table 1.

Table 1 Summary statistics of geometric data

	R (m)	CL (m)	DA (deg)	W (m)	SE (%)	PTL (m)	SW (m)	SD (m)
Mean	161.12	46.26	18.04	6.63	2.56	144.64	2.33	30.45
Med	152.15	38.64	15.50	6.81	2.77	115.00	2.1	30.20
Mode	157.92	20.56	11.00	6.43	0.53	90.00	2.1	40.00
Std. dev.	71.07	20.92	11.19	0.73	1.85	74.76	1.58	12.57
Min	33.38	20.56	7.00	4.94	0.03	60.00	0.00	10.80
Max	303.47	94.63	58.00	7.83	7.77	370.00	8.30	57.50

3.3 Driver Workload Data Collection

As part of the workload study, a device, namely Road Driver Data Acquisition System (RDDAS) consisting of tools to measure driver heart rate, galvanic skin resistance, and eye blinking, was developed. A GPS module was attached to the system to get the real-time positioning of the vehicle. The vehicle was equipped with video camera to capture the view of road ahead of the vehicle.

A driver to be tested is equipped with RDDAS (Fig. 1). A pair of 5 V/3.3 V Grove—GSR sensors/electrodes is attached on to middle and fore fingers of the driver (Fig. 2). Data from the sensors are processed through an Arduino Mega 2560 microcontroller board. Arduino is having USB connectivity, and the data is transferred to a laptop. A monitoring tool was designed in Visual Studio.Net 2012 to monitor and store sensor input and GPS data. Back end is an Excel file which shows date, time, latitude, longitude, speed (Km/h), heart rate, and galvanic skin response of the driver. The latitude and longitude of the study curves are identified among the data. GSR values

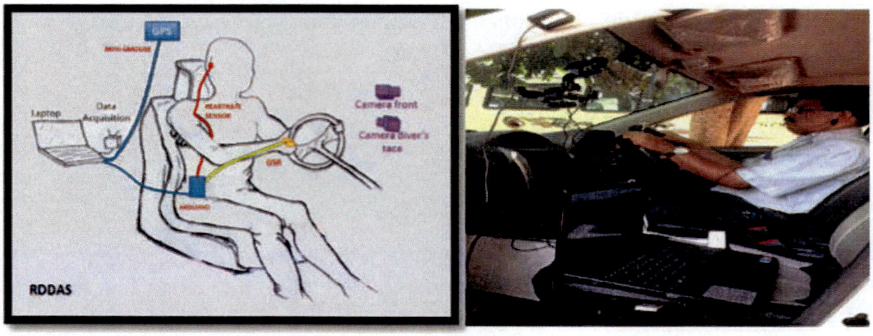

Fig. 1 Driver equipped with RDDAS

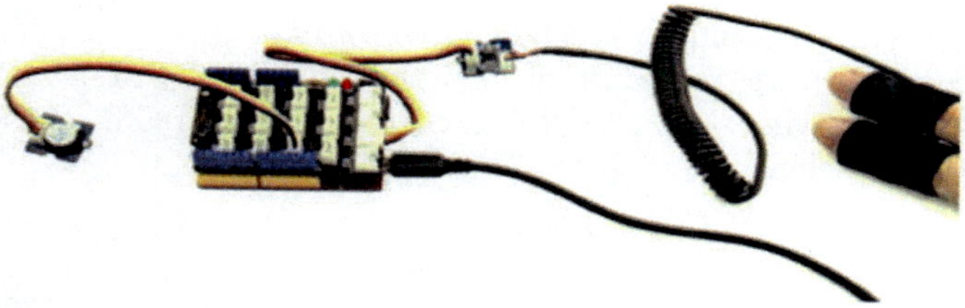

Fig. 2 GSR connected to the fingers

at those locations are recorded in a compact tabular form. In this study, galvanic skin resistance was taken as surrogate measure for driver workload.

The study was done with 30 car and bus drives through each curve section. Altogether 90 drivers took part in the exercise. Age of the drivers varied from to 24 to 60 years and driving experience from 1 to 20 years. Prior to the start of the test drive, each driver underwent laboratory experiments on vision acuity, reaction time, peripheral vision, and depth perception to understand his/her physical characteristics.

Once, the driver and the vehicle are ready with the RDDAS, a ten minute rest period is given during which the base GSR value of the subject is recorded. The base data will act as a control for the GSR measurements and is used to determine the deviation of GSR (DGSR) from the base condition during driving as given in Eq. 1.

$$\text{DGSR} = \text{GSR during driving} - \text{Base GSR} \tag{1}$$

Subsequently, the driver drives along a ten kilometer stretch before reaching the study stretch, so that the driver familiarizes himself with the vehicle and RDDAS. During the drive along the study stretch, GSR data is collected continuously along with GPS coordinates. Continuous recording of video of the road ahead is also carried out.

The data is then retrieved and is verified for any possible errors. Video output from a road facing camera in the RDDAS is used to cross-check the data for any interfering elements during driving. This ensured that only data corresponding to free flow condition of the vehicle are used for analysis. Thus, GSR data of drivers at the identified horizontal curves during free flow driving condition is considered for analysis.

4 Data Analysis

4.1 Preliminary Analysis and Data Exploration

One-way analysis of variance test was conducted seeking answers to the following questions:

1. whether age significantly affects GSR or DGSR of drivers?
2. whether driving experience has any influence on GSR or DGSR while driving?
3. whether there exists significant difference in the GSR or DGSR of a driver by profession and other casual drivers?

K-means clustering was done to group drivers based on their age, experience, occupation, and reaction time. Classification table is given in Table 2.

The results of the ANOVA test are given in Table 3.

Table 2 Classification table for ANOVA

Driver characteristics	Class	Description
Age	Young	< 30
	Middle	30–50
	Old	> 50
Experience	Less	< 5
	Medium	5–20
	High	> 20
Occupation	Driver	–
	Non-driver	–
Reaction time	Fast	< 0.535 s
	Medium	0.535–0.690 s
	Slow	> 0.690 s

Table 3 Results of ANOVA done on car and bus driver characteristics

Workload parameter	GSR			DGSR		
Hypothesis	H_0 (%)	H_1 (%)	Accepted	H_0 (%)	H_1 (%)	Accepted
Car						
Age	60	40	H_0	82	18	H_0
Experience	85	15	H_0	90	10	H_0
Occupation	80	20	H_0	85	15	H_0
Reaction time	100	0	H_0	90	10	H_0
Bus						
Age	100	0	H_0	100	0	H_0
Experience	100	0	H_0	100	0	H_0
Occupation	100	0	H_0	100	0	H_0
Reaction time	100	0	H_0	100	0	H_0

Table 4 Results of ANOVA test conducted on GSR and DGSR of car and bus drivers

Workload measure	*F*-value	*F*-critical	*P*-value	Accepted hypothesis
GSR	235.22	3.915	0.000	H1
DGSR	498.56	3.916	0.000	H1

Table 3 shows the percentage of times H_0 and H_1 which are resulted in the ANOVA test. For example, in the case of car, for 60% of curves studied, H_0 is the ANOVA test result. It means that there is no significant variance in the GSR values among different age groups. Hence, age is not a significantly influencing variable for GSR. Similar tests were conducted for other variables as well. The hypothesis accepted is the one that is accepted more than 50%. From the table, it can be inferred that GSR and DGSR values for car and bus drivers do not vary significantly with age, experience, occupation, and reaction time. Hence, no segmentation of data based on driver characteristics is needed in further analysis.

Another ANOVA test was done to test the following hypothesis:

1. There is no significant difference in the mean GSR values of car and bus drivers.
2. There is no significant difference in the mean DGSR of car and bus drivers.

Results of the test are given in Table 4.

Table 4 shows that the workload measures of car and bus drivers are significantly different. Hence, separate analysis is required for modeling driver workload measures of car and bus drivers.

Further, correlation and scatter plot study was done for the workload measures with geometric variables. Table 5 shows the results of correlation analysis.

Table 5 shows that correlation between driver workload, traffic and geometric variables is having less linearity except for correlation of AVG GSR and sight distance. Later, scatter plot study is done to further understand the data. In many cases, availability of absolute minimum sight distance at curves seems to have the highest linear correlation with the dependent variables. Interestingly, the correlation is found to be negative for car drivers and positive for bus drivers. Figures 3, 4, 5, and 6 show the scatter plots of average GSR and DGSR of car and bus drivers with explanatory variables.

Figure 3g, h show the relationship of shoulder width and sight distance with GSR. Both the trends are negative. With increase in shoulder width or sight distance, skin resistance decreases which imply that workload of car drivers increases. Similar kinds of plots were obtained for DGSR also and are shown in Fig. 4.

Following are the major observations from the scatterplots:

Table 5 Correlation matrix for GSR and DGSR of car and bus drivers

Dependent variable	Traffic volume (ADT)	Radius of curve (m)	Curve length (m)	Deflection angle (deg)	Width of road (m)	Super elevation (m)	Tangent length (m)	Shoulder width (m)	Sight distance (m)
Car									
AVG GSR	0.11	−0.08	−0.27	−0.07	0.18	−0.10	−0.14	−0.30	−0.71*
AVG DGSR	0.14	−0.09	−0.32*	0.01	0.31*	−0.07	−0.10	−0.27	−0.54*
Bus									
AVG GSR	−0.19	0.12	0.23	0.02	−0.11	−0.23	0.10	0.17	0.35*
AVG DGSR	0.07	0.12	0.13	−0.02	0.04	−0.03	−0.01	−0.17	0.24*

The * shows the correlation coefficient at 5 percentage significance level

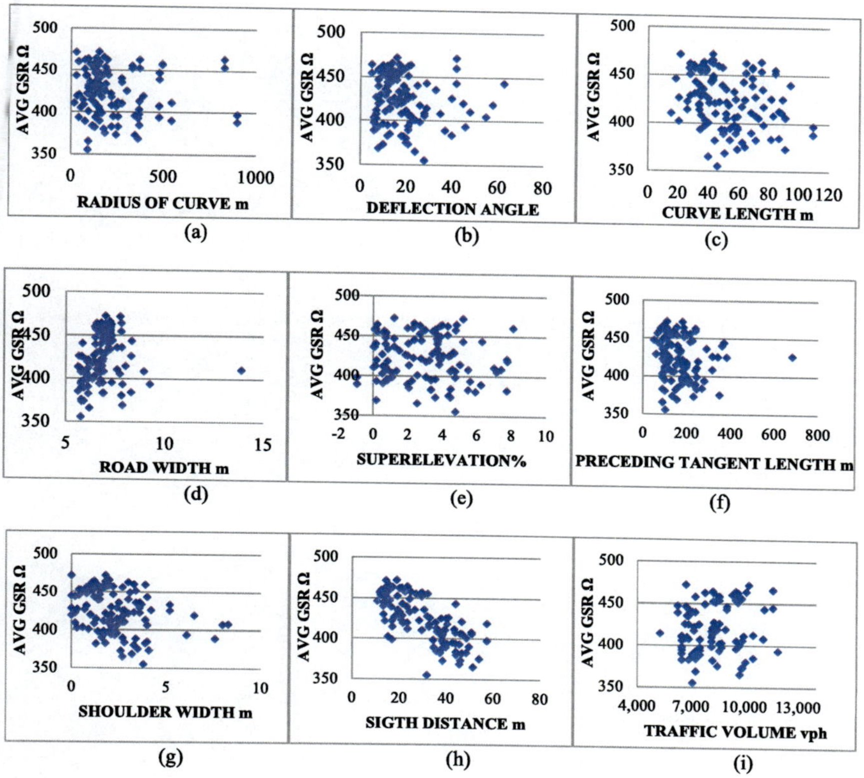

Fig. 3 Variation of average GSR with geometry and traffic volume—Car

1. Heteroscadasticity of workload measures decreases or stabilizes when radius of curve becomes flatter.
2. The average GSR of bus drivers increases with increase in curve length and shoulder width, reaches a maximum, and then decreases. It shows that workload decreases, reaches a minimum, and then increases. A curve length of 60 m and shoulder width of 2.5 m are optimum for minimizing the driver workload of bus drivers.
3. GSR of car drivers is having negative trend with sight distance, whereas for bus drivers, the trend is positive.

4.2 Workload Analysis and Modeling

The dispersion in the data is reduced by adopting interval estimate of workload data. Normality test was done to verify the feasibility of regression analysis on the data. Regression analysis was then done with various geometric and traffic variable as

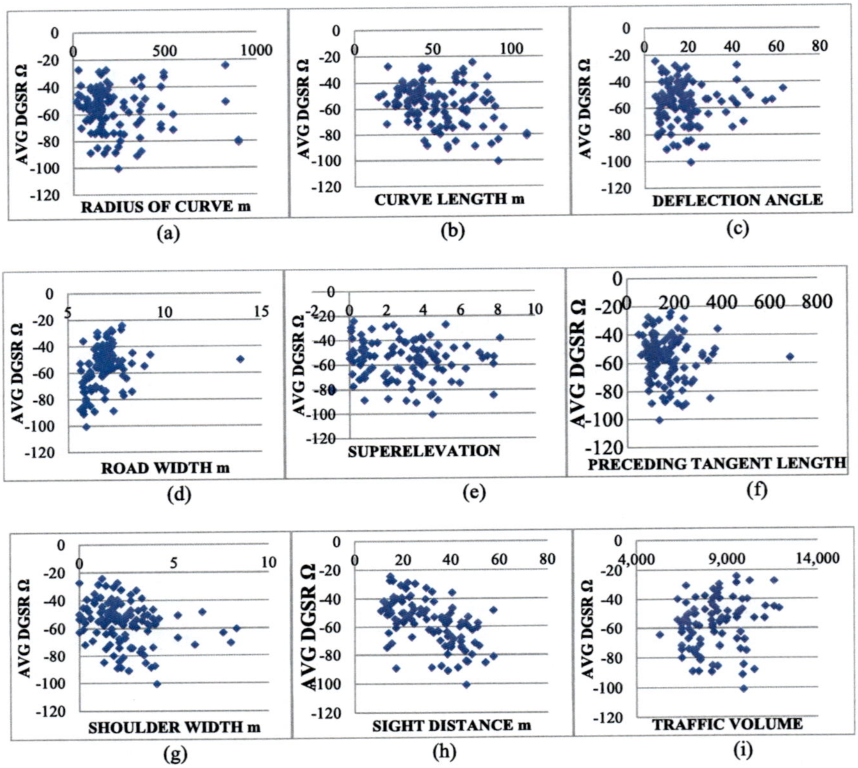

Fig. 4 Variation of average DGSR with geometry and traffic volume—Car

influencing variables. Models with sight distance as the independent variable were found to be giving the best coefficient of determination and minimum root mean square error.

The results of the normality test and the best regression models are given in Table 6. Table 6 summarizes the skewness and kurtosis of the geometric data. They are the measures for checking the normality of distribution. Skewness is a measure of the asymmetry of the probability distribution of a random variable about its mean.

If skewness is less than −1 or greater than 1, the distribution is highly skewed.

If skewness is between −1 and −0.5 or between 0.5 and 1, the distribution is moderately skewed.

If skewness is between −0.5 and 0.5, the distribution is approximately symmetric or normal.

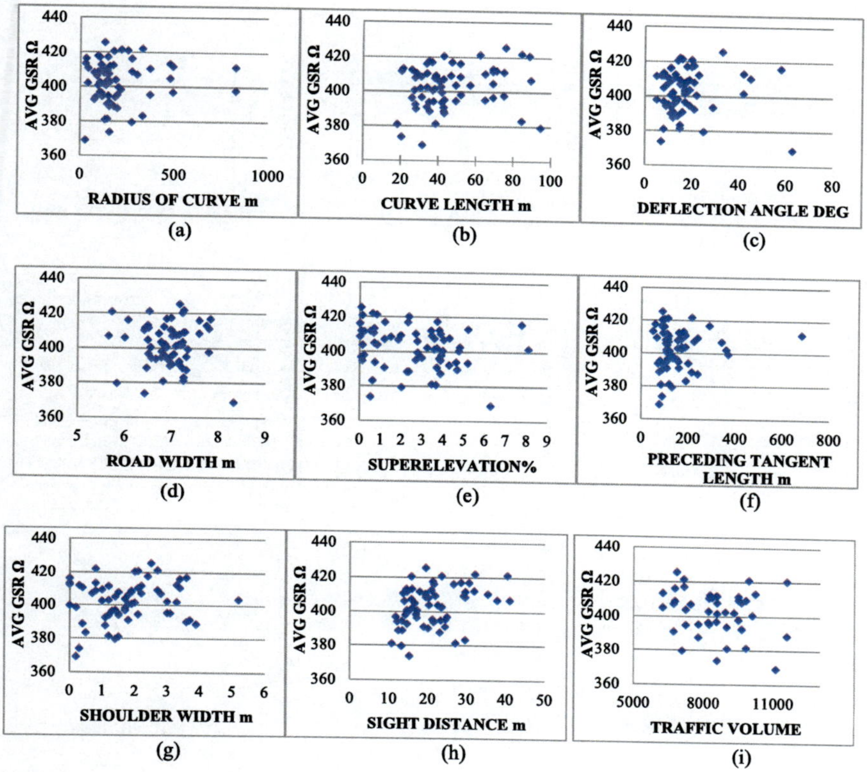

Fig. 5 Variation of average GSR with geometry and traffic volume—bus

Kurtosis gives an idea of the height and sharpness of the central peak, relative to that of a standard normal curve. A value of 3 is the perfect normal curve. A range of + or − 2 is accepted as approximately normal [16]. In the study for all cases, the normality test is satisfied. Hence, the regression analysis that was conducted holds good. Table 6 shows the best models that could be developed for GSR and DGSR along with model statistics.

The models are graphically represented in Figs. 7 and 8.

GSR decreases with increase in sight distance for car. This implies that skin conductivity increases with sight distance for a car driver. But skin conductivity varies with sweating or workload of drivers. Thus, for a car, workload increases with increase in sight distance. The effect of temperature within car could be controlled during the driving exercise. A limitation of the study was that the temperature within the bus could not be controlled during the test. This may be the reason for unexpected trend in the case of bus. Similar trend is observed for DGSR also, as shown in Fig. 8.

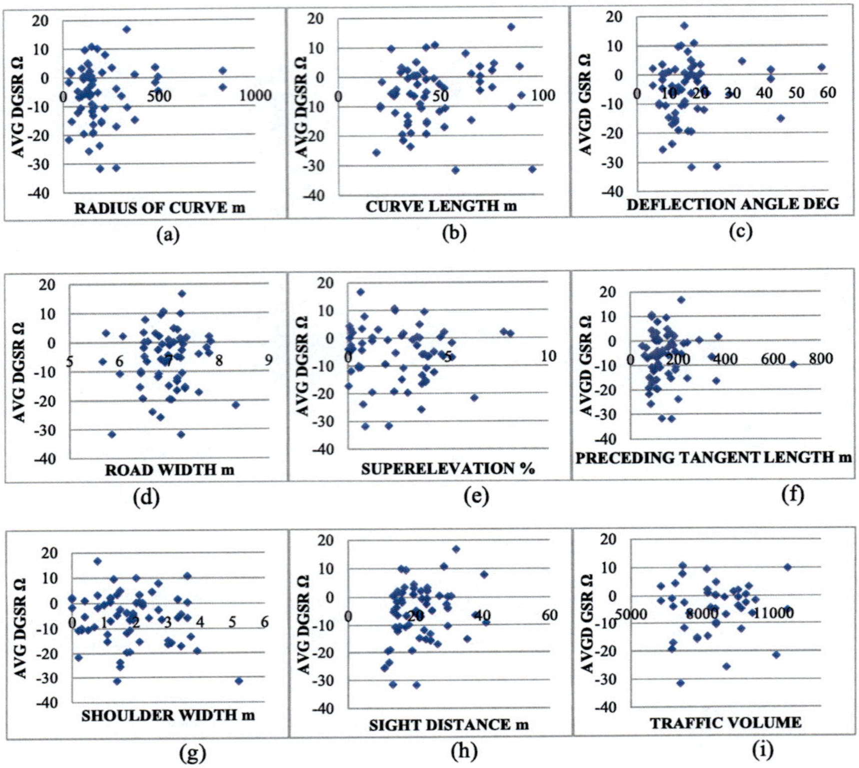

Fig. 6 Variation of average DGSR with geometry and traffic volume—bus

5 Results and Discussions

Galvanic skin resistance of a driving person is a direct indicator of one's workload. Study could reveal the effect of geometry of horizontal curves on driver workload. Sight distance and shoulder width are the two significant variables that had linear correlation with workload of car drivers. This study comes up with the following results.

- Physical characteristics such as vision acuity, peripheral vision, and depth perception do not have any effect on GSR values. A logical explanation for the result can be given as that the driver naturally adapts his driving behavior to his physical constraints. Hence, these parameters do not have much effect on GSR.
- There is significant difference in the GSR values of a car and a bus driver.
- Sight distance is the most significantly influencing variable for bus and car driving.
- GSR decreases with increase in sight distance for car drivers. This indicates that workload increases with increase in sight distance. With more visual input to a

Table 6 GSR and DGSR models for car and bus driver workload

Workload measure	Model	Value of X	R^2	RMSE	Skewness of residuals	Kurtosis of residuals
CAR						
GSR	464.7–16.9X	$X = 1$ for SD < 20 $X = 2$ for 20 < SD < 30 $X = 3$ for 30 < SD < 40 $X = 4$ for 40 < SD < 50	0.87	19.99	0.20	0.06
DGSR	−37.69–7.526X	$X = 1$ for SD < 20 $X = 2$ for 20 < SD < 30 $X = 3$ for 30 < SD < 40 $X = 4$ for 40 < SD < 50	0.85	13.07	0.38	0.59
BUS						
GSR	394.6 + 5.15X	$X = 1$ for SD < 20 $X = 2$ for 20 < SD < 30 $X = 3$ for 30 < SD < 40 $X = 4$ for 40 < SD < 50	0.87	9.38	0.42	−0.02
DGSR	−9.04 + 2.224X	$X = 1$ for SD < 20 $X = 2$ for 20 < SD < 30 $X = 3$ for 30 < SD < 40 $X = 4$ for 40 < SD < 50	0.85	9.68	0.46	−0.01

driver, the mental thought process increases thereby increasing the brain activity. This increases the workload or reduces the skin resistance.

- From the scatter plot observations, it was found that for a minimum workload, curve length and shoulder width are to be 60 and 2.5 m, respectively.

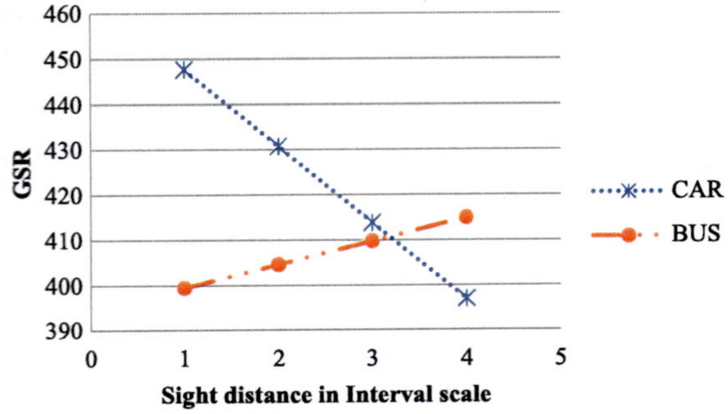

Fig. 7 Variation of GSR with sight distance

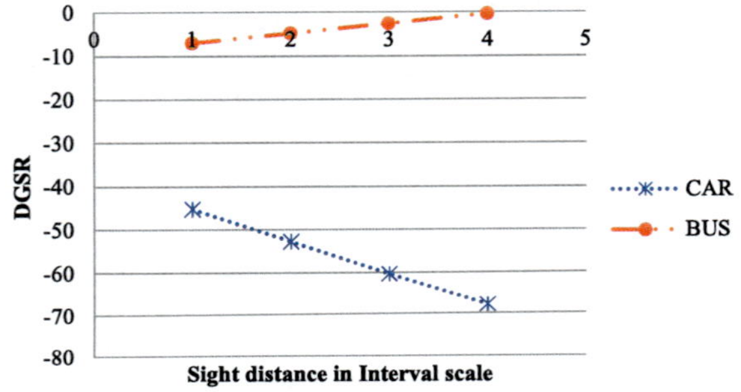

Fig. 8 Variation of DGSR with sight distance

Acknowledgements Authors acknowledge Kerala State Council for Science, Technology and Environment, Thiruvananthapuram, Kerala, for funding this project. The manpower and financial support given by faculty and management of Jyothi Engineering College, Cheruthuruthy, Thrissur, is greatly acknowledged.

References

1. https://imotions.com/wp-content/uploads/Guides/iMotions_Guide_GSR_2015.pdf
2. Paxion J, Galy E, Berthelon C (2014) Mental workload and driving. Front Psychol 5:1344
3. Lamm R, Choueiri EM, Mailaender T (1990) Comparison of operating speed on dry and wet pavement of two lane rural highways. Transportation Research Board, Washington, D.C., pp 199–207
4. Fitzpatrick K, Wooldridge MD, Tsimhoni O, Collins JM, Green P, Bauer KM, Parma KD, Koppa R, Harwood DW, Krammes RA, Poggioli B (2000) Alternative design consistency rating methods for two-lane rural highways. Federal Highway of Administration, FHWA-RD-99-172. U.S. Department of Transportation, Washington D.C.
5. Essam Mohamed SA, Dabbour EA (2003) Truck stability on different types of horizontal curves combined with vertical alignments. A thesis Presented to Ryerson University, Toronto
6. Misaghi P, Hassan Y (2005) Modeling operating speed and speed differential on two-lane rural roads. J Transp Eng 131(6):408–417

7. Anderson IB, Bauer KM, Douglas WH, Fitzpatrick K (1999) Relationship to safety on geometric design consistency measures for rural two-lane highways. Transportation Research Record, Washington, D.C., Paper No. 99-0921
8. Al-Masaeid HR, Hamed M, Aboul-Ela M, Ghannam AG (1995) Consistency of horizontal alignment for different vehicle classes. Transportation Research Record 1500, Transportation Research Board, Washington, DC, pp 178–183
9. Senders JW (1970) Estimation of operator workload in complex systems. In: DeGreen KB (ed) Systems psychology. McGraw Hill, New York
10. Messer CJ (1980) Methodology for evaluating geometric design consistency. Facility Design and Operational Effects. Transportation Research Record, 757, Transportation Research Board, National Research Council, Washington D.C., pp 7–14
11. Wooldridge MD (1994) Design consistency and driver error. Transportation Research Record, 1445, Transportation Research Board, Washington, DC, pp 148–155
12. Messer CJ, Mounce JM, Brackett RQ (1981) Highway geometric design consistency related to driver expectancy. FHWA-RD-81-035. Federal Highway Administration, Washington, DC
13. Wooldridge MD, Fitzpatrick K, Koppa R, Bauer K (2000) Effects of horizontal curvature on driver visual demand. Transp Res Rec 1737:71–77
14. Wu DY, Chu F, Zhang Q, Hsu DF (2013) Sensor feature selection and combination for stress identification using combinatorial fusion. Int J Adv Rob Syst 10:306
15. Singh M, Queyam AB (2013) A novel method of stress detection using physiological measurements of automobile drivers. Int J Electron Eng 5(2):13–20
16. Gravetter F, Wallnau L (2014) Essentials of statistics for the behavioral sciences, 8th edn. Wadsworth, Belmont

Structural Equation Model for Sustainable Transport System Performance Enhancement

Rupali Zope, N. Vasudevan, Shriniwas Arkatkar, and Gaurang Joshi

1 Introduction

Transportation is forms an integral part of human life. It has become important for national, regional, and local economy. It plays crucial role in booting up common user life through facilities and accessibility to them. However, the growing economy of India in recent years has contributed in terms of increase in vehicle population. It further led to different issue like congestion, pollution, fuel consumption, accidents, etc. India is aiming for rapid economic growth. The role of secondary and tertiary sector would be more prominent for such growth. The involvement of these sectors in urban areas is a crucial factor for economic growth. The transport sector is a major contributor to the GDP also. In 2012–2013, the sector contributed about 5.2% to the nation's GDP, with road transportation having a major share of it [1]. Census 2011 reveals growth of 2774 towns comprising 242 Statutory and 2532 Census towns over the decade. Growth rate of population in urban areas was 31.8% [2]. This resulted into immense pressure on urban infrastructure, particularly more due to diverse issues already prevailing in Transport domain. The issues have become more complex because of rapid growth of private vehicles; with no improvement in city bus service over the years. India has experienced a tremendous increase in the total number of registered motor vehicles as shown in Fig. 1. From March 31, 1951 to March 31, 2012, exponential growth of 0.3 to 159.5 million is observed in total number of registered vehicles in India. The total registered vehicles in the country grew at a Compound Annual Growth Rate (CAGR) of 10.5% between 2002 and 2012. It may be noted that there is a rampant growth in registered motor vehicles and among total number of registered vehicles, growth of motorized two-wheelers is predominantly high, indicating flawed transport system.

R. Zope (✉) · N. Vasudevan · S. Arkatkar · G. Joshi
Transportation Engineering and Planning Section, Department of Civil Engineering, SVNIT, Surat, Gujarat 395007, India

M. Parida et al. (eds.), *Proceedings of the Fifth International Conference of Transportation Research Group of India* , Lecture Notes in Civil Engineering 220,
https://doi.org/10.1007/978-981-16-9925-2_8

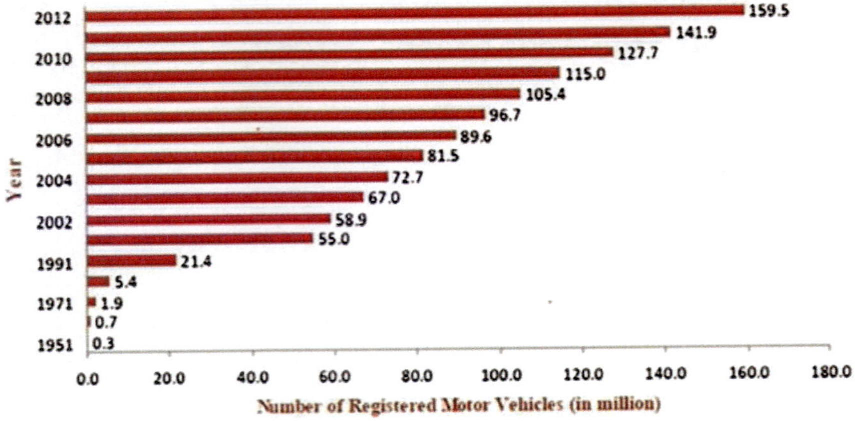

Fig. 1 Number of registered motor vehicles: 1951–2012, *Source* MORTH, 2012

All these factors have contributed in increase in travel demand. The increased use of private vehicles has also resulted into various issues like road congestion, air pollution, road fatalities, and social issues of equity and security.

Therefore, sustainability should be seen as answer to these issues. Sustainable development is defined as "development that meets the needs of the current generation without undermining the ability of future generation to meet their own needs" [3]. The entire concept of sustainability works on three pillars: economic, environmental, and social. So, studies need to incorporate sustainability into transport planning [4]. Inclusion of these pillars involves different variables of different scales, nature (qualitative and quantitative). Interaction of these variables makes the analysis process more complex. There are different methods used to assess these complexities of variables and their cumulative impact on transport system. Apart from cumulative impact, variable also have causal effect on transport system. So, if we compare with traditional evaluation methods, there are different aspects which need to address like (i) identification of measurable exogenous and endogenous variables of transport system (ii) a single model assessment framework for causal effects of variables (iii) minimization of potential errors in measurement process. SEM is one such approach, which help to address all above listed aspects. SEM is a statistical method widely used in scientific studies in recent years. The most important reason for the wide use of SEM is that is it helps to measure the direct and indirect relationships among observed and unobserved variables with a single model [5]. It also facilitates the accounting of measurement errors and the relationships between errors in the observed variables. Therefore, with scope of evaluating the causal impact of exogenous and endogenous variables of transport on sustainability, a second order SEM model is developed. For the second order model developed, hypotheses are formed which represents the hypothesis that these seemingly distinct, but related constructs can be accounted by one or more common underlying higher order constructs. A second order confirmatory factor analysis (CFA) is performed to validate model. The model is tested for their goodness-of-fit criteria, p-value, relative chi-square, goodness-of-fit index (GFI), and root mean square error of approximation and is found satisfactory. Regression weights obtained from SEM analysis provide a clear understanding of effect of users' acceptance and expectation of transport system on performance improvement. Results obtained from the study highlights significant variables for better performance of transport system from sustainability point of view. Paper consists of five sections. First section is all about introduction while second section covers the earlier research and studies done in the same area through literature review. Third section of the paper explains SEM developed for sustainable transport system evaluation and its assessment. Fourth section briefs about result and conclusion of the study.

2 Literature Review

Various researchers and organizations have explored application of different analytical approaches. The complexities, interdependencies, nonlinearities, vagueness, and incomplete information associated with data collected for indicators has been studied effectively with the use of Multi-Criteria Decision methods (MCDM) by many researchers. Different MCDM methods like Fuzzy, AHP, ANN, etc. are widely used for evaluation sustainable transport index. Other than the use MCDM methods, other analytical approaches are also used in analysis of sustainable transport system like composite index, Multi-view Black box (MVBB) framework, multiattribute utility theory, etc. Different research efforts have been devoted in the past to evaluate the effectiveness of sustainable transport policies. They could be classified into few approaches [6] such as life cycle analysis [7], cost–benefit analysis and cost-effectiveness analysis [8], environmental impact assessment [8, 9], optimization models [10], system dynamic models [11], assessment indicator models [12, 13], data analysis [14], and multi-criteria decision analysis methods [15]. These approaches are particularly useful in evaluating the impacts of physical policies. They are also useful to identify the areas that are required for further improvements of a sustainable transport system. However, the effect of exogenous variables on endogenous variables and on other endogenous variables of transport system is an important aspect of research. So, there is a need of an approach which will help to (i) capture the causal influences of the exogenous variables on the endogenous variables (ii) causal influence of endogenous variable upon one another (iii) measurement of direct and total effect of variables on transport system.

Structural equation modeling (SEM) is one such technique that can handle a large number of endogenous and exogenous variables, as well as latent (unobserved) variable specified as linear combinations (weighted averages) of the observed variables. SEM was adopted in several fields of research and generalized by Joreskog and Wiley [16, 17]. SEM was applied in Psychology and Social Science, Natural Science, and especially in the field of Economy and Statistics. SEM is founded on three primary analytical developments: (1) path analysis, (2) latent variable modeling, and (3) general covariance estimation methods. While presenting literature about SEM, major focus is kept on use of SEM in transport and travel behavior system [18]. In the field of transportation research some applications of SEM were proposed to analyze land-use and transport interactions [19–21]. Also in public transport some authors proposed SEM applications [22–24]. More specifically, SEM was also adopted for investigating on customer satisfaction on public transport services [25–29]. SEM is also widely used in travel behavior studies. The earliest known applications of SEM to travel behavior are a joint model of vehicle ownership and usage [30] and a dynamic model of mode choice and attitudes [31, 32]. The dependence of SEM on the homogeneity of a causal travel behavior process across the population of interest was also investigated [33]. Results are presented from models estimated on simulated data generated from competing causal structures. These estimates are shown to perform poorly in the presence of structural heterogeneity. Similarly, models are developed for trip chain generation. As these models encompass activity duration in addition to conventional travel measures of trip generation and travel time [34–36]. So, the effectiveness of SEM can be better understood with the widespread of it in various studies. In the current study, an attempt is to evaluate the transport system sustainability and role of user's perception in its performance enhancement. A second order SEM model is developed for the same as detailed explained in next section.

3 Structural Equation Modeling for Sustainable Transport Planning

There is growing interest in the concepts of sustainability, sustainable development, and sustainable transportation. Sustainability is generally evaluated using various indicators, which are specific variables suitable for quantification (measurement). Such indicators are useful for establishing baselines, identifying trends, predicting problems, assessing options, setting performance targets, and evaluating a particular jurisdiction or organization. Various studies done on indicator selection for urban transport system are referred for deciding indicators used for analysis [37, 38]. Set of indicators selected for the study and use of it for model developed are discussed in next section.

3.1 Model Development

For the current study, a second order SEM is developed. The second order mode indicates apparently dissimilar hypothesis. The mode also represents constructs that are formed with one or more common primary higher order constructs. The model is developed in two parts:

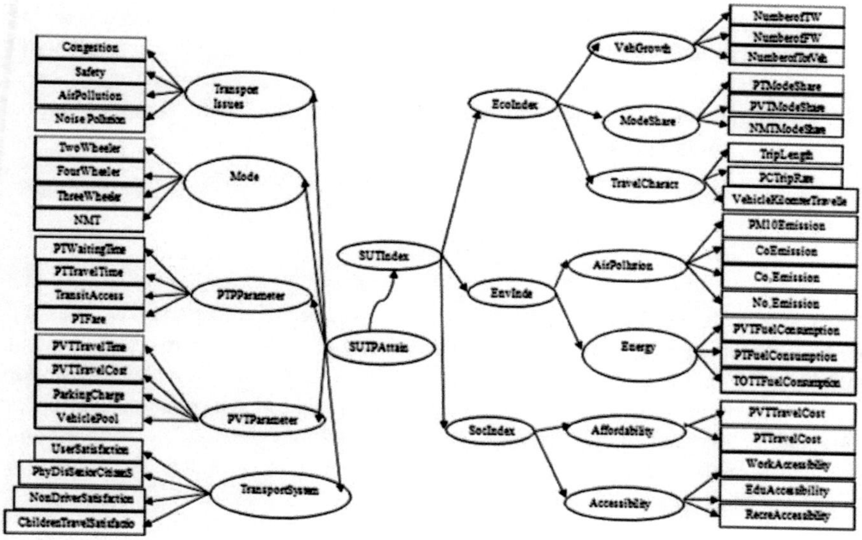

Fig. 2 Conceptual SEM model for sustainable transport system

Part I- Structural Equation Model for Sustainable Urban Transport Index (SUTIndex).

Part II- Structural Equation Model for Sustainable Urban Transport Attainment (SUTAttain).

Variables obtained from PCA analysis are used in the development of model. Figure 2 shows the SEM model developed for Sustainable Transport System. The effect of different variables on sustainability dimensions and sustainable transport index is examined using the SEM method. For the same, the maximum likelihood (ML) approach is used. The ML approach provides unbiased, efficient, and consistent estimates (Kline 2011). A combination of confirmatory factor analysis and path analysis is used to investigate the relationship between observed variables and latent variables. Five point Likert scale is used to measure the observed variables. A model is constructed to measure the latent variables. Latent variables in the study are comprised of two types: (i) endogenous latent variables (ii) exogenous latent variables. Endogenous latent variables are influenced by exogenous variables and exogenous variables are independent variables and they cause fluctuations in endogenous variables. The SEM model considered for the study is comprised of two components: (i) measurement model (ii) a structural model. The measurement of latent variables using relevant observed variables is studied through measurement model. For the investigation of relationship between latent and observed variables, structural model is used. The structural model uses regression analysis and path analysis for the same. "The measurement models for endogenous and exogenous latent variables can be written as follows (Bollen 1987):

$$x = \lambda_x \xi + \delta$$
$$y = \lambda_y \eta + \varepsilon$$

where x = vector of the exogenous variables; λ_x = coefficient matrix that links the observed variables and latent exogenous variables; ξ = vector of the exogenous latent variables; δ = vector of the measurement errors in the exogenous variables; y = vector of the endogenous variables; λ_y = coefficient matrix that links the observed variables and latent endogenous variables; η = vector of the endogenous latent variables; and ε = vector of the measurement errors in the endogenous variables. The structural model can be expressed as follows (Bollen 1987):

$$\eta = \beta\eta + \Gamma\xi + \zeta$$

where η = vector of the endogenous latent variables; ξ = vector of the exogenous latent variables; ζ = vector of the latent errors in the equations; β = coefficient matrix that maps the endogenous latent variables to the endogenous latent variables; and Γ = coefficient matrix that maps the exogenous latent variables to the endogenous latent variables". In the current paper, hypothesis testing is used to study causal effect of variables. It is also used to revels dependency of variables. Hypothesis is an assumption made about a phenomenon and used to test the relationships between independent and dependent variable. The probability statement about population parameters can be tested using hypothesis testing. It is also useful for generalization of hypothesis as it helps to analysis sample data fit. It will also help to build a theory for expressing the causal relation, individual and cumulative effect. Using all these advantages of hypothesis testing, sixteen hypotheses are formed. For hypotheses formulation, a conceptual model with latent and direct variables is formed. The conceptual model of sustainable transport index measurement consists of ten latent variables: EcoIndex, SocIndex, EnvIndex, VehGrowth, ModeShare, TravelCharact, Air Pollution, Energy Consumption, Affordability, and Accessibility.

SUTAttain consist of five latent variables: Transport Issues, mode Preference, PTPParameters, PVTParameters, Transport system satisfaction. Figure 2 shows SEM conceptual model developed.

H1: Performance Improvement parameters will help to improve SUTIndex

H2: Higher SUTIndex considers higher Economic Index

H3: Higher SUTIndex considers higher Social Index

H4: Higher SUTIndex considers higher Environmental index

H5: Higher Economical Index considers lower vehicle growth

H6: Higher Economical Index considers better mode share

H7: Higher Economical Index considers better travel characteristics

H8: Higher Environmental Index considers less Air Pollution.

H9: Higher Environmental Index considers less Energy Consumption

H10: Higher Social Index considers better Affordability

H11: Higher Social Index considers better Accessibility

H12: High degree of SUTAttain consider less transport issues

H13: High degree of SUTAttain consider better use of sustainable modes

H14: High degree of SUTAttain consider better PTP performance parameters

H15: High degree of SUTAttain consider better PVT performance parameters

H16: High degree of SUTAttain consider higher transport system satisfaction of all users

For all above listed hypotheses, data collection for calculated sample size is done and explained in next section.

3.2 Data Collection

The sample size is an important factor in the sample design. It affects the precision, cost, and duration of the survey more than other factors (like the method of survey, duration, etc.). Three factors significantly affect the determination of sample size, variability of a population, degree of precision, and population size. The sample size is calculated using Eq. 1 (Krejcie and Morgan, 1970) as given below:

$$S = \frac{x^2 NP(P-1)}{d^2(N-1) + x^2 P(P-1)} \tag{1}$$

where S = required sample size; χ^2 = Table value of chi-square (3.83); N = population size; d = degree of accuracy (0.05). Therefore, for the need of 980 samples, data for 1080 samples is collected. All the responses are collected randomly. All the responses received and collected data are converted into Likert scale 1–5. The data reliability is verified using Cronbach's alpha, which returned values of 0.76 and 0.89 for SUTIndex and SUTAttain. Cronbach's alpha value is used to assess the reliability or internal consistency. The obtained are 0.76 for SUTIndex data and 0.89

Table 1 Socio-economic characteristics of respondents

Sr. No	Characteristics	Category	Percentage
1	Age	0 to 18	8.89
		18 to 35	38.56
		35 to 60	41.89
		>60	10.66
2	Gender	Male	81.67
		Female	18.33
3	Occupation	Govt. Job	5.00
		Semi-Govt. Job	9.00
		Private service	30.00
		Private business	18
		Education	21
		House-wife	17

for SUTAttain revealing the strength of data consistency. Socio-economic details of respondents are listed in Table 1.

With this set of data and developed conceptual SEM model, analysis is done. Results obtained are discussed in next section.

4 Results

The very first step of analysis is the verification of conceptual model. It is done using IBM SPSS Amos (version 24) software using confirmatory factor analysis (CFA). Validation of model is done for the collected data. The conceptual model is tested and modified. The process is continued till the goodness-of-fit criteria met the recommend index. Finally obtained p-value, relative chi-square ($\chi^2 = df$), goodness-of-fit index (GFI), root mean square error of approximation (RMSEA) values are listed in Table 2. Values obtained indicate that the measurement model fit the data well.

Next step of the analysis is to validate the model by testing above formulated hypotheses H1-H16. First hypothesized relation is depicting the relation between SUTIndex enhancements through SUTAttain. The coefficient obtained for the relation is 0.69 with R^2 value of 0.53. In addition, p-value is observed as 0.088, which is in the range of $0.05 < p \leq 1.00$. It indicates that hypothesis H1 is statistically significant and 53% variance in SUTAttain will help to improve SUTIndex. Therefore, the hypothesis that SUTAttain have direct effect on improvement of SUTIndex is accepted. Results for rest of hypotheses are listed in Table 3. All hypotheses considered for the study are found statistically significant and been accepted for the current study. Thus, obtained squared multiple correlation shows relative variance of values to enhance SUTAttain for better SUTIndex.

Table 2 Goodness-of-fit criteria and indices

Sr No	Goodness-of-fit criteria	Recommended index values	SUTIndex	SUTP performance improvement	SEM model
1	*p*-Value	$0.05 < p \leq 1.00$	0.0056	0.078	0.261
2	Relative chi-square ($\chi^2 = df$)	$0 \leq \chi^2 = \text{df} \leq 2$	0.98	0.86	1.423
3	Goodness-of-fit index (GFI)	$0.09 \leq \text{GFI} \leq 1.00$	1.102	0.67	0.716
4	Root mean square error of approximation (RMSEA)	$0 \leq \text{RMSEA} \leq 0.08$	0.017	0.022	0.010

Thus, the results of the study help to understand existing sustainable transport index and use of user's perception for performance improvement.

5 Conclusion

From last few decades, planning of sustainable transport has become one of the major concerns of transport planning and policy. It is considered as one of the major contributing factor toward sustainable development. It involves the assessment of transport system and its economic, environmental, and social impact on urban areas. Assessment process helps in understanding the existing behavior. It also helps in finding areas of improvement. There are various methods of assessment. These methods facilitate to understand the relationships between variables of transport system. But the complexity and dynamism has causal effect and it becomes crucial to analyze them accordingly. SEM is a data driven multivariate statistical modeling technique facilitating factor analysis of data to evaluate the validity of test and addresses multicollinearity of variables. In the current study, a second order confirmatory factor analysis (CFA) is performed to validate model. The model is tested for their goodness-of-fit criteria, p-value, relative chi-square, goodness-of-fit index (GFI), and root mean square error of approximation and is found satisfactory. A second order confirmatory factor analysis (CFA) is performed to validate model. The model is tested for their goodness-of-fit criteria, p-value, relative chi-square, goodness-of-fit index (GFI), and root mean square error of approximation and is found satisfactory. SEM model analysis is performed and generates regression weights for selected latent variables. From the results obtained for developed SEM model following conclusions can be drawn:

1. The final SEM model for SUTIndex revealed that out three latent variables EcoIndex, EnvIndex and SocIndex, EcoIndex is observed with lowest coefficient value as 0.51. Increase in private vehicular growth in Pune city and lower mode share of public transport and non-motorized transport affect negatively for

Table 3 Results of the hypotheses testing

Sr No	Hypotheses	Regression weight	Squared Multiple correlation (R^2)	Remark
1	H1: SUTAttain have direct effect on improvement of SUTIndex	0.69	0.53	Statistically significant
2	H2: Higher SUTIndex considers higher economic index	0.54	0.78	Statistically significant
3	H3: Higher SUTIndex considers higher social index	0.65	0.81	Statistically significant
4	H4: Higher SUTIndex considers higher environmental index	0.60	0.67	Statistically significant
5	H5: Higher economical index considers lower vehicle growth	0.51	0.69	Statistically significant
6	H6: Higher economical index considers better mode share	0.54	0.74	Statistically significant
7	H7: Higher economical index considers better travel characteristics	0.58	0.67	Statistically significant
8	H8: Higher environmental index considers less air pollution	0.55	0.71	Statistically significant
9	H9: Higher environmental index considers less energy consumption	0.65	0.64	Statistically significant
10	H10: Higher social index considers better affordability	0.60	0.76	Statistically significant
11	H11: Higher social index considers better accessibility	0.70	0.69	Statistically significant
12	H12: High degree of SUTAttain consider less transport issues	0.84	0.84	Statistically significant
13	H13: High degree of SUTAttain consider better use of sustainable modes	0.68	0.73	Statistically significant

(continued)

Table 3 (continued)

Sr No	Hypotheses	Regression weight	Squared Multiple correlation (R^2)	Remark
14	H14: High degree of SUTAttain consider better PTP performance parameters	0.75	0.79	Statistically significant
15	H15:High degree of SUTAttain consider better PVT performance parameters	0.58	0.63	Statistically significant
16	H16: High degree of SUTAttain consider higher transport system satisfaction of all users	0.66	0.59	Statistically significant

SUTIndex. Whereas SocIndex is found as maximum out of, three index values. Better performance of affordability and accessibility helps to achieve a better coefficient value.

2. Latent variables for EnvIndex with lower coefficient values indicate higher air pollution and energy consumption further reducing performance of transport system on environmental parameters. As per user's perception, in case of SUTAttain all variables are observed with coefficient more than 0.5 reflecting higher influence of variables of performance enhancement.

The regression weight found between SUTIndex and SUTPAttain proved that user's perception for improvement of transport system directly help to improve index value. SUTIndex has strong relation with social and environmental variables of transport system. Better performance of social and environmental factors will definitely help to raise the transport system sustainability. Regression weights obtained under social and environmental dimension revels that better accessibility would help to enhance the performance. Thus, the study facilitates (i) emphasis on the ranking of significant criteria (ii) identification of the low performers (iii) select appropriate strategies for improving performance. Considering all this, SEM will definitely help to ensure a methodology for sustainability of transport system and its performance improvement.

References

1. US Department of Energy (2014) Sensors and controls department of energy. Retrieved 22 May 2014, Available at: http://energy.gov/eere/buildings/sensors-controls
2. Ministry of Urban Development (2011) National mission on sustainable habitat policy. Rural urban distribution of population, Provisional Population Totals –Data Highlights (Urban). Retrieved from http://moud.gov.in

3. World Commission on Environment and Development (1987) Report of the world commission on environment and development: *our common future*. UN Documents: Gathering a Body of Global Agreements
4. Marsden G, Kimble M, Nellthorp J, Kelly C (2010) Sustainability assessment: the definition deficit. Int J Sustain Transp 4(4):189–211
5. Meydan CH, Şen H (2011) Structural equation modeling AMOS applications. Detail Publishing, Ankara
6. Awasthi A, Chauhan SS, Omrani O (2011) Application of fuzzy TOPSIS in evaluating sustainable transportation systems. Expert Syst Appl 38:12270–12280
7. Goedkoop MJ, Spriensma R (2000) The Eco-indicator 99: A damage oriented method for life cycle impact assessment. Methodology Report, 2nd ed. 82–132 (2000)
8. El-Diraby TE, Abdulhai B, Pramod KC (2005) The application of knowledge management to support the sustainable analysis of urban transportation infrastructure. Can J Civ Eng 32:58–71
9. Bond R, Curran J, Kirkpatrick C, Lee N (2001) Integrated impact assessment for sustainable development: a case study approach. World Dev 29(6):1011–1024
10. Wood C (2002) Environmental impact assessment: a comparative review, 2nd edn. Prentice-Hall, UK
11. Zuidgeest MHP (2005) Sustainable urban transport development: A dynamic optimization approach. Ph.D. thesis, University of Twenty
12. Tao CC, Hung CC (2003) A comparative approach of the quantitative models for sustainable transportation. J Eastern Asia Soc Transp Stud 5:3329–3344
13. Browne D, Regan B, Moles R (2008) Use of ecological foot printing to explore alternative policy scenarios in an Irish city region. Transp Res Part D 13(5):315–322
14. Ulengin F, Kabak O, Onsel S, Ulengin B, Aktas E (2010) A problem-structuring model for analyzing transportation—environment relationships. Eur J Oper Res 200(3):844–859
15. Yedla S, Shrestha PM (2003) Multi-criteria approach for the selection of alternative options for environmentally sustainable transport system in Delphi. Transp Res Part A 8:717–729
16. Jöreskog KG (1973) A general method for estimating a linear structural equation system. In: Goldberger AS, Duncan OD (eds) Structural equations (1973)
17. Wiley DE (1973) The identification problem for structural equations with unmeasured variables. In: Goldberger AS, Duncan OD (eds) Structural equation models in the social sciences. Academic Press, New York, pp 69–83
18. Bollen KA (1989) Structural equations with latent variables. Wiley, New York
19. Tschopp M, Axhausen KW (2007) Transport infrastructure and spatial development in Switzerland between 1950 and 2000. In: Proceedings of 86th annual meeting of the transportation research board, Jan 2007, Washington, D.C.
20. Van Acker V., Witlox F, Van Wee B (2007) The effects of the land use system on travel behavior: a structural equation modeling approach. Transp Plann Technol 30(4):331–353
21. Abreu e Silva J, Goulias KG (2009) A structural equations model of land use pat-terns, location choice, and travel behavior in seattle and comparison with Lisbon. In: Proceedings of the 88th annual transportation research board meeting, Washington D.C., pp 11–15
22. Bamberg S, Schmidt P (1998) Changing travel-mode choice as rational choice: results from a longitudinal intervention study. Ration Soc 10(2):223–252
23. Fillone AM, Montalbo CM, Tiglao NC (2005) Assessing urban travel: a structural equations modeling (SEM) approach. In: Proceedings of the Eastern Asia society for transportation studies, vol. 5, pp 1050–1064
24. Tam ML, Tam ML, Lam WHK (2005) Analysis of airport access mode choice: a case study in Hong Kong. J Eastern Asia Soc Transp Stud 6:708–723
25. Andreassen TW (1995) (Dis)satisfaction with public services: the case of public transportation. J Serv Mark 9:30–41
26. Stuart KR, Mednick M, Bockman J (2000) Structural equation model of customer satisfaction for the New York City subway system. Transp Res Rec 1735:133–137
27. Karlaftis MG, Golias J, Papadimitriou E (2001) Transit quality as an integrated traffic management strategy: measuring perceived service. J Public Transp

28. Ngatia GJ, Okamura T, Nakamura F (2010) The structure of users' satisfaction on urban public transport service in developing country: the case of Nairobi. J Eastern Asia Soc Trans Stud 8
29. Muhammad IS, Daisy MHK, Saman S (2011) Service quality in rail transport of Pakistan: a passenger perspective. In: Proceedings of 3rd SAICON: international conference on management, business ethics and economics (ICMBEE). Lahore, Pakistan, pp 28–29
30. Den Boon AK (1980) Opvattingen over Autogebruik en Milieuvervuiling. Baschwitz Institute for Public Opinion and Mass Psychology, University of Amsterdam
31. Lyon PK (1981a) Time-dependent structural equations modelling of the relationships between attitudes and discrete choices behavior of transportation consumers. Dissertation, Northwestern University and Technical Report IL- 11- 0012, Office of Policy Research, Urban Mass Transportation Administration, Washington, DC.
32. Lyon PK (1981b) Dynamic analysis of attitude–behavior response to transportation service innovation, Report 423- II-2. Prepared for the Office of Policy Research, Urban Mass Transportation Administration, Transportation Center, Northwestern University, Evanston, IL
33. Pendyala R (2001) Causal analysis in travel behavior research: a cautionary note. In: de Ortuzar JD (ed)
34. Fujii S, Kitamura R (2000) Anticipated travel time, information acquisition and actual experience: the case of the Hanshin expressway route closure. Transp Res Rec 1725:79–85
35. Fujii S, Kitamura R (2000) Evaluation of trip-inducing effects of new freeways using a structural equations model system of commuters' time use and travel. Transp Res B 34:339–354
36. Golob TF (2000) A simultaneous model of household activity participation and trip chain generation. Transp Res B 34:355–376
37. Liman T (2008) Sustainable transportation indicators. Sustainable Transportation Indicators Subcommittee of the Transportation Research Board, Washington, DC
38. Litman T (2009) Sustainable transportation indicator data quality and availability. Victoria Transport Policy Institute, Victoria

Performance Improvement of Bus Transport System from User Perspective

Vijay Singh Solanki and P. K. Agarwal

1 Introduction

Public transport plays a social role in the urban environment; it improves access to workplaces and service infrastructure and, at the same time, reduces travel expenses. Public transportation is defined as transportation by a conveyance that provides continuing general or special transportation to the public Belwal et al. [1]. The public transport system is an essential key factor in improving a city's social and economic welfare in growing countries, including India. In the past few years, India has experienced high population growth. Increasing urbanization in Indian cities of all types face problems related to urban transport. Presently, there are different public transport systems like bus rapid transit (BRT) system, light rail transit (LRT) system, mass rapid transit (MRT) system and various others like non-motorized transport. Many types of intermediate transport systems are being maintained and operated in Indian cities.

Despite huge investments in the development of the public transport system, all cities experience ever-growing congestion, accidents, air and noise pollution, and many hazards to our environment. These days' people in large cities have started using their private vehicles (two-wheelers and cars), and in small cities, different forms of intermediate public transport are used. The intermediate public transport provided either by the private or by the informal economy fight to meet the demands of the public like the minimum cost of travelling for many kilometres daily. The decrease in public transport shares in the cities over the past few years. In which, the cities are categorized under 1 to 6 in which the city category 1 has the lowest population and city category 6 has a maximum population which indicates that if the same situation continues to do so for years, it will affect the public transport share in the cities hardly [2]. Thus, it is necessary to promote the use of public transport system.

V. S. Solanki (✉) · P. K. Agarwal
Department of Civil Engineering, MANIT, Bhopal, Madhya Pradesh 462003, India

M. Parida et al. (eds.), *Proceedings of the Fifth International Conference of Transportation Research Group of India* , Lecture Notes in Civil Engineering 220,
https://doi.org/10.1007/978-981-16-9925-2_9

User satisfaction plays a vital role in evaluating the performance of public transport systems based on travel cost, travel time, safety, comfort, reliability, accessibility, etc. However, an inexpertly planned system causes trouble to the users, loses ridership, encourages private vehicles instead of public buses, metros, etc. and imposes a burden on the operator for his daily earnings. An accessible, flexible, safe, comfortable, economical and dependable service encourages a user to shift from personal vehicle to public transport. Hence, it is required to identify the factors on which users lag to use the public transport systems and implement some strategies that can improve existing system performance, so that people will shift from private to public modes of transportation. Hence simple, logical and scientific strategies are required for improving the performance of the public transport system from a user point of view. This study gives simple strategies for improvement in various factors of bus transport in Bhopal cities as per user satisfaction level.

This paper consists of five sections. Section one presents an introduction in this section problem and strategies. Section two presents literature review to analyse the problem and to identify the works for the state capital Bhopal, section three presents the development of user satisfaction index for bus transport system in Bhopal city, section four presents analysis result for state capital Bhopal based on the analysis of data and developing some indices for satisfaction level, and this section also presents strategies for improving the performance of Bhopal bus transport system which are identified. Section five presents the conclusion and outcomes of the study.

2 Literature Review

The performance improvement of the public transport system is an important measure in determining the level of success of the national transportation policies and knowing the impact of a policy and the operation of transportation services effectively and efficiently. Hence, a systematic literature review is an important and useful approach to identifying and analysing all relevant research on performance improvement of public transport systems in developing and developed countries from a user perspective. This section reviews the literature on performance improvement of public transport systems.

Chowdhury et al. [3] considered three performance indicators such as travel time, travel cost and comfort for improving the performance of the public transport system. This study shows travel time, travel cost and comfort on commuters' decision to travel on public transport system involving transfers and concluded that the travel time and travel costs are the most important parameters affecting the commuter's decisions. Singh et al. [4] examined vast literature about performance indicators that influence the public transport system from a user perspective. They identified eleven performance indicators to compare the customer satisfaction perceived by the Mini bus system and the Low Floor bus system in Bhopal city. These are journey time, frequency, fare, safety, seat availability, reliability, staff behaviour, speed, internal aesthetic, entry and exit. According to this study, the most influential parameters are fare, journey time, safety and reliability.

Javid et al. [5] considered thirteen performance indicators to compare commuters' satisfaction from different modes of transport services and alternate public transport services in Lahore, Pakistan. These are comfort, safety, vehicle's physical condition, staff attitude, schedule information, punctuality, route coverage, service frequency, convenience, travel cost, walking time, waiting time and travel time.

Castillo and Francisco [6] considered thirty-five performance indicators to evaluate the performance of the public transport system in Bilbao, Spain. These thirty-five performance indicators are grouped into eight major categories, i.e. connectivity, accessibility, information, time satisfaction, user attendance, comfort, security/safety and environmental impact. The study has identified five performance indicators that are very relevant from a set of 35 different performance indicators affecting the overall performance of the public transport system.

Dell'Olio et al. [7] analysed six performance indicators for the evaluation of the public transport system. The identified performance indicators are waiting time at the bus stop, journey time in the bus, vehicle occupancy, cleanliness of the vehicle, driver's kindness and bus comfort. This study concluded that the most important to least important indicators are waiting time at the bus stop, cleanliness of the vehicle, bus comfort and vehicle occupancy. Some of the major findings of the literature review are summarized as follows:

- A critical literature review indicated that limited studies are available on performance improvement of public transport system from a user perspective in the Indian context. The performance improvement from a user perspective is a complex process. This complexity is raised due to the multiple user performance indicators, and objectives from user satisfaction are considered concurrently. Additionally, the user performance indicators have no systematic categorization under various aspects from a user perspective. Further, it is more complicated to evaluate some of these indicators due to the absence of a database, or the data collection process is time-consuming, difficult and expensive.
- Therefore, most of the studies could not provide meaningful information to identify simple strategies that can be easily used for performance improvement. Hence, there is an urgent need to develop some simple strategies for improving the performance of the bus transport system in Indian cities, which can be implemented easily.

3 Performance Improvement of Bus System from User Perspective

In this proposed study, some parameters are identified for determining user satisfaction level: travel time, travel cost and quality. The performance of the bus transport system from a quality aspect is affected by safety, user comfort and user reliability.

This section presents a framework of the proposed methodology for the determination of user satisfaction level. Figure 1 presents a framework of the proposed methodology.

The methodological framework comprises four major stages for performance improvements of the bus transport system which are as follows:

- Stage I: Development of a hierarchical structure for identification of performance indicator
- Stage II: Determination of condition indices
- Stage III: Determination of relative weightage of identified performance indicators
- Stage IV: Determination of user satisfaction level of the bus transport system.

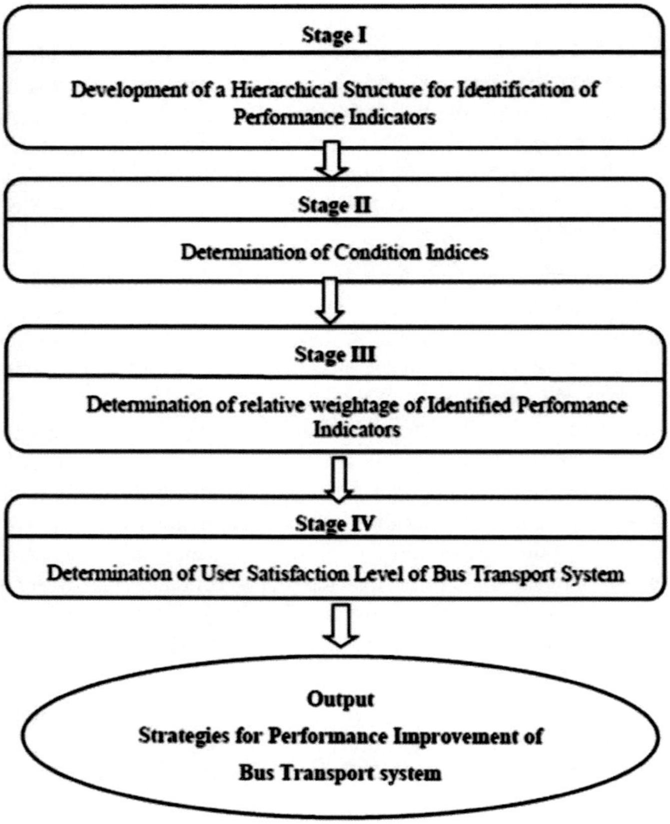

Fig. 1 Framework of proposed methodology for performance improvement of bus transport system

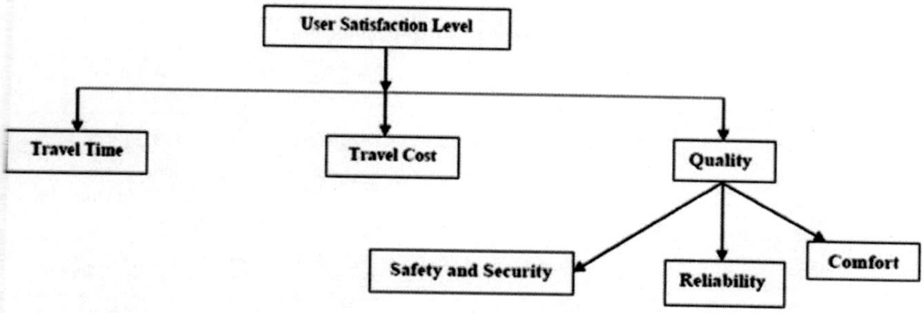

Fig.2 Hierarchy structure for identification of performance indicator for bus transport system

Stage I: Development of a hierarchical structure for identification of performance indicator

The first stage of this proposed methodology is developing a hierarchical structure for identifying performance indicators. A comprehensive hierarchical structure has been developed according to the conditions of the Indian public transport system. The major criteria selected for bus transport route are total travel time, travel cost, quality, comfort, safety and security and reliability for bus transport routes. These criteria are based on literature, primary data collection and discussion with the passengers and field studies. Figure 2 shows the hierarchy structure for the identification of performance indicator for the bus transport system.

Stage II: Determination of condition indices

This stage aims to determine condition indices for identified performance indicators to evaluate the user satisfaction level of the bus transport system of Bhopal cities. Identified parameters are as user travel time, user cost and user quality. The three indices developed to evaluate user satisfaction are user travel time index, user cost index and user quality index. The user satisfaction index (USIs) is developed in this study which can be evaluated using Eq. 1.

$$\mathrm{USI_s} = W_{\mathrm{UTs}} * \mathrm{UTI_s} + W_{\mathrm{UCs}} * \mathrm{UCI_s} + W_{\mathrm{UQs}} * \mathrm{UQI_s} \tag{1}$$

where

USIs	User satisfaction index of system;
W_{UTs}	Relative weight of user travel time in overall user satisfaction level;
$\mathrm{UTI_s}$	User travel time index for system;
W_{UCs}	Relative weight of user cost in overall user satisfaction level;
$\mathrm{UCI_s}$	User cost index for system;
W_{UQs}	Relative weight of user quality in overall user satisfaction level;
$\mathrm{UQI_s}$	User quality index for system.

Table 1 Details of condition indices for user performance indicators

S. No	Notation	Performance indicators	Methodology for evaluating the performance indicators
1	UTIs	User travel time index for system	$\text{Usertraveltimeindex(UTIs)} = \frac{\text{Triplengthbetweenoriginanddestination(TLs)}}{\text{Traveltimebetweenoriginanddestination(TTs)}}$
2	UCIs	User cost index for system	$\text{Usercostindex(UCIs)} = \frac{\text{Triplengthbetweenoriginanddestination(TLs)}}{\text{Travelcostbetweenoriginanddestination(TCs)}}$
3	UQIs	User quality index for system	UQIs = WUSs *USIs + WUCFs *UCFIs + WURs *URIs
3A	USIs	User safety index for system	USIs is determined by average safety rating given by a passenger
3B	UCFIs	User comfort index for system	UCFIs is determined by average comfort rating given by a passenger
3C	URIs	user Reliability index for system	URIs is determined by average reliability rating given by a passenger

The details of condition indices for identified performance indicators for evaluating the user satisfaction level of the bus transport system of Bhopal cities are presented in Table 1.

where

W_{USs} Relative weight of user safety in overall user quality for system;
W_{UCFs} Relative weight of user comfort in overall user quality for system;
W_{URs} Relative weight of user reliability in overall user quality for system.

Stage III: Determination of relative weightage of identified performance indicators

The performance indicators may not equally affect the satisfaction level of the bus transport system. The relative weights of the performance indicators are determined using passenger's opinion survey and their rating. A total of 250 responses are collected, and the relative weight of identified performance indicators is determined using the fuzzy AHP technique relative weights. Therefore, the objective of this stage is to the determination of relative importance of identified performance indicators. Table 2 presents details of relative weights obtained for various performance indicators.

Stage IV: User satisfaction level of the bus transport system

This stage aims to develop strategies for performance improvement of public transport to enhance user satisfaction level. Now putting the value of the relative weight of user performance indictors in Eq. (1), overall user satisfaction level can be evaluated, which can be written as Eq. (2):

Table 2 Details of relative weights obtained for various performance indicators

S. no	Notation	Performance indicators	Relative weight
1	WUTIs	Relative weight of user travel time in overall user satisfaction level	0.334
2	WUCIs	Relative weight of user cost in overall user satisfaction level	0.402
3	WUQIs	Relative weight of user quality in overall user satisfaction level	0.264
		Total =	**1.000**

$$USI_s = 0.334 * UTI_s + 0.402 * UCI_s + 0.264 * UQI_s \quad (2)$$

Thus, from the above equation, user satisfaction level has determined. Further with USI_s value, some rational and simple strategies can be identified for performance improvement of the Bhopal bus transport system.

4 Analysis Results for Bhopal Bus Transport System

To illustrate the proposed study, the city has three main services under Bhopal's bus transport system: City Link Limited, Mini bus service and Magic service. Three routes have been selected for each service connecting important areas of Bhopal city. The satisfaction level index of bus service is evaluated during peak hour. Details of analysis results for performance improvement of the bus transport system are given below.

4.1 Analysis Results for Performance Improvement of Bhopal Bus Transport System

The data is collected for three routes that have been selected for each service selected for analysis. The data has been collected by a survey conducted during peak hours. A team of four members is required to collect data first member note down the waiting time, boarding and alighting time, the second members note down the number of passengers boarding and alighting at each bus stop. The third and fourth members were engaged in collecting data from passengers about safety level, comfort level and service level. The information about the number of trips per day and travel cost is collected by conductors. The data has analysed for identified bus routes in Bhopal city, and indexes have been developed. These indexes are travel time index, travel cost index, comfort index, safety index and service reliability index. The travel cost index measures the ticket cost per kilometre of route length, and the travel time index measures the time required to travel per kilometre of route length. The safety index, comfort index and reliability index are purely based on passenger rating. Table 3 shows details of input data for performance improvement of the Bhopal bus transport system.

It indicates that index for user travel time, cost, safety, comfort, realibility, quality calculated with input data. Table 4 presents the analysis result of performance improvement of the Bhopal bus transport system.

4.2 *Strategies for Performance Improvement of Bus Transport System*

Based on the analysis results, strategies have been developed for performance improvement from a user perspective. Further, strategies developed in this study are illustrated using a case study of the Bhopal bus transport system. The salient finding of the study is the development of some rational and simple strategies for performance improvement of bus transport system from a user perspective which can be implemented easily. Some of the strategies identified are as follow:

- **Strategies to improve travel time performance (Reduction in number of bus stops/km in a route)**

From input data, it is clear that the travel time index increases with reducing bus stops. Hence, strategy may be developed to reduce the number of bus stops to increase the travel time index. Thus, user satisfaction level has increased. Figure 3 presents the effect of reduction in the number of bus stops/km in a route on the user satisfaction index.

From Fig. 3, it is clear that the value of USI_s is increasing with a reduction in bus stop/km in a route, and at the 0.5 value of USI_s, the value obtained for bus stop/km by interpolation is 0.94. Hence, strategy is proposed to reduce bus stop/km up to 0.94 bus stop/km to achieve user satisfaction level, i.e. user satisfaction index to 0.5.

- **Strategies to improve travel cost performance (Increase passenger kilometre travel per litres of fuel)**

From input data, it is clear that travel cost index increases with increasing passenger kilometres per litres of fuel. Hence, strategy may be developed to increase total passenger-km/litre to increase the travel cost index. That travel cost may be reduced, and further user satisfaction level has increased. Figure 4 presents the effect of passenger-km/litre on user satisfaction index.

From Fig. 4, it is clear that the value of USI_s is increasing with an increase in passenger-km/litre, and at the 0.5 value of USI_s, the value obtained for passenger-km/litre by interpolation is 10.91 by interpolation. Hence, strategy is proposed to increase the passenger-km/litre up to 10.91 to achieve user satisfaction level, i.e. user satisfaction index to 0.5.

Table 3 Details of input data for performance improvement for Bhopal bus transport system

Route name	Bhopal city link bus routes			Mini bus routes			Magic routes		
Particulars	Nehru nagar to Katara hills (SR2)	Coach factory to Nayapura-chichli (SR8)	Chirayu hospital to Awadhpuri (SR5)	Nehru nagar to Board office	Railway station to Trilanga	Habibganj Naka to Katara hills	Board office to Katara hills	Piplani to Board office	New market to Lal Ghati
Trip length (Km)	15	20.5	24.5	12	12.6	8.0	10	8	8
Travel time(Min.)	75	75	70	35–40	35–45	27	20	15	22
Travel cost(Rs.)	24	22	24	12	15	8	10	10	12
Vehicles per hour (Peak period)	5	6	5	4	6	5	7	10	11
Passengers per trip (Peak period)	120–160	120–140	140–160	25–55	28–60	25–50	20–24	15–22	20–32
Total no. of trips	4–5 trips/bus/day	5–6 trips/bus/day	5–6 trips/bus/day	5–6 trips/bus/day	5–6 trips/bus/day	3–4 trips/bus/day	– -	35 magic × 6–7	7
No. of stops/Km	1.87	1.43	1.37	1.33	1.30	1.27	1.25	1.13	1.00
Passenger-km/lit	52	48	48	32	30	28	20	18	18
Inconvenient bus stops (%)	29	32	34	41	42	49	51	54	56

Table 4 Analysis result of performance improvement for Bhopal bus transport system

Index	City link bus route				Mini bus routes				Magic routes			
	Nehru nagar to Katara hills (SR2)	Chirayu hospital to Awadhpuri (SR5)	Coach factory to Trilanga (SR8)	Average	Nehru to Board office	Railway station to Trilanga	Habibganj Naka to Katara hills	Average	Board office to Katara hills	Piplani to Board office	New Market to Lal Ghati	Average
UTIs	0.5	0.53	0.36	0.46	0.33	0.36	0.33	0.34	0.30	0.28	0.29	0.29
UCIs	1.00	0.84	1.00	0.95	0.96	0.96	0.88	0.93	1.00	0.8	0.67	0.82
USIs	0.8	0.75	0.85	0.8	0.70	0.92	0.8	0.8	0.55	0.5	0.52	0.52
UCFIs	0.8	0.71	0.6	0.71	0.6	0.55	0.57	0.57	0.4	0.42	0.4	0.41
URIs	1.00	0.75	0.6	0.75	0.52	0.70	0.55	0.58	0.44	0.46	0.48	0.46
UQIs	0.86	0.74	0.68	0.75	0.61	0.72	0.64	0.65	0.46	0.46	0.46	0.46
USIs	**0.81**	**0.74**	**0.71**	**0.75**	**0.65**	**0.69**	**0.63**	**0.66**	**0.59**	**0.55**	**0.49**	**0.54**

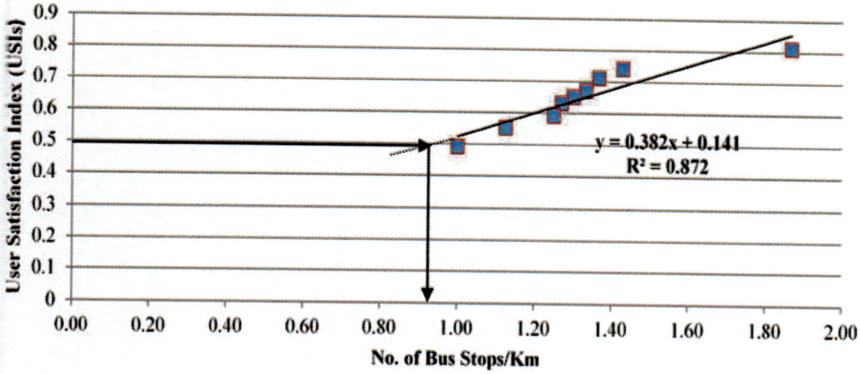

Fig. 3 Effect of reduction in number of bus stops/Km in a route on user satisfaction index

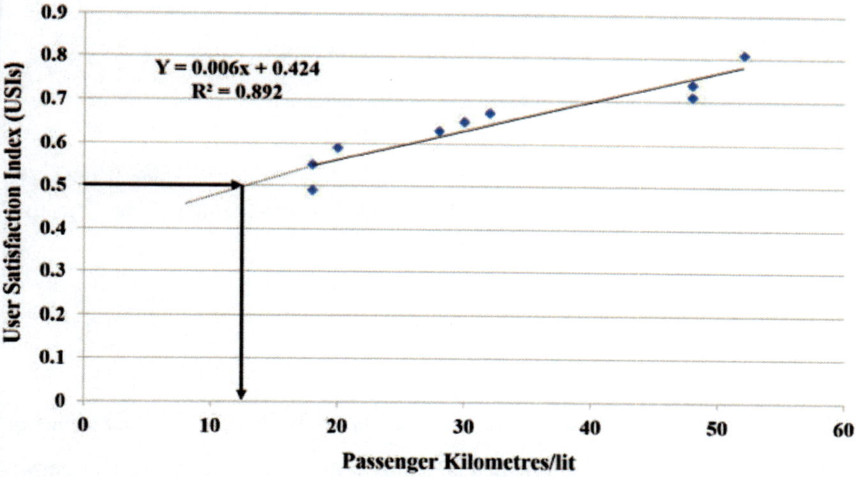

Fig. 4 Effect of passenger-km/litre on user satisfaction index

- **Strategy to improve quality performance (Increase convenient locations of the bus stop)**

From input data, it is clear that the user comfort index is reducing with an increase in the percentage of bus stops with an inconvenient location. Hence, strategy may be developed to reduce the percentage of inconvenient location bus stops, to increase user safety index, so that user safety may be increased further user satisfaction level has increased. Figure 5 presents the effect of % of the inconvenient bus stop on the user satisfaction index.

From Fig. 5, it is clear that the value of USI_s is increasing with a decrease in % of inconvenient bus stops and at the 0.5 value of USI_s. The value obtained for % of

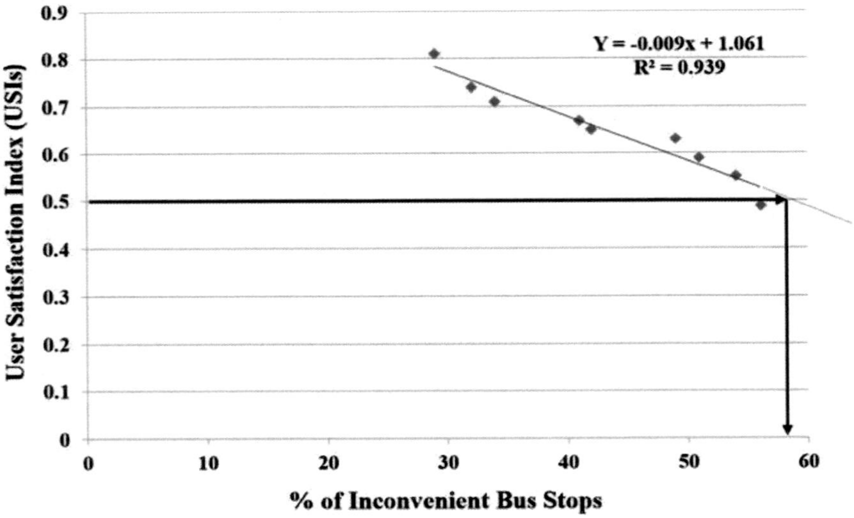

Fig. 5 Effect of % of inconvenient bus stop on user satisfaction index

inconvenient bus by interpolation is 59% interpolation. Hence, strategy is proposed to reduce inconvenient bus stop up to 59% to achieve user satisfaction level, i.e. user satisfaction index to 0.5.

5 Conclusions

Some of the important conclusions drawn from this study are summarized as follows:

- A critical literature review indicated that limited studies are available on the performance improvement of the public transport system from the user perspective in the Indian context. The performance improvement from a user perspective is a complex process. Additionally, the user performance indicators have no systematic categorization under various aspects from a user perspective. Most of the studies could not provide meaningful information to identify simple strategies that can be easily used for performance improvement. Hence, there is an urgent need to identify some simple strategies for improving the performance of the bus transport system in Indian cities, which can be implemented easily.
- In this study, the methodology consists of four stages. In the first stage, a hierarchical structure is developed to identify the critical factors affecting the user satisfaction level. The second step explains the methodology for the evaluation of condition indices, which has been developed. In the third stage, the relative weightage of identified performance indicators has been determined. Overall, user satisfaction level has been determined in the fourth stage.

- In this study, a user satisfaction level index is developed for the existing bus transport system; also different index, i.e. user travel time index, user cost index and user quality index, has developed. Further, the user quality index, user safety index, user comfort index and user reliability index are also developed. Based on this analysis results, some strategies are identified for improving the performance of the bus transport system in Bhopal city.
- In this study, the application of the proposed methodology is illustrated using BCLL bus system, Mini bus system and Magic system of Bhopal city in the state of Madhya Pradesh, and some strategies for improving user satisfaction level of bus transport system in Bhopal city are identified.
- The salient finding of this study is the identification of some rational and simple strategies for performance improvement of bus transport system from a user perspective which can be implemented easily. Some of the strategies identified are reducing the number of bus stops/km in a route, increasing passenger kilometre travel per litres of fuel, increasing convenient bus stop locations, etc.

It is expected that strategies developed in this study will be useful to improve the performance of the bus transport system in state capital Bhopal city. Thus, this study will be useful in improving the user satisfaction level of the bus system and hence will be useful in enhancing ridership of the city bus transport system. It is recommended that the removal of exact stops for a particular route be incorporated in the future study.

References

1. Belwal R, Belwal S (2010) Public transportation services in oman: a study of public perceptions, J Pub Transp 13(4):1–21
2. Associates WS (2008) Study on traffic and transportation policies and strategies in urban areas in India. Technical Report, Ministry of Urban Development, India
3. Chowdhury S, Ceder AA, Schwalger B (2015) The effects of travel time and cost savings on commuters' decision to travel on public transport routes involving transfers. J Trans Geogr 43:151–1592
4. Singh AP, Sharma A, Sharma AK, Singh VK (2014) "Analysis of customer's satisfaction in public transport using fuzzy logic for bhopal city", Int J Eng Tech Res 2(12):272–275
5. Javid MA, Okamura T, Nakamura F, Wang R (2013) Comparison of commuters' satisfaction and preferences with public transport: a case of wagon service in Lahore. Jordan J. Civil Eng. 7:461–472
6. Castillo JMD, Francisco GB (2012) A methodology for modeling and identifying user's satisfaction issues in public transport systems based on users surveys. Procedia Soc Behav Sci 54:1104–1114
7. Dell'Olio L, Ibeas A, Cecf́n P (2010) Modeling user perception of bus transit quality. Transp Policy 17:388–397 (2010)

Construction Safety Management of Chirwa Ghat and Kuthiran Twin Tunnel

Ashok Kumar Sharma and Kamlesh Kumar Ahirwar

1 Introduction

The main aim of the road tunnel is to provide safe means of access for traffic in difficult terrain [1]. To provide safe access in the difficult terrain construction of the tunnel is an emerging and viable solution for the rapid development of Indian highways. Tunnel reduces social, economic, and environmental costs significantly. Working in underground structures such as tunnels is inherently risk prone. The risk element increases with the lengths of the tunnels, poor rock conditions, and tunneling by conventional methods. The conventional methods of tunneling are drill and blast method (DBM) and the New Austrian Tunneling Machine (NATM). Along with uncertain and complex environment of the tunnel, frequent safety violation at the tunnel construction site imposes huge latent danger to all workmen [2]. From the literature review, we have observed that many accidents are reported in local media during the tunnel construction, but unfortunately, classified data of these accidents are rarely available. We find a few studies [3–7] of tunnel construction and its safety management in India. Certainly, we have to acknowledge that our understanding of safety management is incomplete in tunneling with the conventional methods in poor rock condition. This arises the need to understand tunnel construction safety management in India. The first step will be to understand current safety practices and identify the gaps for improvements in tunnel construction. This study presents the construction safety management of two long road tunnels. The first tunnel is the Chirwa Ghat twin tunnel with a length of 910 m near Udaipur, Rajasthan, India, completed in 2016. The second tunnel is the Kuthiran twin tunnel with a length of 1980 m in Kerala, India, completed in 2018. First, we discuss the basic aspects of safety during construction and then the methods used for risk assessment in working

A. K. Sharma
Lion Engineering Consultants, Bhopal 462046, India

K. K. Ahirwar (✉)
Department of Civil Engineering, Indian Institute of Technology Madras, Chennai 600036, India

M. Parida et al. (eds.), *Proceedings of the Fifth International Conference of Transportation Research Group of India* , Lecture Notes in Civil Engineering 220,
https://doi.org/10.1007/978-981-16-9925-2_10

conditions. After that, we also focus on other allied services necessary for tunnel safety during construction.

2 Literature Review

Earlier it was a common assumption, and road tunnel accidents claimed one life for every kilometer of tunnel construction in India. Later with the increased concern over construction safety and its management during construction, it led to the improvement in working conditions, with a subsequent reduction in the frequency of deaths and disabilities in accidents. We tried to find out the data of incidents and accidents during road tunnel construction from the National Safety Council of India, (NSCI), Ministry of Road Transport and Highways (MoRTH), and National Crime Record Bureau, (NCRB) Web sites, but we did not get the classified data. Mostly, the record of incidents and accidents is available in local newspapers publicly [8, 9]. A literature review indicates that a reliable record of accidents in tunnels during construction is not available in India because of the absence of a sophisticated framework for the incident and accidents record collection. So, in the absence of incidents and accidents record, we are unable to find out the causes of accidents and suggest adequate remedial measures in road tunnel construction.

The drill and blast method (DBM) used for excavation and NATM method works on the principle of mobilization of rock mass strength. In all Himalayan road tunnels construction such as Rothang road tunnel, Zojila road tunnel, and Z-Morh road tunnel, the DBM and NATM are being used [10]. NATM method is considered a better option than tunnel boring machine (TBM) method in heavy loading, soft clay, and massive mudflow conditions. So, we can say conventional methods (DBM and NATM) are extensive in current practice. But the safety of workmen remains a question using conventional methods for tunnel construction in India.

Since each road tunnels project has its own peculiarities and special features in given topography, rock features, etc., therefore, each road tunnel project needs to carry out a comprehensive risk analysis of the particular project and evolve a Project Safety Plan (PSP) [1]. The PSP addresses all site-specific issues that may be encountered during construction and also elaborates how to tackle all the risk elements identified. In this continuation, two tunnels were taken for study to assess the ground reality regarding the construction safety management taken place/or not, following the requirement, good engineering practices, and codal provisions.

3 Project Profile

Because of the steep gradient, sharp curves, and forest on either side of the existing road, the National Highway Authority of India (NHAI) decided to provide a six-lane tunnel in the Kuthiran malai area, Kerala. This tunnel is a part of NH544 and connects the Palghat to Kochi. The construction was started in 2014 and completed in 2018. Similarly, given the ghat section, sharp curves, and forest on either side of the existing road, NHAI decided to provide a six-lane twin tunnel in the Chirwa Ghat area near Udaipur, Rajasthan. This tunnel is a part of NH58 and connects Udaipur city and Nathdwara city in Rajasthan. The Chirwa Ghat tunnel started in 2014 and was completed in 2016. The project's site locations are shown in Figs. 1 and 2.

M/s Sadhbav Engineering Limited, Ahmedabad, was the contractor for Chirwa Ghat tunnel construction, and M/s Pragati Engineering, Hyderabad, was the contractor for Kuthiran tunnel construction. The first author of this study was General Manager (Technical) in both the tunnel's construction.

(i) top view (source: google earth image) (ii) Project site

Fig. 1 Kuthiran twin tunnel (1980 m)

(i) top view (source: google earth image) (ii) Project site

Fig. 2 Chirwa Ghat twin tunnel (910 m)

3.1 Technical Features of Twin Tunnels

Kuthiran twin tunnel is a medium unidirectional tunnel, (with cross passage), and Chirwa Ghat twin tunnel is a small unidirectional tunnel, as per IRC: SP:91–2010 [1] guidelines. The detailed technical features of tunnels are presented in Table 1.

4 Safety Management

This section of the paper will present the general safety requirement, details of incident and accidents occurred, and field observations with a different sequence of activities during road tunnel construction.

Table 1 Technical features of twin tunnels

S.No	Particular	Kuthiran twin tunnel	Chirwa Ghat twin tunnel
1	General geology and rock-type	Igneous rock and poor to medium/fair granite	Metamorphic and very poor to fair quartz arenite. With phyllite and schist, clay bends in-between
2	Length of twin tunnel	1980 m	910 m
3	Number of lanes in each tube	3	3
4	Geometry of tunnel	D-shape (wall and arch)	D-shape (wall and arch)
5	Construction methodology	Drill blast method	Drill blast method/NATM
6	Geological challenges encountered	Shear zones, portal collapse, and wall collapses	Shear zones, portal collapse, and wall collapse
7	Finished width of the tunnel	14 m	14.5 m
8	Finished height of the tunnel	10 m	10 m
9	Width of the foot path	1.25 m	1.25 m + 2.5 m
10	Lighting and ventilation	Provided	Provided
11	Distance between two tunnels	23 m	26 m
12	Cross passages	Two passages, 300 m each from the portal	Not applicable

4.1 General Requirement of Safety During Construction

The general requirement of safety during construction includes personal protective equipment, access control systems, signage, and safety systems. These general requirements of safety must follow during construction without any excuse.

4.2 Risk Assessment

The construction of the tunnel evolves uncertainties. To tackle uncertainties, risk assessment is required. The risk assessment helps to find the potential weak point and suggests possible safety measures. This risk assessment approach can be applied in tunnel construction safety management. The competent authority of tunnel construction followed the guidelines of the Road Tunnels Manual, World Road Association (PIARC), [8] for risk assessment in both tunnels. The risk assessment has three steps: First step is risk analysis, the second step is risk evaluation, and the third is planning safety measures (see Fig. 3).

Risk analysis identifies the probable hazard and estimates the risk from the hazard. This can be performed using a qualitative or quantitative and a combination of both. A systematic risk evaluation is based on safety criteria. If the risk evaluation value is higher than the acceptable limit, then we have proposed additional safety measures. It is lower than safety criteria, then we will not propose safety measures.

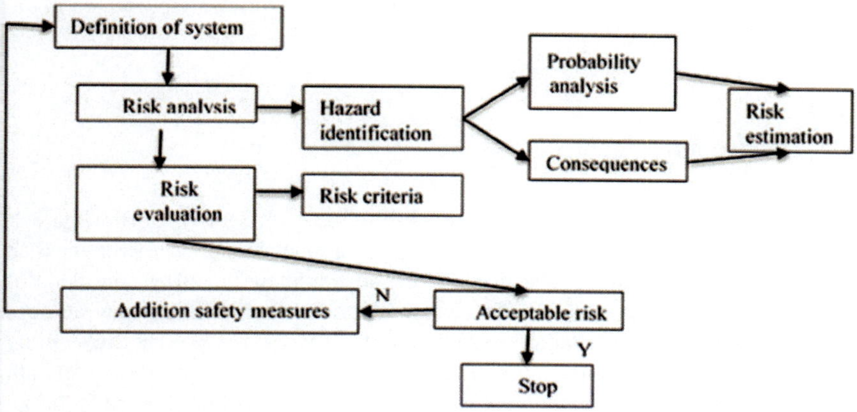

Fig. 3 Flowchart procedure for risk assessment

4.3 Field Observations During Tunnel Construction

This subsection presents incidents and accidents that happened during tunnel construction and the compliance of safety provisions based on field observations at the tunnels.

Incidents and accidents happened during tunnel construction

We observed many incidents and accidents in both the tunnels. The incident is an undesired event that may result in property damage and does not cause injury or illness to workmen, and the accident may result in property damage as well as injury or illness to workmen. The details of incidents and accidents that happened in both tunnels are presented in Table 2.This data is a compilation of details from daily progress reports and hindrance registers of both tunnels.

From Table 2, we have observed that several incidents and accidents have happened and workmen got injured and hospitalized also; therefore, actual implementation part of PSP must be on the topmost priority during the construction of tunnels to avoid the reoccurrence of such incidents/accidents.

Field Observations

Table 3 shows the safety provisions made in Project Safety Plan (PSP) and its actual implementation, during the construction of both tunnels.

From Table 3, the gaps in the sequence of various construction activities and their actual implementation at the site are very explicit. Even the basic requirement of safety during construction is not fully compiled in both tunnel construction. Table 3 will help us draw the conclusions and recommendations, which are presented in the last section of the study.

4.4 Discussion

We can take various safety aspects into the discussion from Sect. 4.3. From Table 3, we can say that safety management at the Chirwa Ghat tunnel seems better than at the Kuthiran road tunnel. The reasons behind this may be safety conscious of management and resources availability. Many shortcomings like emergency services were not as per prescribed standards, safety training to workmen, poor housekeeping, and no use of noise controlling devices which we can figure out from Table 3 in both the tunnels. Poor ventilation and poor lighting cause many incidents in the Kuthiran tunnel (refer to Table 2). It is also observed that workmen are not wearing safety masks and glasses while drilling the hole, rock bolting, and grouting activities. But the main concern is many workmen get injured and hospitalized because of the fall of rocks that happened when workers were set off explosive charges, installing steel arch supports in the cutting face. The shotcreting, bolting, and removal of rock masses in the cutting face are basic safety measures. We can also do drilling for drainage,

Table 2 Incidents and accidents that happened in both tunnels during construction

S. No	Activities inside tunnel	Location	Nature of incident and accident inside the tunnel
1	Drilling hole by hydraulic jumbo on tunnel face	Chirwa Ghat tunnel	Drilling water went into the eyes of workmen, because of the non-wearing of safety goggles, during the change of drilling rods
2	Loading and charging	Chirwa Ghat tunnel	Because of the mishandling of explosives some detonators missing
3	Blasting (Controlled)	Chirwa Ghat/Kuthiran tunnel	Misfire occurred damaged, a bucket of excavator and minor injuries to the operators/helpers
4	Mucking to disposal areas	Kuthiran tunnel	The dumper front crashed on the sidewall due to a poor ventilation system
5	Trimming of tunnel face	Kuthiran tunnel	The collapse of rock pieces on the head and other parts of the body, due to adequate headgear and trimming tools
6	Scaling of tunnel cutting face	Chirwa Ghat tunnel	The collapse of rock pieces and dust on the head and over the body of workmen, because of inadequate headgear and mechanized scaling tool with the bucket
7	Cutting tunnel face and inspection	Kuthiran tunnel	Sudden fall of rock pieces with soil from wall and crown, injured surveyor, foremen and geologist and one person hospitalized for a week
8	First spraying of shotcrete layer (shotcrete against the cutting face and around)	Chirwa Ghat tunnel	The giddiness felt by workmen and nozzlemen because of poor arrangement of ventilation, by fumes of admixture in the mix
9	Erecting a steel arch support on walls and arches	Chirwa Ghat tunnel	A segment of steel ribs fallen on workmen because of the poor condition of wire rope in lifting machine admitted to hospital for four days

(continued)

Table 2 (continued)

S. No	Activities inside tunnel	Location	Nature of incident and accident inside the tunnel
10	Second spraying of shotcrete layer	Chirwa Ghat/Kuthiran tunnel	The giddiness felt by workmen and nozzlemen due to poor arrangement of ventilation, by fumes of admixture in the mix
11	Driving rock bolts on the crown and springing level	Chirwa Ghat tunnel	Drilling water went into the eyes of workmen, because of non-wearing of safety goggles; medical treatment for seven days
12	Formwork and reinforcement binding and fixing along with the tunnel profile	Chirwa Ghat tunnel	Workmen got electrical shock due to inadequate earthing for electric lines and ELCB, medical treatment given for a week to two persons
13	RCC/CC lining work with gantry formwork	Chirwa Ghat tunnel	The workmen, got fingers injured, because of inadequate illumination at the workplace, hospitalized for four days
14	Grouting of concrete lining	Chirwa Ghat/Kuthiran tunnel	Grout mix went into the eyes of workmen, because of non-wearing of full-face safety mask with, during the grouting operation
15	Portal development at both exit and entry ends	Chirwa Ghat/Kuthiran tunnel	Five workmen got slipped and injured from the slope because monkeys disturbed the surface soil since proper rope fencing with caution tape was not provided

Table 3 Field observations during tunnel construction

S.No	Required safety provision	Chirwa Ghat tunnel	Kuthiran tunnel	Explanatory remarks
1	Personnel protective equipment	Partially used	Partially used	Helmet, safety shoes, reflective jackets, gloves, goggles, dust masks, earplugs/muffs, safety harness
2	Access control system	Partially used	Sparingly used	Exit/entry control system for workmen, material and equipment, no-entry for unauthorized persons
4	Signage—display at prominent locations	Ordinary boards used	Ordinary boards used	Well-illuminated boards, yellow danger, blue mandatory, red prohibition, green safe condition
5	Training—better safety management	Conducted some times	Not conducted	Safety training—initial, basic, and routine, pep talks and toolbox talks, mock drills
6	Medical/first aid facilities	Available daytime only	Rarely available	Availability of medical attendant/first aid attendant and ambulance
7	Ventilation system used	Mechanical ventilation	Natural but insufficient	Natural/mechanical ventilation system used to force fresh air
8	Noise control	Used earplugs/muffs	Rarely used	Protective device to be used to maintain noise level <85 dB
9	Lighting	Adequate lighting arrangements	Inadequate/poor lighting arrangements	50/30/10 lx to be maintained at heading, inside tunnel, portal
10	Communication system—warning signs and notice boards, telephone system, and CCTV system	Adequate warning signs and display boards, telephone	Adequate warning signs, telephone	Hooter/siren/display of construction activities inside the tunnel, CCTV system for better monitoring of construction operation inside the tunnel
11	Fire protection—general items, fire system, electrical installations	Adequate and satisfactory arrangements made	Poor and inadequate arrangements made	Removal of flammable liquid/combustible materials, water-type fire extinguisher and sand bucket, fire alarm and phone no. of nearest fire brigade station
12	Housekeeping—general, traffic control, pipes, and cables, water control	Adequate and satisfactory arrangements made	Poor and inadequate arrangements made	Keep required material only, safe walkways for workmen/passage for vehicles, proper installation of pipelines, dewatering system for water ingress in the tunnel
13	Protection against insects, leeches, vermins, and snakes	Insufficient arrangements	No arrangements	Spray, drainage of breeding area, netting, elimination of unsanitary conditions
14	Emergency management system	In place, as per PSP	Not available	Emergency rescue measures, described in EMP, as part of approved PSP
15	Drilling operations	Wet drilling with hydraulic/robotic jumbo	Wet drilling with hydraulic/robotic jumbo	Wet drilling is permitted to keep the safety/visibility at heading and inside of the tunnel

(continued)

Table 3 (continued)

S.No	Required safety provision	Chirwa Ghat tunnel	Kuthiran tunnel	Explanatory remarks
16	Blasting operations	Adequate arrangements were in place	Poor arrangements were in place	Nonel/explosives/detonators/330 m distance/no cell phone are used, B.O. by experienced blaster/shot firer licensed—PESO
17	Inspection after blasting	Proper arrangements were in place	Proper arrangements were in place	After 5 min and an all-clear signal, the shot firer checks the exploded face
18	Checking of misfire	Proper arrangements were in place	Proper arrangements were in place	After 15 min of explosion—all-clear signal—face inspection check for misfires status carried out with multi-meter
19	Scaling and Mucking	Adequate arrangements were made	Adequate arrangements were made	Scaling ladder/equipment used to remove the loose/distressed surface; dumping/disposal yard identified for muck disposal
20	Explosive's storage and handling	All necessary arrangements were in place	Necessary arrangements were in place	Fenced portable magazines under watch by gun-guard, handling to the workplace by the mobile magazine
21	Explosives disposal and accounting	Necessary arrangements were made	Necessary arrangements were made	Safe disposal of packaging of explosives/maintain approved register of received and consumed explosive material
22	Installation of supports as approved	Necessary supports provided	Necessary supports provided	Safely supported within the stand-up time of rock, to avert uncontrolled collapses
23	Erection-fabricated structural steel section	Safe practices used in erection works	Safe practices used in erection works	Signalmen, equipment check for hooks, cables, ropes, slings, hoists, and protection from falling objects
24	Scaffolds and working platforms	Safe workmanship practiced	Safe workmanship practiced	Use of safety belt, suitable guard rails as safety fence on-slippery work area
25	Concreting works—mixing and pumping	Safe workmanship practiced	Safe workmanship practiced	RMC—use dust mask, pipeline, and pumping operation coordinated by signalmen
26	Grouting and shotcreting	Safe workmanship practiced	Safe workmanship practiced	Safe condition maintained for shotcrete pump, grouting machine, experienced nozzlemen engaged for safety/quality work
27	Instrumentation for monitoring—during construction	Strain gage, and load cell for shotcrete and rock bolts used	Monitoring carried out by visual observation only	Safety and stability of excavation supports, warning of potential ground failure, safety and serviceability of adjacent structures
28	Welding and cutting works by gases, hoses, and torches	Safe workmanship practiced	Safe workmanship practiced	Safety goggles, mask, gas lighter, well-ventilated work area

(continued)

Table 3 (continued)

S.No	Required safety provision	Chirwa Ghat tunnel	Kuthiran tunnel	Explanatory remarks
29	Electric arc welding and cutting	Safe workmanship practiced	Unsafe workmanship practiced	Proper earthling of welding sets, and armored cable, ELCB fitted
30	Competency test of drivers and operators	Conducted as per PSP	Randomly conducted	For safe and secure—pre-employment training to work in the tunnel
31	Competency test for workmen	Conducted as per PSP	Randomly conducted	For safe and secure—pre-employment training to work in the tunnel

measure the displacement, and adequate lighting at cutting face chances of reducing the incident. The shotcreting in combination with the above safety measures at cutting face is very effective because it seals the weak points in the rocks and avoids air exposure. However, even after shotcreting, some incidents of fall of rocks happened because the thickness of the shotcrete was inadequate and upwelling of groundwater.

Even after applying the above safety measures, the cases of the fall of rocks remain high, then the client, contractor, and construction management team need to discuss the issues and find out safety measures which may be from the design and use of advanced instrumentations.

5 Conclusion

Road tunneling is a viable and environment-friendly solution for rapid highway development in difficult terrain. The safety management of tunnel construction is crucial when conventional methods, i.e., drill and blast method (DBM) and NATM method, are used for tunneling in poor rock conditions. From the study, following conclusions can be drawn:

- The literature review indicates that classified record of incidents and accidents during road tunnels construction is not available.
- The guidelines of the Road Tunnels Manual, World Road Association (PIARC), were used for risk assessment in both tunnels.
- We observed the partial non-compliance of safety provisions mentioned in the Project Safety Plan (PSP) observed, thus many incidents and accidents happened in both road tunnels.
- The proper arrangements are made for inspection after blasting, checking of the misfire, scaling and mucking, explosives storage and handling, explosives disposal and accounting installation of supports, and erection of fabricated structural steel section in both tunnels.

- The safe workmanship practices were adopted in scaffolds, concreting works, instrumentation, welding, and cutting works by gases, hoses, and torches activities in both tunnels.
- Rock collapse and falls are a major concern in poor rock conditions. So, to reduce the incidences and accidents, extensive use of instrumentation during tunnel construction is necessary.

Recommendations

The following recommendations can be made based on observation presented in the above sections

- The communication and alarm system for evacuation should be placed together with the fire water supply, and emergency vehicles must be available for effective evacuation.
- The combustion material storage underground should be limited to what is required in one shift of work.
- Consider the escape masks as personal protective equipment in tunnel construction.
- A comprehensive plan should be available about working conditions, organizational factors, and societal factors for fire and rescue emergencies.
- Collect and analyze the incidents and accidents that happened in road tunnels. This will help us to find out causes and analyze the effectiveness of taken measures.

Thus, the study provides detailed insight into road tunnel construction safety management and identified the gaps for improvement in current field practice. This study is limited to conventional methods used for tunneling in poor rock conditions. Additionally, case studies of new road tunnels in India will further demonstrate the strategies and upgrading measures.

Acknowledgements We acknowledge M/s Sadhbav Engineering Limited, Ahmedabad, for their technical support in the Chirwa Ghat tunnel study, and M/s Pragati Engineering, Hyderabad, for their technical support in the Kuthiran tunnel study.

We also acknowledge that this study was presented at the fifth conference of the Transportation Research Group of India held at Bhopal (India) from December 18–21, 2019.

References

1. IRC:SP:91–2010 (2010) Guidelines for road tunnels. New Delhi (2010)
2. Ding L, Zhang L, Wu X, Skibniewski MJ, Qunzhou Y (2014) Safety management in tunnel construction: case study of Wuhan metro construction in China. Saf Sci 62:8–15. https://doi.org/10.1016/j.ssci.2013.07.021
3. Goel RK, Singh B, Zhao J (2012) Underground Infrastructures (2012). https://doi.org/10.1016/C2010-0-67210-5

4. Palomba M, Russo G, Amadini F, Carrieri G, Jain A (2013) Chenani-Nashri Tunnel, the longest road tunnel in India: a challenging case for design-optimization during construction. Underground. W Futur 964–971 (2013). https://doi.org/10.1201/b14769-132
5. Goel RK (2001) Status of tunnelling and underground construction activities and technologies in India. Tunn Undergr Sp Technol 16:63–75. https://doi.org/10.1016/S0886-7798(01)00035-9
6. Sharma AK (2018) Construction Methodology of Twin Tunnel at Chirwa Ghat—A Review. 6:518–523 (2018)
7. Sharma AK, Sharma MR, Garg MD, General A (2018) Safety audit during tunnel construction. 6:688–700
8. Road Tunnels Manual, W.R.A. (PIARC): risk-assessment @ tunnelsmanual.piarc.org, https://tunnelsmanual.piarc.org/en/safety-tools-safety-management/risk-assessment
9. Probe indicts 2 companies for death of 10 workers, https://www.tribuneindia.com/news/archive/features/probe-indicts-2-companies-for-death-of-10-workers-188521
10. Srb M (2017) Long transport tunnels in Indian Himalaya finished, progressing. In: Preparation. Tunnel (2017), pp 23–29

Development of Risk Assessment Model for Waterway Safety

N. M. Sabitha, A. S. Athulya, B. G. Sreedevi, and V. S. Sanjay Kumar

1 Introduction

Risk can be considered an essential part of the life of all human beings. Every day we used to make judgments which balance our goals against the probability that something undesirable (a risk) will happen on the way. Risk is defined as the product of two factors—the probability (or likelihood) of an undesirable incident occurring and if it does occur, the severity of its potential long and short-term impact (or consequence).

Most of the activities in our highly complex society are not entirely in the control of the individuals. We regularly rely on organizations to manage risks for us and always expect risks to be addressed for us to reduce to lower levels, thereby increasing the perceived level of safety. Reducing or avoiding risk may cause inconveniences that can be expressed, not only in terms of money but also in terms of loss of amenity, restrictions of access, and denial of opportunity. The use of inland waterways is also subjected to these trends and changes related to risk.

Inland Water Transport (IWT) is an economically and environmentally sustainable mode of transportation, which happens to be one of the oldest modes available to mankind [1]. Though IWT has many advantages over other modes of transport, it faces many challenging risks, such as an increasing rate of accidents and collisions. The need of the hour is to mitigate the risks involved in passenger vessel operations to minimize accidents, with the development of technology and public awareness. To ascertain a sustainable and safest mode of Inland Water Transportation, it is necessary to develop risk assessment methods by incorporating the factors and components that would cause hazards in waterways [2]. The process of risk assessment is a systematic approach to identify major hazards of waterway-related safety and to evaluate the potential mitigation measures.

N. M. Sabitha (✉) · A. S. Athulya · B. G. Sreedevi · V. S. S. Kumar
National Transportation Planning and Research Centre, Thiruvananthapuram, India

M. Parida et al. (eds.), *Proceedings of the Fifth International Conference of Transportation Research Group of India* , Lecture Notes in Civil Engineering 220,
https://doi.org/10.1007/978-981-16-9925-2_11

Kerala has 1687-km-long waterways. It includes 590 km of West Coast Canal (WCC) from Neeleswaram in the north of the state to Kovalam in the south [1]. The remaining portion comprises feeder canals/rivers. The unique feature of WCC is that it flows parallel to the Arabian Sea with several sea openings. A part of WCC, from Kollam to Kottappuram (168 km) and two other canals in Kochi (Champakkara Canal—14 km and Udyogamandal Canal—23 km), constituting 12.15% of the total IWT in Kerala [3], was declared by the Government of India (GoI) as National Waterway-3 (NW-3) in the year 1993 [4]. This stretch is developed and maintained by the Inland Waterways Authority of India (IWAI). The backwaters of Kerala, such as Ashtamudi Lake and Vembanadu Lake, provide a good means of inland navigation. Kerala Public Canals and Ferries Act regulated the canal transportation for cargo and passenger movement in Kerala [4]. Before the advent of faster modes of transport, the inland waterways were used for communication and connectivity in Kerala, mainly due to the geographic and demographic pattern of the state as there are no network of roads existed in that era. Though IWT sector is economically viable, this important mode of transport is affected by tragic accidents every year, incurring a heavy toll of human lives and properties. The climatic conditions, nature of operations, human error (crew and passengers), lack of safety standards, etc., pose a number of risks to safety of passengers and boats. The accident rates are high and needed to be taken care of to minimize the accidents.

1.1 Waterway Risk Model

The risk may be defined as the product of the probability of a casualty and its consequences [5]. The Waterway Risk Model includes various variables dealing with the causes of waterway accidents and their effects. The various risk categories addressed in the model are as follows:

- Vessel conditions: the quality of vessels and the crews that works in a vessel.
- Traffic conditions: the number of vessels available in a waterway and their interactions with other vessels.
- Navigational conditions: the conditions related to the environment that vessels must deal in a waterway relating to wind, water movement (i.e., currents), and weather.
- Waterway conditions: the physical properties of the waterway that affect the maneuverability.
- Immediate consequences: the immediate impacts of a waterway accident: injury or fatality to people, spilling of petroleum and hazardous materials requiring response resources, and the waterway transportation system disruption.

Risk management is the systematic approach of identifying, analyzing, and action on risk(s). It incorporates many steps, from the preliminary identification of risks to the final decision taken on risk-reducing actions and risk monitoring. Risk assessment is done based on risk values assigned for each of the above parameters such as vessel conditions, traffic conditions, and waterway conditions.

A risk management system mainly includes the following processes:

- Risk identification
- Risk analysis
- Risk estimation
- Risk assessment
- Risk mitigation
- Risk monitoring.

Risk management system must be formulated by understanding the manner through which each hazard occurs, their initiating factors, and their consequences that may affect waterway safety. It is the evaluation of alternative risk reduction measures and the implementation of such measures.

2 Methodology

A measure of the probability of an unwanted event and its impacts or consequences can be termed as a risk [6]. The attributes that make up risk are therefore those that affect the accident probability and the impacts or consequences of potential accidents. So the attributes determined are grouped into a tree-like hierarchy with six main groups; four of these groups affect accident frequency, and the other two affect the consequences of potential accidents. Each of these must be minimized to achieve the overall objective of maximizing safety in a given waterway. The factors considered in the present model include condition of vessels, traffic conditions, environmental conditions, waterway configuration, short-term consequences, and long-term impacts. Scales to measure each variable were established from the available data. Figure 1 shows the waterway risk assessment model with hierarchy of waterway attributes that affect its safety. The model consists of linear combination of the attributes, like a linear-additive value function, where a single-dimensional value function that essentially converts each attribute is multiplied with a weight reflecting the importance of each attribute. After the parameters were established for each risk-inducing factor, waterway-specific risk is estimated by putting those values into the model for the specific values for that waterway for each variable. The model allows comparison of relative risk and the potential efficacy of various improvements between different waterways. Thus, the model is based on a multi-attribute decision analysis technique and the effects safety alternatives would have on these levels.

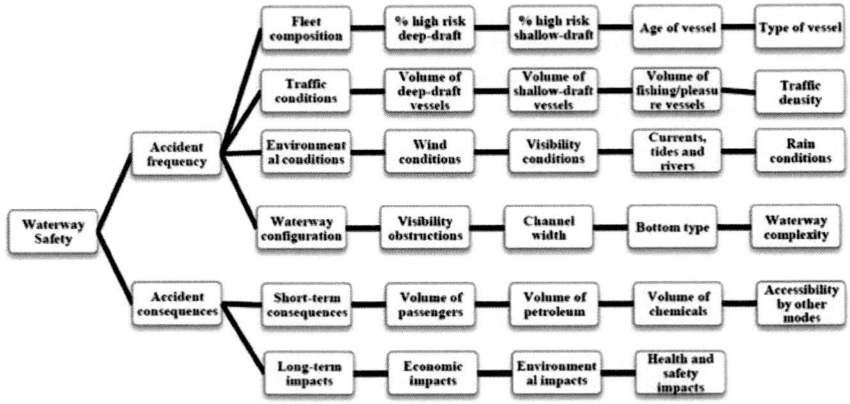

Fig. 1 Waterway risk assessment model with hierarchy of attributes that affect safety

3 Boat Accident Data Collection and Analysis

Tragic accidents happen in IWT sector every year, causing a heavy toll of human lives and properties [7]. The details of boat accidents occurring in the waterways of Kerala are registered in the respective Port Offices. The accident data were collected and analyzed for the causes of boat accidents and safety aspects. Boat accidents have a wide range of causes. The best way to prevent these accidents is to start learning from these accidents and understand the causes of injuries and fatalities, a major of which are avoidable. The cause of accident can be due to the ignorance of passengers or due to inattention of operator or may be due to other reasons. The main reasons why boat accident occurs are as follows:

- Careless operation
- Improper lookout (to watch for threats or hazards)
- Operator inexperience
- Overloading
- Excessive speed
- Impaired operation (alcohol or drug use)
- Equipment failure
- Violation of navigation rules
- Weather (storm, flood, and wind patterns)
- Force of waves or wakes
- Dangerous waters.

4 Questionnaire Survey and Analysis

Questionnaire survey was conducted for boat passengers and crew at Kollam, Alappuzha, and Ernakulum boat jetties, which are the main active centers of water transport in Kerala. Separate questionnaires were prepared for the boat passengers and boat crew. Samples were chosen randomly, and the experiences about the present boat service system and their expectations were queried. The questionnaire for this purpose was designed for extracting relevant information without seeming to be too intrusive. The questionnaire for passengers had questions which enabled the collection of data regarding the occupation, income, purpose of journey, origin, destination, frequency of travel, awareness about the safety equipment and rules, safety of boat and boat jetties, problems faced by passengers during journey, and suggestions for improving safety and satisfaction level on boat services. The questionnaire for boat crew had questions on educational qualification, designation, trip duration, total working hours in a week, details of training attended regarding boat safety, rescue operations, etc., awareness about rules and regulations, their knowledge on operation of safety equipment, necessity of refresher courses related to safety, precautions taken by them to ensure passenger safety, suggestions for improving passenger safety, the routine checking they do before starting a boat trip, frequency of maintenance of boat, and the safety equipment available in the boat. The data collected from boat passengers and crews were analyzed separately and are discussed in subsequent sections. In Alappuzha, the opinion survey was conducted for 437 boat passengers and 164 boat crew. In Kollam, the opinion survey was collected from 207 boat passengers and 23 boat crew. In Ernakulam, the opinion survey was collected from 381 boat passengers and 30 boat crew.

4.1 *Passenger Data Analysis*

Alappuzha. From the survey, it was clear that majority of people in Alappuzha depends on boat services for their daily travel. 44% of passengers travel daily in the boat and majority of the boat passengers use boat service for traveling to and back home after work, shopping, education, etc. About 15% of the passengers use the boat services for social and recreational purposes. The ferry system plays a very important role in the day-to-day lives of office-goers, students, and housewives. About 46% of the sample was found to be workers in government/private sectors and self-employed businessmen, and another 22% were found to be students making their trip to education, a short distance away from their homes. The poorer sections of the citizens mostly depend on the ferry services. About 38% of the ferry users who responded had less than ten thousand rupees as their monthly family income. Also about 46% of the total sample interviewed had only primary education, and another 23% were graduates. The interviewees were enquired about the causes of

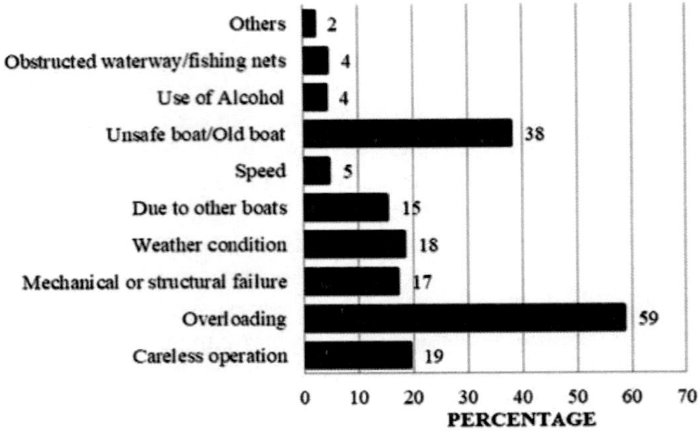

Fig. 2 Factors leading to boat accidents in Alappuzha as per passenger opinion

boat accident and were asked to mark them according to their priorities. The results are given in Fig. 2.

Kollam. In Kollam, majority of the boat passengers use boat service for the purpose of going for work and also for educational purposes. About 12% of the sample uses the boat services for social and recreational purposes. About 63% of the sample was found to be workers in government/private sectors and self-employed businessmen, and another 15% were found to be students, a short boat ride away from their homes. About 38% of the ferry users who responded had less than ten thousand rupees as their monthly family income. Also about 52% of the total sample interviewed had only primary education, and another 19% are graduates. 89% responded that they have never attended any training classes related to passenger safety. The interviewees were enquired about the causes of boat accident and were asked to mark them according to their priorities. The results are given in Fig. 3.

Ernakulam. In Ernakulum, majority of the boat passengers use boat service for reaching their work destinations. Other purposes include shopping, education, back home, social and recreational, etc. The ferry system was used in a daily basis by the office-goers and students. Majority of the people using the boat services are employees. About 48% of the sample was found to be workers in government/private sectors and self-employed businessmen, and another 22% were found to be students. The interviewees were enquired about the causes of boat accident and were asked to mark them according to their priorities. The results are given in Fig. 4.

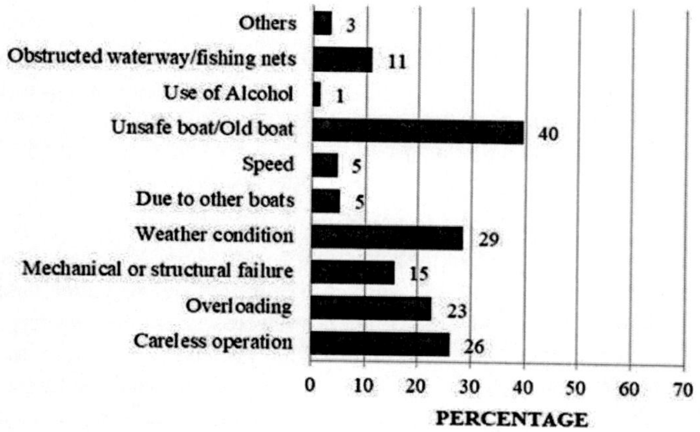

Fig. 3 Factors leading to boat accidents in Kollam as per passenger opinion

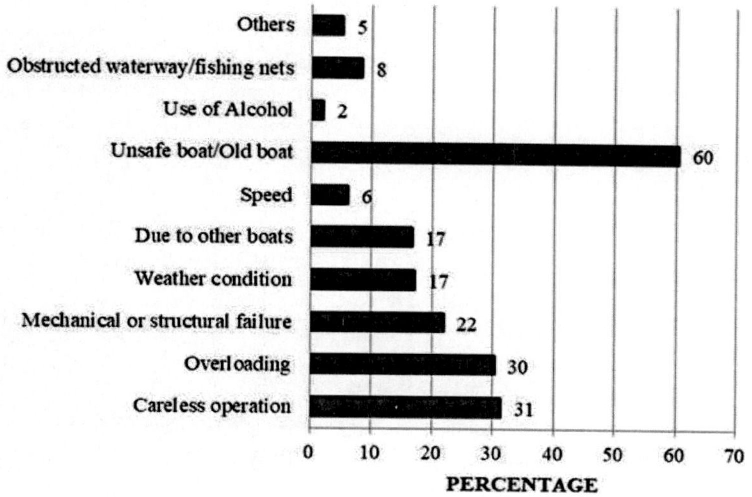

Fig. 4 Factors leading to boat accidents in Ernakulam as per passenger opinion

4.2 Crew Data Analysis

Alappuzha. The survey was conducted in both State Water Transport Department (SWTD) boats and private boats. The questionnaire survey for crew was conducted among Serang, Boat master, Driver, and Lascar in SWTD boats. It was inferred that above 90% of the crew have attended training in the fields of boat safety, first aid, rescue operations, and to understand signals and signs in waterways. The trainings were given at Maritime Institute-Neendakara (Kollam), Tourism department, Alappuzha port, Dock and repair (SWTD), etc. Also above 90% of the crew were aware of Kerala Inland Vessel (KIV) rules and operation of safety equipment available in the boat. 79% of the boat crew suggested that there is a need of refresher courses related to safety in the fields of safety and rescue, use of safety equipment, etc. The boat crews were enquired about the causes of boat accident and were asked to mark them according to their priorities. The results are given in Fig. 5.

Kollam. From survey, it was clear that above 80% of the crew have attended training in the fields of boat safety, first aid, rescue operations, and to understand signals and signs in waterways. About 91% of the crew were aware of operation of safety equipments available in the boat. 83% of the boat crew suggested that there is a need of refresher courses related to safety in the fields of safety and rescue, techniques for propeller cleaning, waste removal, etc. The boat crews were enquired about the causes of boat accident and were asked to mark them according to their priorities. The results are given in Fig. 6.

Ernakulam. From the survey, it is clear that above 90% of the crew have attended training in the fields of boat safety, first aid, and to understand signals and signs in waterways. 100% of the crew had attended training in rescue operations. About 83% of the crew are aware of operation of safety equipments available in the boat. The boat crews were enquired about the causes of boat accident and were asked to mark them according to their priorities. The results are given in Fig. 7.

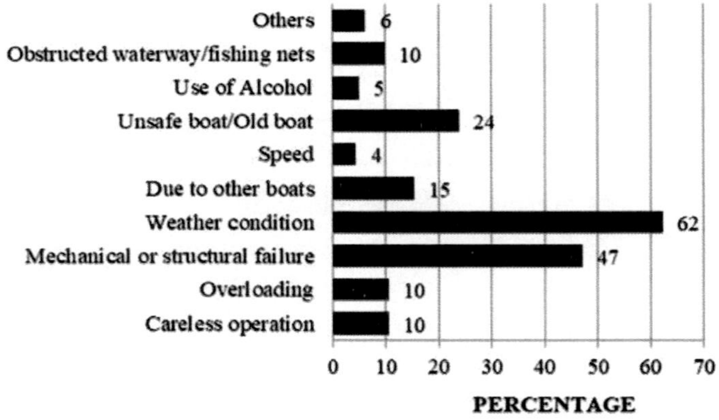

Fig. 5 Factors leading to boat accidents in Alappuzha as per crew

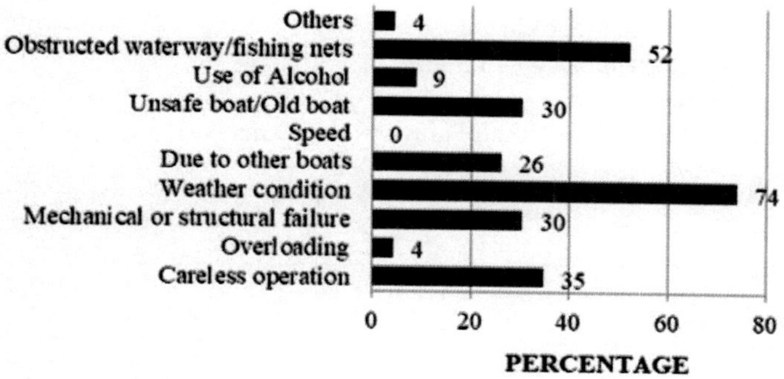

Fig. 6 Factors leading to boat accidents in Kollam as per crew

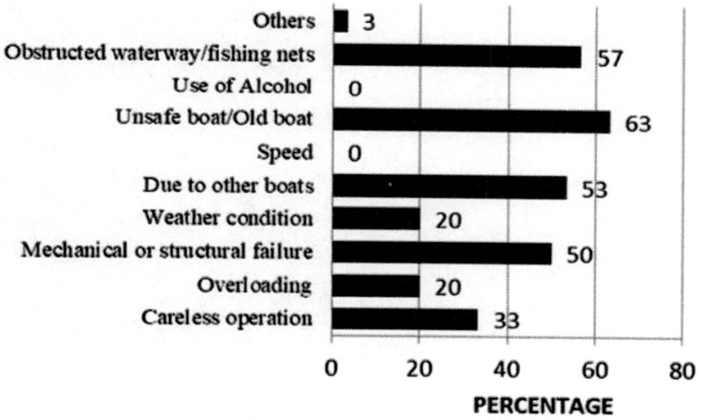

Fig. 7 Factors leading to boat accidents in Ernakulam as per crew

5 Study Area

5.1 Waterway Selection Criteria

In the initial stage of risk assessment, waterways with more users were identified. Passenger traffic is one of the major uses of waterways in Kerala. Also the accident data of the boats in Kerala were collected form the Port Offices and analyzed for trends and peculiarities in the accident pattern. The major accident-prone areas were identified for the study. Areas with busy navigation were also considered. The lists of waterways selected for the study are

1. Alappuzha—Kottayam Canal (AK Canal)—24 km
2. Alappuzha—Changanacherry Canal (AC Canal)—29 km
3. Nedumudi—Edathva—14.8 km

4. Vytila—Kakkanad—5.2 km
5. Ernakulam—Varappuzha—13.3 km
6. Kollam—Sambranikodi—5 km
7. Panavally—Irappuzha—South Paravur—10.6 km.

Map of the selected waterways is given in Fig. 8.

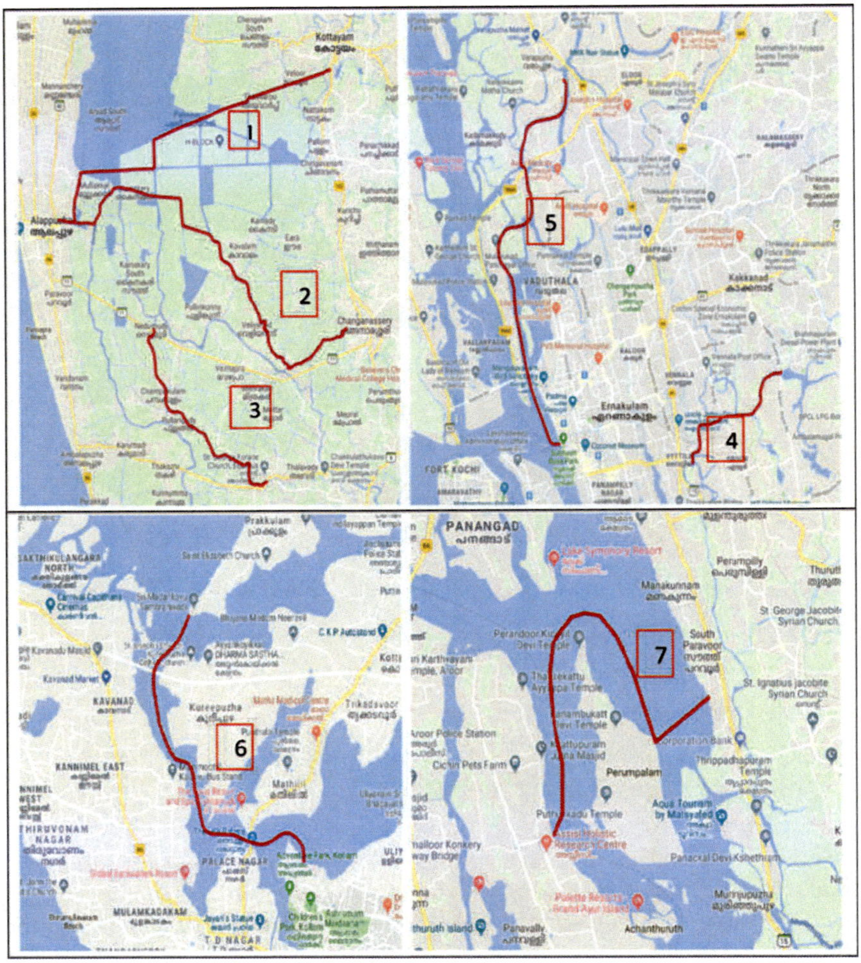

Fig. 8 Map showing study waterways

5.2 Description of the Selected Waterways

A brief description of the selected waterways for the analysis is given in this section. The waterways in Kerala differ in its configuration, use, geography, location, and in importance to its users.

Waterway No. 1—Alappuzha–Kottayam Canal (24 km). Alappuzha–Kottayam Canal also known as AK Canal is a part of Alappuzha–Kottayam–Athirampuzha Canal and was declared as National Waterway No. 9 in the year 2016 [3]. State Water Transport Department (SWTD) is operating regular boat services in this section. SWTD is operating 12 boat services in between Kottayam and Alappuzha [4]. This route starts from Alappuzha boat jetty, and the initial 4 km is highly congested with heavy traffic of passenger and recreational boats. 2.5 km length of this reach is a part of NW-3. After that about 2.75 km of this route passes through Vembanad Kayal. Next 4 km passes between the paddy fields, and then for 5 kms, it passes through the kayal. The remaining 8.25 km stretch passes through narrow canal. This section is shallow and silted with weeds obstructing the waterway. Bridges with low clearance and remains of old bridge and construction waste dumped in the waterway are also problems in this section.

Waterway No. 2—Alappuzha–Changanacherry Canal (29 km). Alappuzha–Changanacherry Canal (AC Canal) is the National Waterway No. 8, and SWTD is operating eight boat services in between Alappuzha and Changanacherry. The section starts from Alappuzha boat jetty, and the initial 1 km is highly congested with heavy traffic of passenger and recreational boats. Next 12 km waterway passes through paddy field area and has enough width and depth for navigation. House boats are also operating in this section. The remaining portion of the waterway passes through narrow canals. Depth of waterway is also low due to silting. Low bridges, weeds, sharp turnings, and branches of trees obstructing vision are threats to navigation.

Waterway No. 3—Nedumudi–Edathva (14.8 km). This section of waterway starts from Nedumudi and passes through Kuttanad area. This canal is wider and deeper, and the curves are gentle with good turning radius. Fourteen trips are operated by SWTD in this route. House boats are also operating in this area.

Waterway No. 4—Vytila–Kakkanad (5.2 km). This is a 5.2 km waterway connecting Vyttila mobility hub and Kakkanad area. This section is wide and deep, and there are no obstructions in this waterway route. Cargo boats are also operated in this section carrying petroleum and chemicals to the industrial area of Ambalamugal. SWTD is operating 16 trips in this route.

Waterway No. 5—Ernakulam–Varappuzha (13.3 km). This section has 13.3 km length, and it starts from Ernakulam boat jetty and ends at Varappuzha jetty. This route crosses many ferry routes, especially at the staring region. Sea opening is very near to the starting point causing disturbance in the waterway. The waterway is wide and deep, and the curves are gentle. SWTD operates four trips in this route.

Waterway No. 6—Kollam–Sambranikodi (5 km). This waterway route passes entirely through backwaters, and the waterway is wide and deep. SWTD is operating around 60 short trips in this region connecting the two shores of the kayal. This section is a part of NW-3.

Waterway No. 7—Panavally–Irappuzha–South Paravur (10.6 km). This route passes through Vembanad Lake and is very wide and deep. It crosses the NW-3.

6 Waterway Risk Assessment Model

The waterway risk assessment model has a linear-additive value function consisting of a linear combination of the attributes [2]. If we denote n factors that should be considered in the decision and denote the level of the attributes for a particular waterway by $x_1, \ldots .x_n$, then we can denote the linear-additive value function as

$$v(x_1, \ldots .x_n) = w_1 v_1(x_1) + \ldots + w_n v_n(x_n), \tag{1}$$

where $v(x_1, \ldots .x_n)$ is a value or preference function that allows us to rank alternatives, the $v(x_i)$ are single-dimensional value functions that essentially convert each attribute x_i to a common scale, and the w_i reflect the importance of each attribute x_i to overall preference over the range of alternatives considered. x_i is the objective attribute that is to be measured on either constructed or natural scales. The weights w_i and single-dimensional value functions $v(x_i)$ can be calculated from the analysis of accident data, questionnaire survey, and inventory of the canals.

For each attribute, weights w_i are calculated based on the collected data. For this, best-case and worst-case waterways were chosen based on field experience and interaction with officials and passengers using the waterway. For worst case, all attributes were set at the levels of the highest, and for the best case, the weights are kept at lowest. The attributes were compared in pairs. For instance, in the first comparison, fleet composition and traffic conditions were compared. The best and worst scenarios were compared with each case and the criterion which would be improved and by how much it is identified and values are assigned accordingly. The weights are given in such a way that the total weight of all factors sum up to 100.

6.1 Risk Categories (Generic Weights Sum to 100)

1 Fleet composition
2 Traffic conditions
3 Navigational conditions
4 Waterway configuration
5 Immediate consequences

5 Subsequent consequences.

Six major categories of risk were identified, and weights were assigned for these major risk categories in such a way that the total value will be 100. For some of the attributes x_i in the model, evaluation measures are constructed because no natural scales were available. Table 1 gives the factor scale list.

Major risk factors in each category were identified and ranked into four categories with one category having ideal condition with scale value 1 and other with scale value 9 as worst condition. Each of the two in-between values is ranked depending upon the severity of the condition. These scale values are used as reference values, and each waterway is ranked for each factor accordingly.

7 Results and Discussion

The selected waterways were assessed for various safety factors, and values were assigned to major risk categories for each waterway as shown in Table 2.

Other factors in each category are given priorities in a scale of 1–9. An illustration of the same (Vyttila Kakkanad section) is shown in Table 3.

The risk factor calculated for each waterway section is given in Table 4.

It can be seen that the Vyttila–Kakkanad section (waterway no. 4) is the risky stretch of water for navigation. This waterway has hazardous cargo operation to the industrial area. Moreover, the sides of the waterway are highly populated as it is near to Kochi city. The waterway has bridges, and width of the waterway is just sufficient for navigation, but these factors make the crossing of boats a bit difficult. The second waterway in the list is Alappuzha–Changanacherry Canal, which is a narrow silted canal with sharp bends and low bridges making navigation risky. The Alappuzha–Kottayam Canal is the third risky canal as some portion of the canal is narrow with low depth and obstructions. Component-wise risks of the waterways are tabulated and shown in Fig. 9.

8 Conclusion

The methodology for developing a risk assessment model for the selected waterways in Kerala is explained in the paper. Based on questionnaire survey and accident data analysis, the risk factors were identified and ranked. The attributes related to risk are those that affect the probability of accident and the impacts or consequences of potential accidents. So the attributes determined are grouped into a tree-like hierarchy with six main groups; four of these groups affect accident frequency, and the other two affect the consequences of potential accidents. The factors considered in the present model included condition of vessels, traffic conditions, environmental conditions, waterway configuration, short-term consequences, and long-term impacts. Scales to

Table 1 Factor scale list

Factor scales-condition list		Scale value	Factor scales-condition list		Scale value
Wind conditions			*Visibility conditions*		
a	Severe wind < 2 days/month	1	a	Poor visibility < 2 days/month	1
b	Severe winds occur in brief periods	2.5	b	Poor visibility occurs in brief period	2.1
c	Severe winds are frequent and anticipated	4.6	c	Poor visibility is frequent and anticipated	4.9
d	Severe winds occur without warning	9	d	Poor visibility occurs without warning	9
Tide and river currents			*Rain conditions*		
a	Tides and currents are negligible	1	a	Rain is not a problem for navigation	1
b	Currents run parallel to the channel	2.1	b	Rain may affect navigation	2.5
c	Transits are timed closely with tide	4.8	c	Severe rain and flooding	5.5
d	Currents cross-channel/turns difficult	9	d	Vessels need to stop during rain	9
Visibility obstructions			*Channel width*		
a	No blind turns or intersections	1	a	Meeting and overtaking are easy	1
b	Good geographic visibility –intersections	2	b	Passing arrangements needed ample room	2.3
c	Visibility obscured, good communications	4.4	c	Meeting and overtaking in specific areas	6.7
d	Distance and communication limited	9	d	Movements restricted to one-way traffic	9
Bottom type			*Waterway complexity*		
a	Deep water or no channel necessary	1	a	Straight run with NO crossing traffic	1
b	Soft bottom, no obstruction	1.6	b	Multiple turns > 15° NO crossing	2.7
c	Mud, sand, and rock outside channel	4.4	c	Converging NO crossing traffic	4.7
d	Hard or rocky bottom at channel edges	9	d	Converging with crossing traffic	9
Number of people on waterways			*Petroleum volume*		

(continued)

Table 1 (continued)

Factor scales-condition list		Scale value	Factor scales-condition list		Scale value
a	Industrial, little recreational boating	1	a	Little or no petroleum cargoes	1
b	Recreational boating and fishing	3.4	b	Petroleum for local use	2.5
c	Cruise and excursion vessels ferries	6	c	Petroleum for transshipment inland	4.6
d	Extensive network of ferries, excursions	9	d	High volume petroleum and LNG/LPG	9
Chemical volume			*Economic impacts*		
a	Little or no hazardous chemicals	1	a	Vulnerable population is small	1
b	Some hazardous chemical cargo	2.3	b	Vulnerable population is large	3.3
c	Hazardous chemicals arrive daily	4.8	c	Vulnerable, dependent, and small	5.3
d	High volume of hazardous chemicals	9	d	Vulnerable, dependent, and large	9
	Environmental Impact			Health and safety impacts	
a	Minimal environmental sensitivity	1	a	Small population around waterway	1
b	Sensitive, wetlands, vulnerable	3.2	b	Medium–large population around waterway	2.3
c	Sensitive, wetlands, endangered	6.1	c	Large population, bridges	5.4
d	Endangered species, fisheries	9	d	Large dependent population	9

Table 2 Category-wise risk of waterways

Waterway no	Fleet composition	Traffic conditions	Navigational conditions	Waterway configuration	Immediate consequences	Subsequent consequences
1	10	15	25	35	5	10
2	5	15	35	25	15	5
3	8	15	22	40	5	10
4	3	10	20	20	42	5
5	15	25	30	5	15	10
6	10	10	35	10	25	10
7	10	20	35	15	15	5

Table 3 Waterway risk assessment for Vyttila–Kakkanad section

Major risk	Fleet composition	Traffic conditions	Navigational conditions	Waterway configuration	Immediate consequences	Subsequent consequences
	3	10	20	20	42	5
Components	Percentage of high-risk deep draft vessel	Volume of deep draft vessels	Wind conditions	Visibility obstructions	Number of people on waterway	Waterway economic impacts
	6	6	2	4	2	7
	Percentage of high-risk shallow draft vessel	Volume of shallow draft vessels	Visibility conditions	Channel width	Volume of petroleum cargoes	Environmental impacts
	2	2	6	6	8	7
	Age of vessel	Volume of fishing and pleasure craft	Tide and river currents	Bottom type	Volume of hazardous chemical cargoes	Health and safety impacts
	2	1	3	3	8.5	7
	Type of vessel	Traffic density	Rain conditions	Waterway complexity	Accessibility by other modes	
	1	4	4	7	3	

Table 4 Calculated risk factor for each waterway sections

Sl No	Waterway route	Risk factor
1	Alappuzha–Kottayam Canal	1720
2	Alappuzha–Changanacherry Canal	1782.5
3	Nedumudi–Edathva	948.5
4	Vytila–Kakkanad	1871
5	Ernakulam–Varappuzha	1415
6	Kollam–Sambranikodi	1340
7	Panavally–Irappuzha—South Paravur	1320

measure each variable were established from the available data. The model is a linear combination of the attributes, like a linear-additive value function, where single-dimensional value functions that essentially convert each attribute are multiplied with a weight reflecting the importance of each attribute. The model allows comparison of relative risk and the potential efficacy of various improvements between different waterways. Thus, the model is based on multi-attribute decision analysis techniques and the effects of safety alternatives would have on these levels. This model can be used to categorize each waterway according to the safety and risks involved and thus to prioritize the waterway improvement decisions. Such a model will be very useful

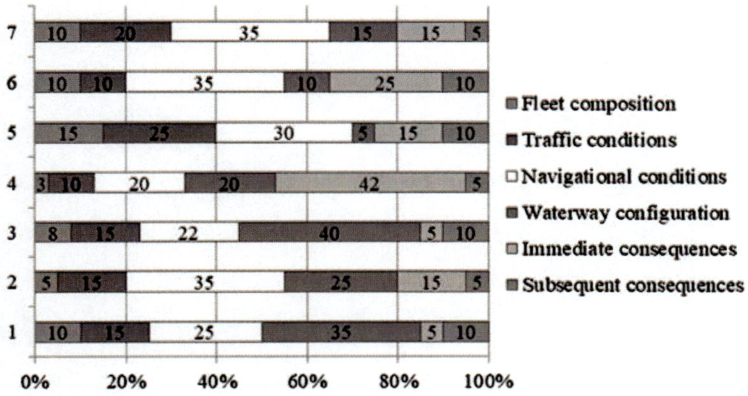

Fig. 9 Component-wise risk factors in each waterway

to decision-makers who otherwise would make judgments by informal comparison of various factors.

References

1. Performance audit of Inland Water Transport, CAG, Govt. of India, https://cag.gov.in/sites/default/files/audit_report_files/Chapter_2_Performance_Audit_on_Inland_Water_Transport.pdf
2. Keeney RL, Raiffa H (1976) Decisions with multiple objectives: preferences and value tradeoffs. Cambridge University Press, New York
3. IWAI website, http://iwai.nic.in/index1.php?lang=1&level=1&sublinkid=124&lid=143, Last accessed 15 June 2019
4. SWTD Website, https://www.swtd.kerala.gov.in/pages-en-IN/boatschedules-index.php, Last accessed 15 June 2019
5. Merrick JRW, Harrald JR (2007) Making decisions about safety in US ports and waterways. Interfaces 37(3):240–252
6. Hazard Identification (2012) Risk assessment and control procedure. University of Western Sydney
7. Kalyani T, Vidyasagar DSP, Srinivas VSJ (2015) Accident analysis of river boats capsize in Indian Inland waters and safety aspects related to passenger transportation. Int J Innovative Res Dev 4(7) (2015)
8. Azad AK (2009) Riverine passenger vessel disaster in Bangladesh.: Options for mitigation and safety, BRAC University, Dhaka Bangladesh
9. Zhang D, Yan X, Yang Z, Wang J (2014) An accident data–based approach for congestion risk assessment of inland, waterways: AYangtze River case, Proc IMechE Part O J Risk Reliab 228(2)

A Review Study on Utilization of Jarosite Waste Produced During Zinc Extraction in Construction of Rigid Pavements

Surendra Kumar Saini, R. N. G. D. Ransinchung, and Praveen Kumar

1 Introduction

Jarosite is a waste material generated as a by-product in the extraction of zinc and lead from the ore concentrate during the hydro-metallurgical leaching of concentrate in the lead–zinc smelter [1]. Due to toxic ingredients like lead, zinc, copper, cadmium, etc., jarosite is universally considered hazardous waste [2–7]. About fifty percent of zinc is present in the ore concentrate, which is roasted at nine-hundred-degree centigrade temperature, and after that, the leaching process is carried out where jarosite gets produced as waste material [1]. Due to the acidic nature of jarosite in its fresh state, the addition of lime is done before its dumping to neutralize it [8, 9]. Leaching from stored jarosite can contaminate groundwater. Jarosite, when mixed with 2% lime, 10% cement and water, is called "Jarofix" [10]. Throughout the world, more than 85% of zinc extraction industries use the roast-leach-electrowinning process. In producing one-ton zinc metal through this process, a half ton of jarosite waste is typically generated [11]. At Hindustan Zinc limited, Chittorgarh, Rajasthan, about five lacs metric tons Jarosite is generating yearly, and the stored part is about fifteen lac metric tons [10] (Fig. 1).

S. K. Saini · R. N. G. D. Ransinchung (✉) · P. Kumar
Department of Civil Engineering, Indian Institute of Technology, Roorkee 247667, India
e-mail: gdranfce@iitr.ac.in

S. K. Saini
e-mail: ssaini2@ce.iitr.ac.in

P. Kumar
e-mail: pkaerfce@iitr.ac.in

M. Parida et al. (eds.), *Proceedings of the Fifth International Conference of Transportation Research Group of India* , Lecture Notes in Civil Engineering 220,
https://doi.org/10.1007/978-981-16-9925-2_12

Fig. 1 A photograph of jarosite [12]

Table 1 Physical properties and composition of jarosite [13–15]

Physical properties		Composition (%)	
Chemical name	Potassium iron sulfate hydroxide	Oxygen	44.73
Chemical formula	$KFe_3(SO_4)_2(OH)_6$	Iron	33.45
Color	Amber yellow or dark brown	Sulfur	12.81
Streak	Pale-yellow	Potassium	7.81
Density	2.9–3.26 g/cm3	Hydrogen	1.21
Transparency	Translucent to opaque		
Luster	Vitreous to dull		

1.1 General Composition and Physical Properties of Jarosite

Jarosite has a trigonal crystal structure with a molecular weight of 500.81 gm, specific gravity 3.15–3.26, and hardness of 2.5–3.5 Mohs [13]. The general composition and physical properties of the jarosite are given in Table 1.

1.2 Particle Size Distribution of Jarosite

The gradation or particle size distribution refers to the relative amount, typically by mass, of particles present according to size. Jarosite contains about sixty-four percent silt-sized and about thirty-two clay-sized particles [16].

1.3 Chemical and Mineralogical Characterization

Rathore et al. [16] and some other authors have listed in detail the chemical analysis of jarosite refer Table 2. As per IS 3812 2003 Part-I, minimum ($SiO_2 + Al_2O_3 + Fe_2O_3$) content is 70% for using fly ash as pozzolanic materials in concrete. This total is

Table 2 Chemical analysis of jarosite

S. no	Parameters	Concentration in Jarosite wt (%) [16]	Concentration in Jarosite wt (%) [17]	Concentration in Jarosite wt (%) [18]
1	SiO_2	6.75±0.412	6.31±0.7	8.09
2	Al_2O_3	6.75±0.152	1.42±0.02	5.29
3	MgO	1.86±0.067	–	0.3
4	Fe_2O_3	32.12±0.436	44.94±0.05	36.75
5	CaO	6.87±0.151	–	8.02
6	ZnO	9.18±0.175	10.91±0.02	–
7	Na_2O	0.61±0.424	–	3.06
8	K_2O	0.74±0.023	–	–
9	PbO	1.95±0.132	–	–
10	SO_3	31.19±0.246	–	31.25

Table 3 Heavy metals trace elements in jarosite

Parameters	Concentration in Jarosite (approx.) [15] (ppm)	Concentration in Jarosite (approx.) [17]
Copper	1040	–
Manganese	1970	–
Cadmium	320	–
Chromium	180	416.59±14.35 ppm
Nickel	90	93.50±1.10 ppm
Cobalt	39	30.38±0.27 ppm

about 50% for jarosite which is not up to the mark but indicates that it may have the possibility to be used in combination with fly ash in concrete mixes.

The available amount of heavy metals trace in jarosite reported by different researchers is given in Table 3.

1.4 Radioactivity Level of Jarosite

Asokan et al. [15] have listed in detail the radioactive characteristics of jarosite refer to Table 4.

Table 4 Radioactivity level of jarosite [15]

Radionuclides	Activity level of radionuclides (Bq/kg)	
	Jarosite	Upper limit
40 K (β emitters)	294.66±17.68	925
226 Ra (α emitters	55.40±0.86	370
228 Ac (α emitters)	72.00±3.47	259

Table 5 Physicochemical characteristics of jarosite [16]

Parameter	Range
Electrical conductivity	13.26±0.437 dS/m
Humidity	44 to 47%
pH	2.6–4.0

1.5 Physicochemical Characteristics

Rathore et al. [16] have listed in detail the physicochemical characteristics of jarosite refer to Table 5.

2 Literature Review

A little work has been done on zinc industry waste. But the various reviewed studies reveal that the strength and other vital properties of the concrete can be enhanced up to a significant amount by using jarosite with Portland cement. Further, related to the use of jarosite in PQC mixes are rare.

Arora et al. [14], in their study on PQC mixes, used jarosite and fly ash together as part replacement of cement. In the study, different doses of fly ash were used with a constant 10% replacement level of the jarosite dose. The results show that substitution with jarosite has positive effects on fresh concrete properties like workability and hardened concrete properties compressive and flexural strength. For the concrete mixes given in Table 6, results shown in Figs. 2, 3, Tables 7, and 8 have been reported for compressive strength and flexural strength in the study.

It can be predicted from Figs. 2, 3, Tables 7, and 8 that the strength increases with the increase in jarosite dose, and the maximum value is achieved at 10% dose of jarosite. Further, for higher cement replacement levels, the strength starts decreasing with the rise in fly ash dose. Moreover, the increment in flexural strength of the control mix is 51.24% which is much higher compared to the 39.11% increment of the compressive strength. While for all other mixes, the increment in flexural strength is almost the same as compressive strength. For the role of supplementary cementing materials in the pozzolanic reaction, the total of $SiO_2 + Al_2O_3 + Fe_2O_3$ contents is considered [19]. Hence, the observed strength increase may be attributed to the same

Table 6 Mix composition considered by Arora et al. [14] for cement (C) with jarosite (J) and fly ash (F) [14]

Mix	Cement substitution level (%)	Cementitious material		
		Cement (%)	Jarosite (%)	Fly ash (%)
CM	0	100	0	0
J10F0	10	90	10	0
J10F10	20	80	10	10
J10F20	30	70	10	20
J10F30	40	60	10	30

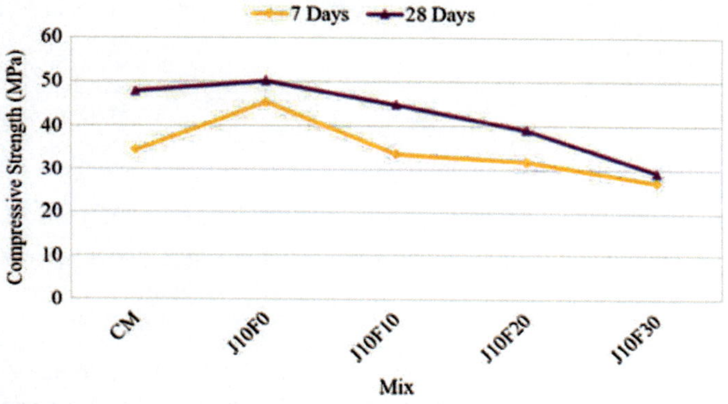

Fig. 2. 7 and 28 days compressive strength reported by Arora et al. [14]

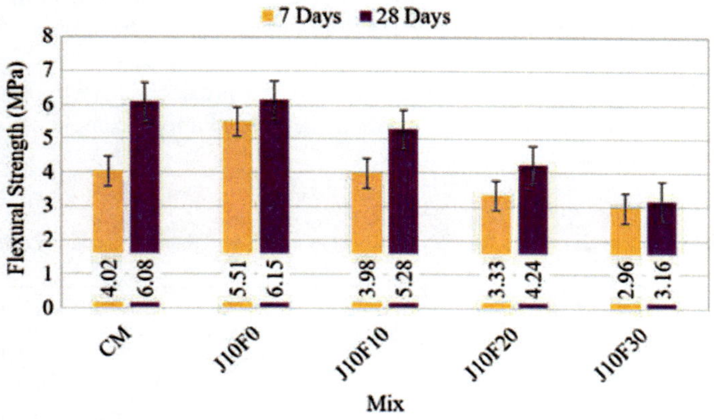

Fig. 3 Flexural strength at 7 and 28 days reported by Arora et al. [14]

Table 7 Percentage increase in compressive strength between 7 and 28 days for the results reported by Arora et al. [14]

Mix	Compressive strength (MPa)		
	7 days	28 days	Percentage increase
CM	34.39	47.84	39.11
J10F0	45.24	50.17	10.9
J10F10	33.47	44.8	33.85
J10F20	31.74	39	22.87
J10F30	26.993	29.33	8.66

Table 8 Percentage change in compressive strength with respect to the control mix for the results reported by Arora et al. [14]

Mix	Percentage replacement	Change in strength w.r.t. CM (%)	
		7 days	28 days
J10F0	10	31.55↑	4.87↑
J10F10	20	2.68↓	6.35↓
J10F20	30	7.71↓	18.48↓
J10F30	40	21.51↓	38.69↓

(Here, ↓ indicates % reduction in the strength and ↑ indicates % rise in the strength)

trend as in the case of various other materials like fly ash, silica fume, copper or zinc slag, etc., due to the pozzolanic content ($SiO_2 + Al_2O_3 + Fe_2O_3$) present in the jarosite or by the finer particle size of the jarosite making the mix denser. Here, it should be noted that the study attempts a constant replacement level of 10% for jarosite with different fly ash doses. Concrete mixes with higher doses of jarosite with fly ash have not been investigated.

Vyas [20] revealed for the concrete mixes that there is a decrease in the unconfined compressive strength of concrete when jarosite is used as a replacement of sand. The results are given in Table 9.

Table 9 Unconfined compressive strength results of concrete with jarosite as part replacement of sand reported Vyas [20]

Cement (kg.)	Fine aggregate (kg.)		Coarse aggregate (kg)	Total weight (kg)	28 days UCS (kg./cm^2)
	Sand	Jarosite			
1.0	2.00	0.00	4.0	7.00	170
1.0	1.60	0.40	4.0	7.00	155
1.0	1.20	0.80	4.0	7.00	140
1.0	0.80	1.20	4.0	7.00	125
1.0	0.40	1.60	4.0	7.00	100
1.0	0.00	2.00	4.0	7.00	65

Table 10 Mixes considered by Katsioti et al. [21] for different ratios of clinker, gypsum, and jarosite [21]

Sample No	Clinker (%)	Gypsum (%)	Jarosite (%)	Substitution of gypsum by Jarosite (%)
1	95.00	5.00	0.00	0
2	95.00	4.75	0.25	5
3	95.00	4.50	0.50	10
4	95.00	4.25	0.75	15
5	95.00	4.00	1.00	20
6	95.00	3.75	1.25	25
7	95.00	2.50	2.50	50
8	95.00	1.25	3.75	75
9	95.00	0.00	5.00	100

Table 11 Setting times and normal consistency results reported for the mixes considered by Katsioti et al. [21]

Sample no	IST (min.)	FST (min.)	Standard consistency (%)
1	145	225	23.2
2	145	220	23.2
3	155	215	23.4
4	150	205	23.7
5	150	215	23.6
6	90	165	24.6
7	60	105	25.6
8	20	32	36
9	15	35	35

Katsioti et al. [21] used jarosite as replacement of gypsum in cement manufacturing, and the following results were observed (refer to Tables 10, 11, and 12) regarding initial setting time (IST), final setting time (FST), standard consistency, and compressive strength for the mixes with different ratios of clinker, gypsum, and jarosite.

The study suggests a 20% dose of jarosite as optimal for substitution of gypsum. However, from the results given in Tables 10, 11, and 12, it can be concluded that cement manufactured with up to 50% replacement of gypsum with jarosite is also achieving the minimum stipulated strength values specified in IRC44:2008 [22] and IRC 15: 2011 [23] for the PQC mixes.

Sinha et al. [10] found that CBR decreased as the jarosite part for jarosite-soil mixes is increased while it increased in the case of the jarosite-bottom ash mixes. Similar to OMC and MDD, it was also evaluated that the change is predominant up to only 75% jarosite content. It was established that jarosite (100%), jarosite-soil mixes (50–75%), and jarosite-bottom ash mixes (50–75%) have the potential for construction of the embankment. According to IRC 58, 2002 [24], design thickness of

Table 12 Compressive strength results reported for the mixes considered by Katsioti et al. [21]

Sample no	2 days strength (MPa)	7 days strength (MPa)	28 days strength (MPa)
1	27.1	37.7	48.1
2	24.3	37.7	44.2
3	25.1	36.6	44.4
4	25.1	37.7	44.5
5	24.0	36.5	44.4
6	23.3	34.7	44.2
7	19.4	34.1	41.67
8	0.6	2.3	39.26
9	0.7	3.0	31.40

the concrete pavement is directly related to the CBR value and modulus of subgrade reaction. Hence, in the context of Sinha et al. [10], it may be possible to reduce the thickness requirements for PQC and DLC layers for rigid pavements by making use of jarosite-bottom ash embankments below these layers.

Mehra et al. [25] have investigated the strength and durability characteristics of M25 concrete containing jarosite as a partial replacement of sand. In the study, toxicity leaching characteristics test at 25% replacement of fine aggregates with jarosite was also performed, and results were within limits as per USEPA (1992) [26] for zinc and copper, but the limits for lead, cadmium, and iron are exceeding. Up to 15% replacement level of jarosite, improvement in strength and durability properties has been reported. However, the cause of these improvements is not reported. Hence, in this context, a detailed investigation is further needed. Furthermore, to perform the toxicity leaching, characteristics test for lower doses of the jarosite is also needed to be studied.

3 Conclusion

Following are the inferences drawn from the present literature review:

1. Past research suggested that there is an increase in CBR value for embankment/subgrade containing jarosite with bottom ash mixes which indicate an increase in modulus of subgrade reaction, i.e., k value, thus reveals that the thickness requirement for concrete pavement slab could be reduced by using jarosite with bottom ash.
2. A detailed investigation is further needed to know the role of jarosite ingredients in strength development when used as partial replacement of cement in concrete mixes like PQC.
3. As for PQC mixes, the flexural strength is more concerned. Available literature reports enhancement in the flexural strength for the concrete containing jarosite.

It indicates that PQC mixes could be made more economical and environmental friendly by using jarosite as a partial replacement of cement which would in turn also solve the problem of its disposal.

4. Jarosite is a resource of great potential to be used in rigid pavement construction, which needs to be recycled in a technically and environmentally feasible manner.

References

1. US_EPA (2014) 12.7 zinc smelting (Epa.Gov/Ttnchie1/Ap42/Ch12/Final/C12s07.Pdf)
2. González-Ibarra AA, Nava-Alonso F, Fuentes-Aceituno JC, Uribe-Salas A (2016) Hydrothermal decomposition of industrial jarosite in alkaline media: the rate determining step of the process kinetics. J Min Metall Sect B-Metal 52(2):135–142 (2016)
3. Shukla A (2017) Head—R&D, Hindustan zinc limited best practices in utilization of alternate raw materials: utilization of Wastes Jarosite and lead zinc slag. In: 3rd international conference on alternate fuels and raw materials in cement industry on 23–24 Mar 2017, Vedanta (2017)
4. Arslan C, Arslan F (2003) Thermochemical review of jarosite and goethite stability regions at 25 °C and 95 °C. Turk J Eng Environ Sci 27:45–52
5. Chen TT, Dutrizac JE (2000) A mineralogical study of Jarofix products for the stabilization of jarosite for disposal. Miner Met Mater Soc (TMS) 917–933
6. Chen T, Dutrizac J, Seyer S (2001) Jarofix: Addressing iron disposal in the zinc industry. JOM 53(12):32–35
7. Pappu A, Saxena M, Shyam RA (2011) Waste to wealth—cross sector waste recycling opportunity and challenges. Can J Environ Constr Civil Eng 2(3):(2011)
8. Report CRRI (2010) Feasibility Study of Jarofix waste material for embankment and sub grade. Published by CSIR-Central Road Research Institute, New Delhi
9. Sinha AK, Havanagi VG, Shahu JT (2019) Stabilised jarofix waste material for road construction. Int J Pavement Eng 1–12
10. Sinha AK, Havanagi VG, Arora VK, Ranjan A, Mathur S (2011) Feasibility study of Jarofix waste material for road construction. In: Proceedings of Indian geotechnical conference, 15–17 Dec 2011, Kochi (Paper No.L-041.), pp 685–688
11. Orko I, Kangas P, Pajarre R, Lundström M, Koukkari P (2016) Hydrometallurgical processing of Jarosite waste
12. Chayan G, Arun P (2018) Variables controlling strength of lime stabilized Jarosite waste. Int J Geo-Eng 9:6 (2018)
13. Piyush S (2016) Feasibility study of industrial Jarosite waste as vital material for construction: positive and negative aspects. Malays J Civil Eng 28(1):139–154 (2016)
14. Arora V, Sachdeva SN, Aggarwal P (2015) Effect of use of Jarosite on workability and early age strength of concrete. IJCMS 4 (2015). ISSN 2347–8527
15. Asokan P, Saxena M, Asolekar S (2006) Hazardous Jarosite use in developing nonhazardous product for engineering application. J Hazard Mater 137(3):1589–1599
16. Rathore N, Patilb MP, Dohare D (2014) Utilization of Jarosite generated from lead-zinc smelter for various applications. Int J Civil Eng Technol 5:192–200
17. Kerolli-Mustafa M, Ćurković L, Fajković H, Rončević S (2015) Ecological risk assessment of Jarosite waste disposal. CROATICA CHEMICA ACTA. 88. https://doi.org/10.5562/cca2554
18. Sinha AK, Havanagi VG, Arora VK, Ranjan A, Mathur S (2012) Recycling Jarofix waste as a construction material for embankment and sub grade. J Solid Waste Technol Manage 38(3):169–181
19. IS 3812(Part 1) (2003) Specification for pulverized fuel ash Part 1 for use as pozzolana in cement, cement mortar and concrete

20. Vyas AK (2011) Solidification-stabilization technique for metal bearing solid waste from zinc industry—A case study. Int Conf Environ Comput Sci 19:151–155
21. Katsioti M, Boura P, Agatzini S, Tsakiridis P, Oustadakis P (2005) Use of Jarosite/Alunite precipitates as a substitute for gypsum in Portland cement. Cement Concr Compos 27:3–9
22. IRC44 (2008) Guidelines for cement concrete mix design for pavements. IRC
23. IRC 15 (2011) Standard Specifications and code of practice for construction of concrete roads. IRC
24. IRC: 58 (2002) Guidelines for the design of plain jointed rigid pavements for highways. The Indian Road Congress, New Delhi
25. Mehra P, Gupta RC, Thomas BS (2016) Assessment of durability characteristics of cement concrete containing Jarosite. J Cleaner Prod 119:59e65 (2016)
26. USEPA (1992) Newly identified mineral processing waste characterization data set, Published by United States Environmental Protection Agency (USEPA), USA Office of Solid Waste, vol 3, pp 42–1–42–3

Impact of Advocacy of Infrastructural Development Policies Through a Road Safety Perspective—A Bengaluru Case Study

Ashish Verma, Sajitha Sasidharan, and Prithvi Bhat

1 Introduction

India is ranked the second among world countries, with the longest road length of 5,903,293 km [1]. Rapid urbanization and industrialization in the country have increased the demand and decreased the capacity of the existing roads. This has led to construction of new road lengths. A constant advocacy for infrastructure development like widening of roads, constructing new roads, flyovers, etc., can be evidently seen in the financial budget announcements and policy schemes of the central and state government in India. This is with the notion that additional infrastructure designed as per safety standards and specification will help in decreasing the road crashes. Road crash deaths and injuries are a major concern globally. Road crashes not only incur personal loss but also economic loss. A 3% loss in GDP is estimated associating with road crashes in India. Ministry of Road Transport and Highways (MoRTH) launched a road safety mission, with the aim of decreasing half the road accidents by 2020 than 2010. But the statistics shows that the mission is way behind in achieving the target.

Bengaluru is situated at the southern part of India and is nicknamed as the "silicon valley of India," as it is the major hub for information technology in India. Due to its drastic and rapid transformation into the world's most dynamic city, Bengaluru has been selected as an ideal case for this study. These developments have initiated huge migration to the city burdening the existing infrastructure. Fatalities and injury trends seem to be fluctuating over the years. Statistics show that safety has not improved as aimed by the safety vision by MoRTH.

Making the roads more safe includes improving many parameters like road infrastructure design, vehicle design, driver behavior, road user behavior, demographics,

A. Verma (✉) · S. Sasidharan · P. Bhat
IISc Sustainable Transportation Lab. (IST Lab.), Department of Civil Engineering, Indian Institute of Science, Bangalore, Karnataka 560012, India
e-mail: ashishv@iisc.ac.in

M. Parida et al. (eds.), *Proceedings of the Fifth International Conference of Transportation Research Group of India* , Lecture Notes in Civil Engineering 220,
https://doi.org/10.1007/978-981-16-9925-2_13

trauma care facility, vehicles mix on the road, vehicle numbers, etc. The main objective of the paper is to understand the correlation of road crashes and road length along with other variables. Both injuries and fatalities were modeled. Negative binomial regression modeling was adopted since the data was over dispersed (the mean being very much less than variance). The independent variables were combined in different combinations to understand their effect on road traffic injuries and fatalities.

The structure of the paper is as follows. A brief literature review is followed by data, trends, and methodology. The following section includes analysis and results from the models and followed by conclusion as the last section.

2 Literature Review

Many studies have been done understanding road accidents and infrastructure and most of the studies are about improving road safety consider roadway geometric design as the important parameter [2–4].

Noland [5] conducted a study in different states in UN to understand the impact of infrastructure effect on fatalities and injuries. Different variables indicating infrastructure development was taken like type of road length, divide road length, road width, trauma care facilities, etc. As a proxy variable for improvement in medical technology, infant mortality rates were considered. Other variables like alcohol consumption and traffic rules like wearing seat belts were also considered. The study found less alcohol consumption and usage of seatbelts reduced fatalities. Study showed that infrastructure impacted fatalities and injuries.

Many microscopic studies focusing straight stretch, intersections, etc., are done of which Porter and England [6] studied that red lights at intersections had more probability of road users being in a crash. On a macroscopic level also, many studies have been conducted how engineered road increase the risk. Mahalel and Szternfeld [7] studied the impact of highly designed roads on road user safety. The study concluded that drivers tend to decrease their concentration and underestimate the difficulty in highly safe and engineered roads which ultimately increases the risk on the road users.

Studies have taken different modeling method in analysis of road traffic fatalities and injuries, but had many gaps and limitations regarding the data. Shaw et al. [8]. investigated the ability of different models to be used in modeling road accident. They suggest since the accident data is over-dispersed, using Poisson distribution may understate the occurrence of crashes. Hence, using more basic distributions like double Poisson or negative binomial distribution models is more fitting for the road accident data.

3 Data

The data for the analysis was collected from different secondary sources. The dependent variable was traffic injuries and fatalities. The data was collected from the annual reports published by National Crime Record Bureau (NCRB), which has reported aggregate number city-wise traffic injury and fatality data. The traffic injury data and fatalities data from NCRB were available from 1967 to 2015. The other independent variables taken were population, population density, vehicles registered, and road length. These data were taken from different official government Web sites and official reports. Population data was availed from the national census reports which is conducted every other decade. Population data was available for 1961, 1971, 1981, 1991, 2001, and 2011. Data for missing years was interpolated from the decadal population. Population density was then calculated based on the population. Numbers of vehicles registered were available from NCRB and also from road transport organization (RTO) reports. Vehicles registered before 2013 were taken from NCRB report. From 2013, the data was taken from Bengaluru RTO vehicle statistics. Road length for the city was not very extensively available. Data regarding road length was only available for few years. The data available was for 1960, 2010, 2011, and 2013. This was available from two different sources, namely Bruhat Bengaluru Mahanagara Palike (BBMP) which is responsible for administrative activities in Bangalore and from Bangalore mobility indicators report. It was assumed an exponential growth in the road network in the city since much of data were not available.

3.1 Trends

General trend variation for different variables seen in the data over the study period is shown in the following sections.

Figure 1 shows the population density trend in Bengaluru city over the study time period.

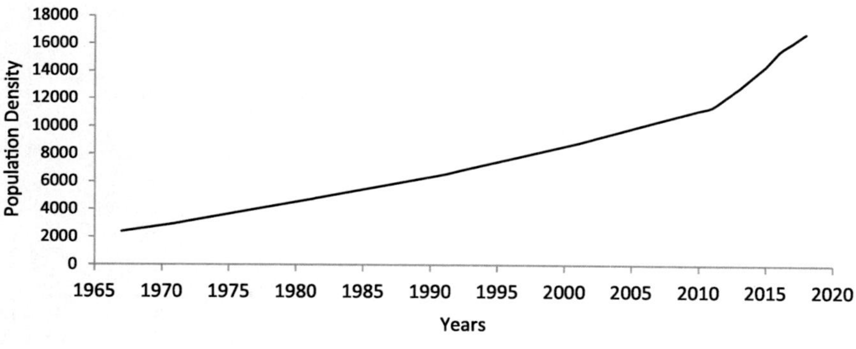

Fig. 1 Population density trend over the years

Figure 2 shows the injury trend over the years. Injury trend shows an overall decreasing trend over the years in the city.

Figure 3 shows the road accident fatality trend over the years in Bengaluru city. Fatality cases over the years show a highly fluctuating trend. But the trend shows an overall increase in the number of fatalities.

Figures 4 and 5 show the male and female fatality trend in the city over the years. Male fatalities are more than female fatalities. More work trips are made by male than females, and this could be the reason of more male fatalities compared to females.

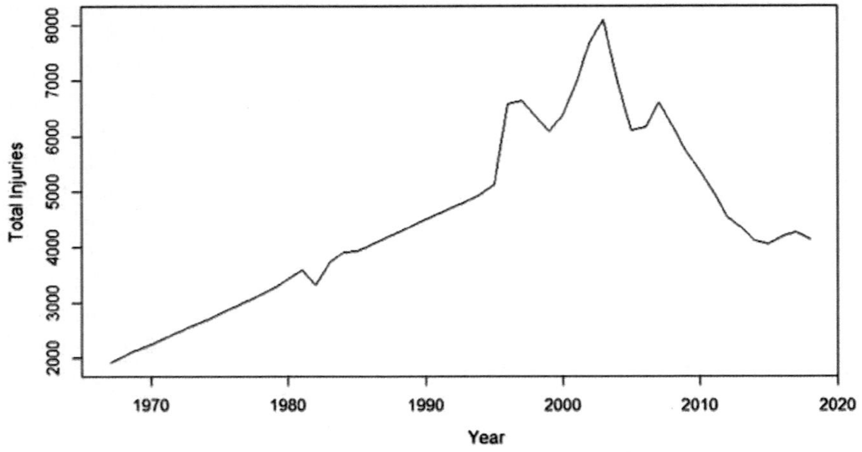

Fig. 2 Injury trend in the city over the years

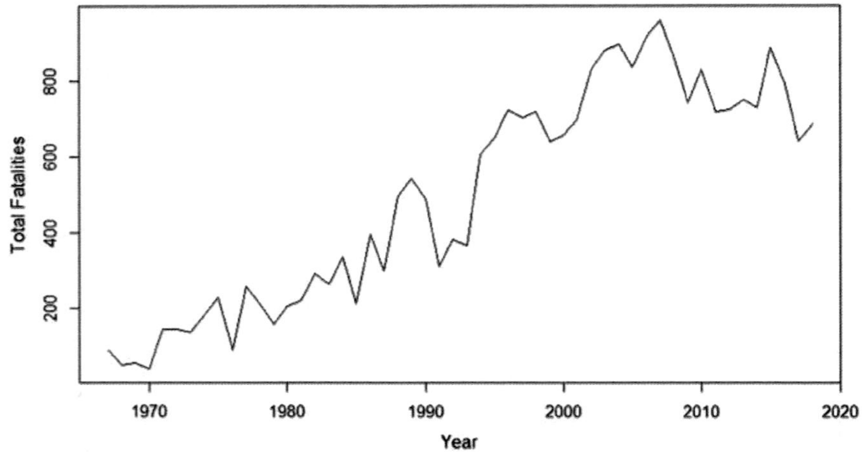

Fig. 3 Road accident fatality trend in the city over the years

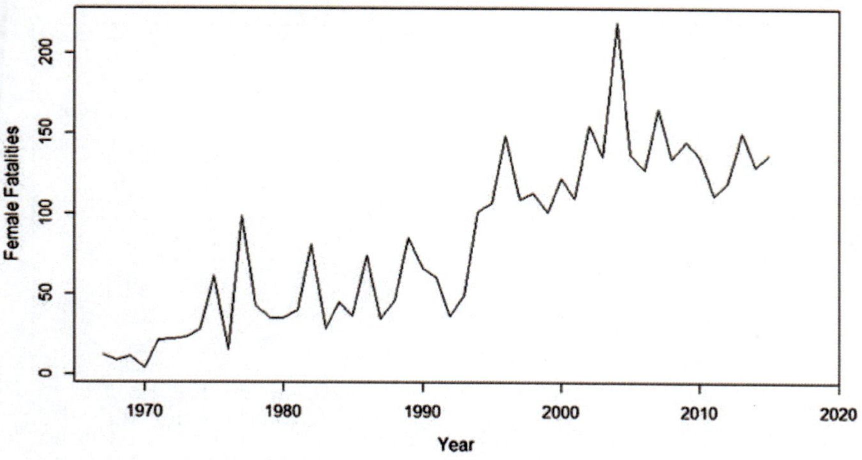

Fig. 4 Female fatality trend over the years

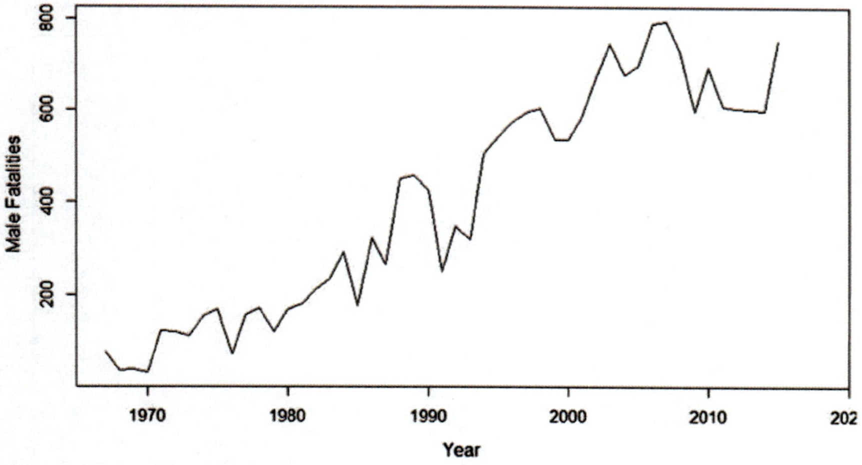

Fig. 5 Male fatality trend over the years

Figure 6 displays the increase in road length over the years taken for analysis.

Figure 7 shows the trend of traffic fatalities rate per length of road over the years. A fluctuating trend is seen. Around 2008, a decrease in the rate is seen compared to 1990 to 2000s.

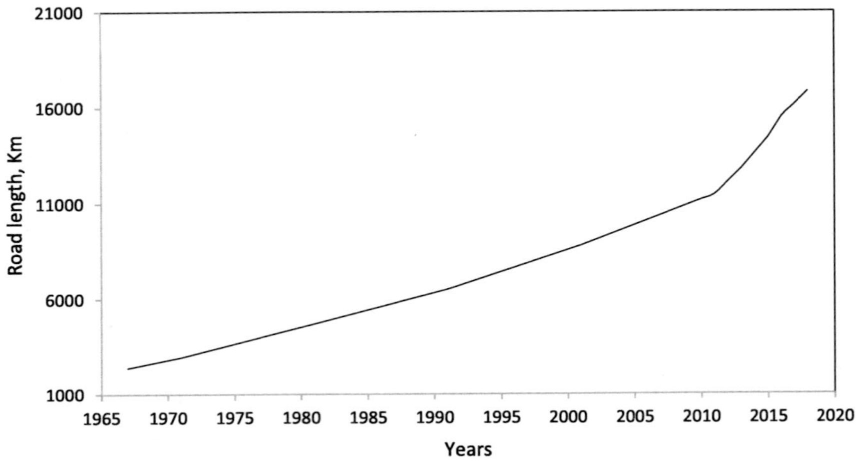

Fig. 6 Road length expansion over the years

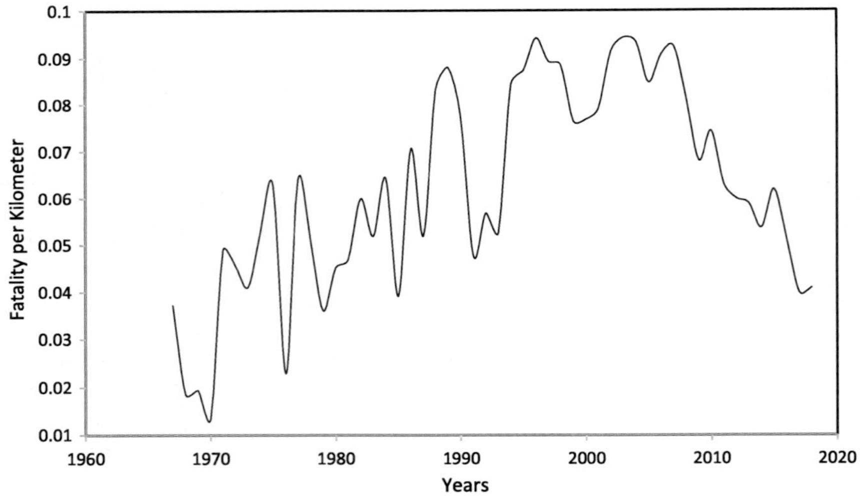

Fig. 7 Fatalities per kilometer over different years

4 Methodology

From the literature for analyzing traffic injuries and fatalities, negative binomial regression modeling is the best method as it accounts for the large variance with respect to the mean. Although results were obtained using Poisson distribution, but the basic assumption of mean equal to variance was not valid. The basic equation for negative binomial regression is as follows:

$$\mu_i = \exp(\ln(t_i) + \beta_1 x_{1i} + \beta_2 x_{2i} + \cdots + \beta_k x_{ki}). \tag{1}$$

where μ is the mean incidence of exposure. β_1, β_2, etc., are the regression coefficient of the parameter estimated.

Equation for negative binomial regression model is:

$$\Pr(Y = y_i | \mu_i, \alpha) = \frac{\Gamma(y_i + \alpha^{-1})}{\Gamma(\alpha^{-1})\Gamma(y_i + 1)} \left(\frac{1}{1 + \alpha\mu_i}\right)^{\alpha^{-1}} \left(\frac{\alpha\mu_i}{1 + \alpha\mu_i}\right)^{y_i}. \tag{2}$$

Freeware *R* software was used for analysis of the data.

5 Results

The analysis was done in *R* using negative binomial regression. The dependent variables taken were total fatalities and total injuries. The independent variables adopted were considering categories like infrastructure development indicated by road length, demographics indicated by city population and population density, and economic development indicated by new vehicle registrations.

For analysis, different combinations of independent variables were combined to understand the effects for both fatalities and injuries. Table 1 shows the model results for total fatalities and road length.

A positive relation between fatalities and road length was obtained as per the model. As the road length increases, the number of fatalities also tends to increase. The intercept value is very in the model indicating unexplained causes other than road length in increasing the number of fatalities.

The model result of the effect of increase in population over the years and its correlation to the number of fatalities is given in Table 2.

As anticipated, a positive relation is obtained between total fatalities and population. As the population increases, the number of accidents also increases. When population increase, the number of trips made increase, and hence, fatalities will also increase as the exposure increases to risk.

Table 1 Model results for total fatalities and road length

Variables	Coefficients	Std. error	*Z* value	Pr(>|z|)
Intercept	4.712e + 00	1.429e-01	32.97	< 2e-16
Road length	1.774e-04	1.692e-05	10.48	< 2e-16

Table 2 Model results for relationship between total fatalities and population

Variables	Coefficients	Std. error	*Z* value	Pr(>|z|)
Intercept	4.712e + 00	1.429e-01	32.97	< 2e-16
Population	2.394e-07	2.284e-08	10.48	< 2e-16

The model result of the effect of population density over the years and its correlation to the number of fatalities is given in Table 3.

The results show negative relation between population density and fatalities. The causal factor might be because, as city become more compact, less number of trips is made. Hence, exposing road users to less risk leading to decreased fatalities.

Model was then run for the relationship between total road traffic fatalities as dependent variable and the independent variables—road length, population, and vehicle numbers. But no significant results were obtained. Table 4 gives the corresponding model result.

Coefficient for density is not defined due to singularities with other variables. Road length is positively correlated. Vehicle numbers is not significant as per the model.

Table 5 shows the model results for total injuries and road length. A positive relation between injuries and road length was obtained as per the model. As the road length increases, the number of injuries also tends to increase. The intercept value is very in the model indicating unexplained causes other than road length in increasing the number of injuries.

Table 3 Model results for relationship between total fatalities and population density

Variables	Coefficients	Std. error	*Z* value	Pr(>|z|)
Intercept	4.712e + 00	1.429e-01	32.97	< 2e-16
Population density	1.774e-04	1.692e-05	10.48	< 2e-16

Table 4 Model result between total fatalities and other variables

Variables	Coefficients	Std. error	*Z* value	Pr(>|z|)
Intercept	4.712e + 00	1.712e-01	28.294	< 2e-16
Road length	1.762e-04	1.681e-05	10.48	< 2e-16
Density	NA	NA	NA	NA
Vehicle numbers	−4.778e-03	4.205e-03	−1.136	0.256

Table 5 Model results for total injuries and road length

Variables	Coefficients	Std. error	*Z* value	Pr(>|z|)
Intercept	7.908e + 00	8.891e-02	88.94	< 2e-16
Road length	6.365e-05	1.692e-05	6.03	1.64e-09

Table 6 Model results for relationship between total injuries and population

Variables	Coefficients	Std. error	*Z* value	Pr(>|z|)
Intercept	7.908e + 00	8.891e-02	88.94	< 2e-16
Population	8.590e-08	1.425e-08	6.03	1.64e-09

The model result of the effect of increase in population over the years and its correlation to the number of injuries is given in Table 6.

As anticipated, a positive relation is obtained between total injuries and population. As the population increased, the number of road traffic injuries also increased. Increase in population leads to equivalent increase in trips made, and hence, traffic injuries will also increase, as more exposure tends to increased risk.

The model result on the effect of increase in population density over the years and its correlation to the number of injuries is given in Table 7.

Contrast to that of in case of fatalities, a positive relation is obtained between total injuries and population density. As the density increases, the number of accidents also increases.

Model was run for the relationship between total injuries and other variables road length, population, and vehicle numbers. The results are given in Table 8. Only vehicle numbers are significant when the variables are considered together.

Model was run for understanding the correlation between total injuries and other parameters road length, population density, and vehicle numbers. The results are given in Table 9. Only road length showed significance. Coefficient for density is not defined due to singularities with other variables. Road length is positively correlated, but vehicle numbers showed a negative correlation.

Table 7 Model results for relationship between total injuries and population density

Variables	Coefficients	Std. error	*Z* value	Pr(>|z|)
Intercept	7.908e + 00	8.891e-02	88.94	< 2e-16
Population density	6.365e-05	1.056e-05	6.03	1.64e-09

Table 8 Model result between total injuries and other variables

Variables	Coefficients	Std. error	*Z* value	Pr(>|z|)
Intercept	8.0734393	0.1009523	79.973	< 2e-16
Road length	−0.1245453	0.1290776	−0.965	0.3346
Population	0.0001678	0.0001742	0.965	0.3346
Vehicle numbers	−0.0065922	0.0024862	−2.652	0.00801

Table 9 Model result between total injuries and other variables

Variables	Coefficients	Std. error	Z value	Pr(>|z|)
Intercept	8.068e + 00	1.018e-01	79.239	< 2e-16
Road length	6.378e-05	1.000e-05	6.375	1.83e-10
Density	NA	NA	NA	NA
Vehicle numbers	−6.253e-03	2.500e-03	−2.136	0.01256

6 Conclusions

Construction of new roads, elevated corridors, etc., are taken as a measure to increase the capacity and also assumed to increase safety by government officials. The study intended to show it otherwise. The main results from the study are as follows:

1. The trends analysis for fatalities and injuries were done for the Bengaluru city. Over the years of the study, period fatalities seemed show an increase, while injuries showed an overall declining trend.
2. Negative binomial regression model was used for the analysis.
3. Both fatalities and injuries showed positive correlation with increase in road length when taken alone. Both fatalities and injuries tend to increase with increase in road length.
4. Fatalities and injuries increase with increase in population. Same was the result seen in case of increasing population density.
5. No significant results were obtained when the variables road length, population, and vehicle numbers were taken together in case of fatalities. In case of total injuries for the same variables, only vehicle numbers showed significance. It showed negative correlation for injuries, i.e.; as vehicle numbers increase injuries decreased. This can be explained using "safety in numbers." As the vehicle number increases, the road users behave with more caution and thus increasing the safety.
6. When variables road length, density and vehicles numbers were taken, only road length showed positive correlation in case of fatalities. No other variables were significant.
7. When variables road length, density, and vehicles numbers were taken in case of injuries, road length showed positive correlation and vehicle numbers showed negative correlation.

To conclude the study in a nutshell, road length do impact road safety. Bengaluru is known for its road congestions and government plans to build more roads flyovers, etc., to alleviate the congestion problem. Impact of providing more infrastructures for increasing capacity for private road users from the study seems to cause more fatalities and injuries. A sustainable solution is what is required for the city, instead of increasing road capacity for the private road users.

7 Study Limitations and Future Scope

1. The variable taken for analysis was not sufficient and did not seem to show significant results as expected.
2. The coefficient of intercept from the model was very high, which indicates that the variables taken were not enough for explain the relation of fatalities or injuries.
3. More rich and diverse indicator has to be selected for future analysis for better interpretation of results from the model.

References

1. Annual Report 2017–2018 Ministry of road transport and highways, Government of India. https://morth.nic.in/
2. Transportation Research Board (1987) Designing safer roads: practices for resurfacing, restoration, and rehabilitation, special report 214. National Research Council, Washington, DC
3. McGee HW, Hughes WE, Daily K (1995) Effect of highway standards on safety. National cooperative highway research program, report no. 374. Transportation Research Board, National Research Council, National Academy Press, Washington, DC
4. Miaou SP, Hu PS, Wright T, Davis SC, Rathi AK (1991) Development of relationships between truck accidents and highway geometric design: phase I. Technical memorandum prepared by the Oak Ridge National Laboratory. Federal Highway Administration, Washington, DC
5. Noland RB (2003) Traffic fatalities and injuries: the effect of changes in infrastructure and other trends. Accid Anal Prev 35(4):599–611. https://doi.org/10.1016/S0001-4575(02)00040-4
6. Porter BE, England KJ (2000) Predicting red-light running behavior: a traffic study in three urban settings. J Safety Res 31:1–8
7. Mahalel D, Szternfeld Z (1986) Safety improvements and driver perception. Accid Anal Prev 18:37–42
8. Shaw-Pin, Miaou Harry, Lum (1993) Modeling vehicle accidents and highway geometric design relationships. Accid Anal Prev 25(6):689–709. https://doi.org/10.1016/0001-4575(93)90034-T.

Impact Evaluation of Exclusive Bus Lanes on Urban Roads

A. R. Arathi and Vincy Verghese

1 Introduction

Transportation aims at safe and efficient movement of goods and passengers. Faster mobility of goods and passengers is the catalyst for economic growth of a country, and this is facilitated by efficient transportation system. In case of road transportation systems, as facility increases, the volume of traffic also increases due to increasing demand for transport, particularly in developing countries like India. Because of the space, financial and material constraints urban road infrastructure cannot be developed beyond a limit, and this leads to increase in congestion, pollution, and reduction in road safety. Hence, there is a need for an appropriate strategy for optimal use of road transport system to reduce congestion and to increase efficiency of road networks. One way to reduce congestion is by encouraging the travelers to use public transport system (buses) instead of private transport modes, because public transport system enables mass transit of passengers in fewer vehicles.

To bring about a shift in the passenger preferences, the public transport system should be highly efficient and relatively less expansive to attract the travelers from private modes of transport. This goal can be attained by encouraging public transport like buses and giving priority to these modes on roads. One of the methods of assigning priority to public transit is by providing exclusive bus lanes. Exclusive bus lanes are the lanes restricted only for buses provided in order to speed up the buses, to reduce the interactions between buses and other modes of vehicles, and thereby reducing the road crashes.

A. R. Arathi (✉) · V. Verghese
Department of Civil Engineering, Jyothi Engineering College, Thrissur, Kerala, India

M. Parida et al. (eds.), *Proceedings of the Fifth International Conference of Transportation Research Group of India* , Lecture Notes in Civil Engineering 220,
https://doi.org/10.1007/978-981-16-9925-2_14

2 Aim and Objectives

Aim of this study is to investigate the impact of provision of exclusive bus lanes on traffic flow characteristics under heterogeneous traffic conditions. To achieve this, the objectives formulated are the following:

1. To develop social criteria based on the proportion of travelers using different modes.
2. To develop economic criteria based on the money value of time of travelers using the different modes.
3. Comparison of performance measures between without and with bus lane conditions.
4. Compare with and without bus lane conditions from VISSIM simulation results.

3 Literature Review

Arasan and Vedagiri [1] used the mode-choice probability curve to determine the probable shift of users from car to bus with the provision of exclusive bus lanes under mixed traffic conditions in Indian city roads. A recently developed simulation model was used to examine the increase in the level of service (LOS). Arasan and Vedagiri [2] developed a micro-simulation model and investigated the impact of the provision of bus lanes on urban roads under heterogeneous traffic conditions. The justification for providing an exclusive traffic lane was also given on the idea of the number of travelers per unit width of the road.

Cervero [3] reviewed experiences with designing and implementing bus rapid transit (BRT) systems worldwide. Siddharth and Ramadurai [4] calibrated VISSIM for Indian mixed traffic flow conditions, and the sensitivity analysis was done using ANOVA and the elementary effects method. Chen et al. [5] conducted a micro-simulation analysis which showed that exclusive bus lanes (XBLs) and transit signal priority (TSP) have a big impact on the operational performance of BRT if both are implemented simultaneously.

Syed et al. [6] used micro-simulation tool VISSIM for simulation and found that the bus priorities are more efficient at high volumes. Abdelfatah and Abdulwahid [7] assessed the impact of XBLs on urban road network performance under different traffic conditions using VISSIM software. Indian Highway Capacity Manual [8] gives the procedures for determining the capacity and level of service (LOS) at signalized and unsignalized intersections. Signalized intersection chapter describes the estimation of capacity and LOS offered by the fixed time isolated signal-controlled intersections. Unsignalized Intersection chapter presents the estimation of capacity and LOS offered by the three-legged and four-legged unsignalized intersections.

4 Study Area

The study area was selected considering the roadway geometry, traffic movement features, and availability of suitable location for mounting the video camera. With this regard, the suitability of the provision of exclusive bus lane is studied along Thrissur–Kechery route, which satisfies the geometric requirements. This road is state highway 69 (SH 69) with a distance of 17 km between Thrissur and Kechery. The selected stretch of road is four-lane divided road and 3.5 km long, including eight intersections such as Punkunnam, Puzhakkal, Muthuvara, Amala, Peramangalam, Mundur, Kaiparambu, and Kechery. Along the identified route, Puzhakkal–Peramangalam road stretch is selected for collecting the traffic data. Within that stretch, four intersections were selected for taking the traffic count, they are, Puzhakkal junction, Muthuvara junction, Amala junction, and Peramangalam junction. The available width of the carriageway is 14.5 m (7.0 m in both directions). The aerial view of this stretch is as given in Fig. 1.

This study area is selected because of the following reasons:

- It is the route which is facing severe traffic congestion.

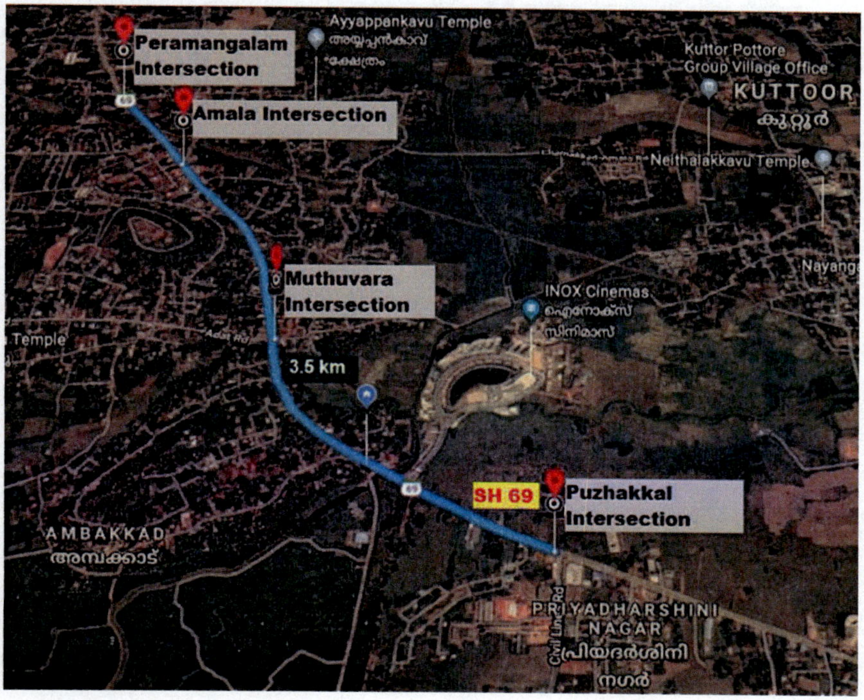

Fig. 1 Aerial view of the study stretch

- Based on five years (2014–2018) accident records (collected from police station), 16% of the total accidents are caused by buses.
- It is the major route through which all the buses including local, limited stops, and KSRTC buses traveling to different place. Thus, it is a representative stretch for the entire Thrissur city.

4.1 Vehicle Composition

The traffic data was collected by conducting a videographic survey at the selected intersections (Puzhakkal, Muthuvara, Amala, and Peramangalam) along the study stretch from Puzhakkal to Peramangalam. The traffic composition on the roads of Thrissur city is highly heterogeneous with vehicles of wide varying static and dynamic characteristics. The observed vehicle types are two wheelers (2 W), three wheelers (3 W), cars, light motor vehicles (LMV), heavy motor vehicles (HMV), and bus. The observed vehicle composition at four intersections is shown in Fig. 2.

Proportion of buses varies from 2–11%. This indicates that the private mode usage is more comparing to public transits. At all the four intersections, the turning volumes (left-turn and right-turn) are small when compared to the through volume on SH 69. Hence, justify the provision of bus lanes.

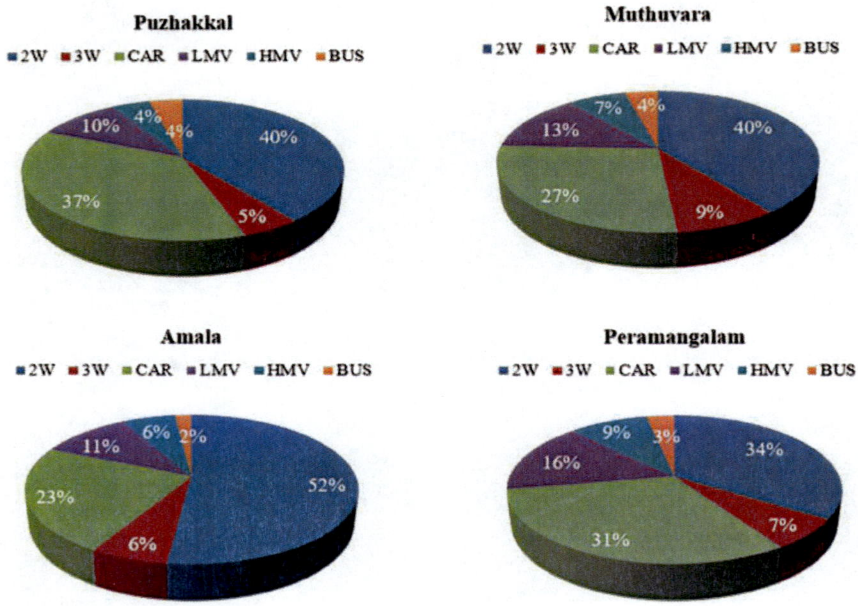

Fig. 2 Observed traffic composition

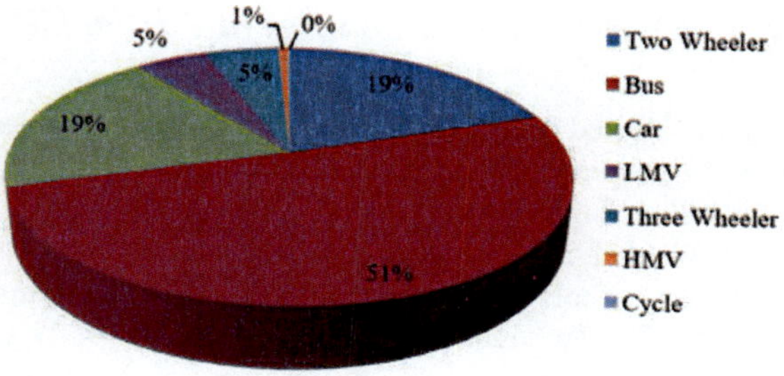

Fig. 3 Proportion of travelers

4.2 Vehicle Occupancy

Occupancy of different modes of vehicles is determined using windshield method. Average occupancies of 1.36, 2.28, 2.32, 2.76, 1.24, and 58.84 are obtained for two wheeler, three wheeler, car, LMV, HMV, and bus, respectively. Proportions of travelers using each mode of vehicle are shown in Fig. 3.

Figure 3 indicates that 51% of the travelers using the buses comparing to the other modes. This states that the buses enables mass transit of passengers.

4.3 Speed

The time taken by different modes of vehicle to pass the selected distance is retrieved from video, and speeds of different modes of vehicles toward Thrissur and Kechery directions are determined. During peak hour, the speeds of vehicles are less, and their journey time is more compared to off peak hour.

4.4 Survey of Income of Travelers

Survey for monthly income of travelers is done at different locations in Thrissur city for different modes of vehicles. Proportion of travelers using different modes of vehicles under each income group is determined. Using Eq. (1), the hourly income of the users of different modes was calculated.

$$\begin{array}{c}\text{Hourly} \\ \text{Income}\end{array} = \frac{\begin{array}{c}\text{Monthly} \\ \text{Income}\end{array} \times 12 \begin{pmatrix}\text{Months} \\ \text{in a Year}\end{pmatrix}}{52 \begin{pmatrix}\text{Weeks} \\ \text{in a Year}\end{pmatrix} \times 6 \begin{pmatrix}\text{Working Days} \\ \text{in a Week}\end{pmatrix} \times 8 \begin{pmatrix}\text{Working Hours} \\ \text{in a day}\end{pmatrix}}. \tag{1}$$

4.5 Existing Signal Timings at Puzhakkal

Out of selected four intersections, the signal is provided only at Puzhakkal intersection. Phase plan and ring diagram are shown in Fig. 4. The observed signal timings at Puzhakkal is shown in Fig. 5.

Puzhakkal is three-legged intersection provided with pre-timed type signal, consisting of total two phases. It includes a fixed time signal controller with a cycle time of 77 s. For signal group 1, a red interval of 49 s, green of 23 s, all-red/amber of 2 s, and amber time of 3 s were observed and recorded. For signal group 2, a red interval of 20 s, green of 52 s, all-red/amber of 2 s, and amber time of 3 s were observed and recorded.

Phase		Ring 1	Ring 2
Phase A			
Phase B	B1		
	B2		

Fig. 4 Phase plan and ring diagram at Puzhakkal intersection

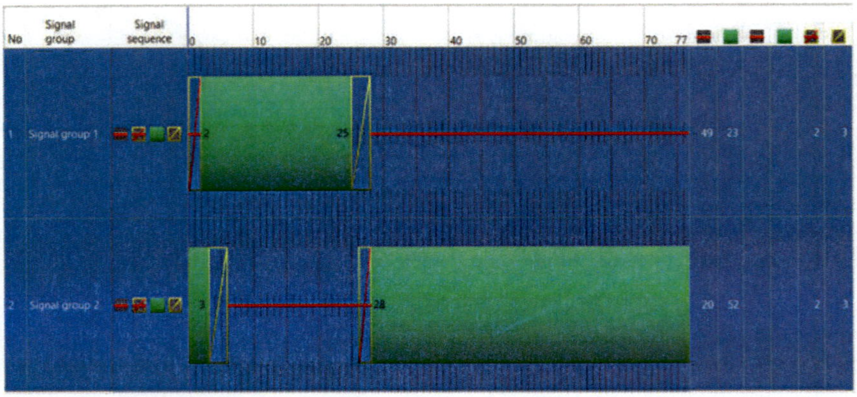

Fig. 5 Existing signal timings at Puzhakkal intersection

5 Data Analysis

5.1 *Journey Time Savings in Terms of Money Value*

According to questionnaire survey, 45% of passengers are ready to shift to bus if the exclusive bus lane system is implemented. The reduced traffic volumes after 45% shift of occupants to bus toward Thrissur and toward Kechery are determined. During peak and off peak hours, the speeds are determined for different modes of vehicles corresponding to actual traffic volume (without bus lane). Then, the speeds of different modes of vehicles except bus for reduced traffic volume (with exclusive bus lane) are determined using interpolation technique. After the exclusive bus lanes are implemented, the bus can flow freely with an increased speed around 70 km/h [6], without any interruption from other vehicles.

Then, from this speeds, journey time of all modes of vehicles with and without bus lanes was calculated for traveling a distance of 17 km road. While analyzing the journey time, it can be observed that the journey time for all modes is reduced with the implementation of exclusive bus lane. Then, the journey time savings for different modes of vehicles are determined. For bus, a saving of around 11 min is obtained after the implementation of exclusive bus lane. The journey time savings are then converted into money value savings by multiplying the journey time savings of each mode with their corresponding hourly incomes, and calculated money savings are given in Table 1.

A total of Rs. 38.5 money savings per hour are obtained when the exclusive bus lanes are provided and out of which saving of Rs. 11 per hour per km is obtained for buses.

Table 1 Money value savings

Vehicle type	Savings in journey time (min)		Average hourly income (Rs.)	Money value savings (Rs.)	
	Toward Thrissur	Toward Kechery		Toward Thrissur	Toward Kechery
Two wheeler	0.37	0.41	132.61	0.82	0.91
Three wheeler	0.68	0.33	115.94	1.31	0.64
Car	0.21	1.29	163.01	0.57	3.50
LMV	0.03	2.26	156.25	0.08	5.89
HMV	0.15	1.18	147.83	0.37	2.91
Bus	10.49	10.58	61.29	10.72	10.81
Total money value savings (Rs.)				38.5	

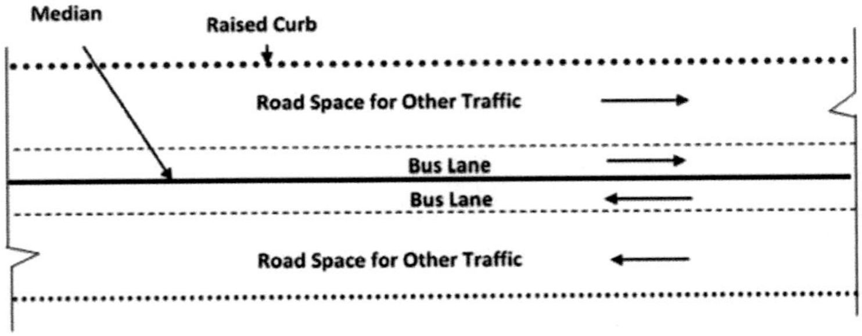

Fig. 6 Pictorial representation of the road stretch with exclusive bus lane

5.2 Road Space Allocation

Total width of the selected road stretch is 7 m (in one direction), an extra 3.5 m is proposed to be provided exclusively for buses on both directions adjacent to the curb so that passengers can easily enter to and exit from the buses. A pictorial representation of the road stretch with 3.5 m exclusive bus lane is shown in Fig. 6.

5.3 Signal Design

Traffic signal is designed at Puzhakkal and Peramangalam intersections using HCM method of signal design. After the implementation of exclusive bus lanes on Puzhakkal to Peramangalam road, turning movements are only allowed at the end intersections. At intermediate intersections (Muthuvara and Amala), turning movements are not allowed. This is done to avoid interruptions to the exclusive bus lane facility provided near the medians on both directions. Both signal design consists of two signal groups; signal group 1—allows turning movement, and signal group 2—allows through movement on SH 69.

At Puzhakkal, a fixed time signal controller with two phases was designed with a cycle time of 120 s. The signal group 1 includes a red interval of 39 s, green of 74 s, all-red/amber of 2 s, and amber time of 5 s. The signal group 2 includes a red interval of 81 s, green of 32 s, all-red/amber of 2 s, and amber time of 5 s. At Peramangalam, a fixed time signal controller with two phases was designed with a cycle time of 120 s. The signal group 1 includes a red interval of 68 s, green of 45 s, all-red/amber of 2 s, and amber time of 5 s. The signal group 2 includes a red interval of 52 s, green of 61 s, red/amber of 2 s and amber time of 5 s.

5.4 *Capacity and Level of Service*

Capacity and level of service (LOS) of Puzhakkal and Peramangalam intersections were determined for both without and with bus lane conditions using Indo-HCM procedure. Presently intersection LOS at Puzhakkal is F, and after the implementation of exclusive bus lanes, intersection LOS is improved to B. At Peramangalam, presently it is LOS C for the right-turn movement from major and minor roads, and after the implementation of exclusive bus lanes, intersection LOS is improved to B. Capacity and LOS at both intersection were improved with the provision of exclusive bus lanes.

5.5 *Traffic Simulation*

Both the scenarios (without and with bus lane conditions) were replicated with the help of VISSIM software. VISSIM was calibrated for Indian heterogeneous traffic conditions by adopting calibration parameters from a literature [9]. The snap shot of simulated traffic from VISSIM (with bus lane condition) is shown in Fig. 7.

The main objective of this study was to evaluate the impact of provision of exclusive bus lanes on the selected urban road stretch. The changes in density and travel time were compared between with and without bus lane conditions. Based on simulation results, with the provision of exclusive bus lanes, density decreases about 40%, and a percentage decrease of travel time of around 6–11% is obtained for the buses.

5.6 *Findings*

The main aim of this study was to evaluate the impact of providing exclusive bus lanes on urban roads. After the analysis, the results obtained are given below:

1. The bus travelers, constituting 51% of the total of the travelers, will use only 33% of the road space, whereas the users of all the other modes (excluding buses) constituting 49% of the total of the travelers will use 67% of the road space. This shows that the provision of exclusive bus lane is justifiable based on the proportion of travelers using different types of vehicles.
2. Total money saving of Rs. 38.5 per hour per km is obtained. Among this, a saving of Rs. 11 per hour per km is obtained for the buses.
3. By introducing an extra bus lane of 3.5 m in both directions near the median, the buses can flow freely with increased speeds without any interruptions from other vehicles.
4. At puzhakkal, after the implementation of exclusive bus lanes, intersection level of service was improved from LOS F to LOS B. At Peramangalam, level of service for the right-turn movements were improved from LOS C to B. Capacity

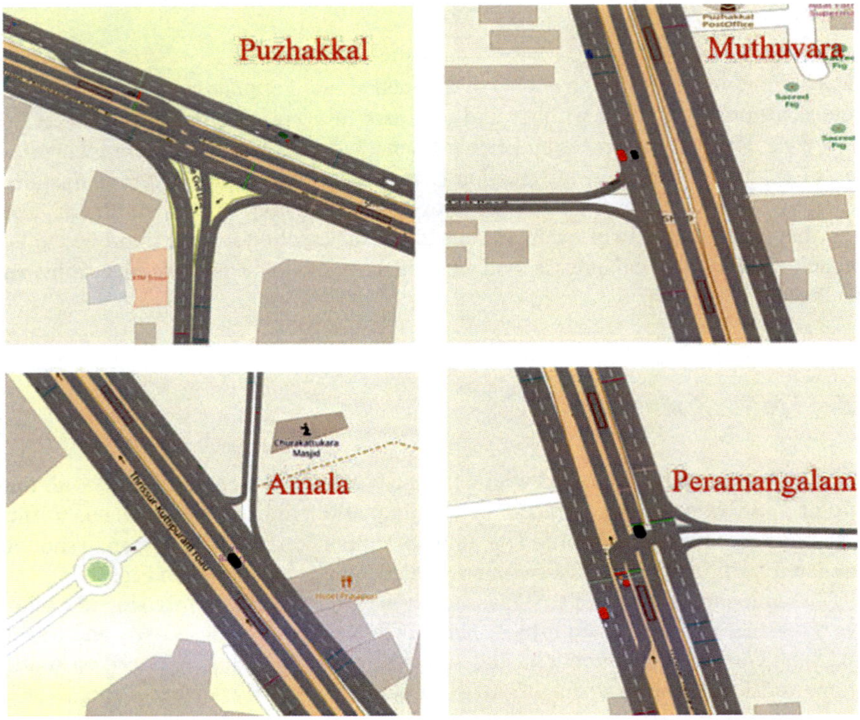

Fig. 7 Snap shot of simulated traffic in 2D (with bus lane condition)

and LOS at both intersection were improved with the provision of exclusive bus lanes.

5. Simulation results shows that density decreases about 40% and a percentage decrease of travel time of around 6–11% is obtained for buses with the provision of exclusive bus lanes.

6 Conclusions

In the study stretch, SH 69 route, only one intersection (at Puzhakkal), is provided with traffic control system such as signal. No other intersections in the study stretch was provided with the signal system. It was also noted that there is no safe crossing facilities which were provided for the road users at these selected intersections, and the bus stops were provided very near to the intersections. It was observed that the selected stretch is the one through which the most of the buses running. The selected study stretch is an area where the buses were going dangerously in very rush manner that leads to many accidents to other vehicle users and to road users as well. The buses were competing in the road, to pick up more passengers, to get more trips, and to reach the destination before the expected time. This makes lot of serious road crashes and loss of life.

With the provision of exclusive bus lanes, capacity and LOS of the study intersections were improved. After the implementation of exclusive bus lanes, density gets reduced by 40%, thereby traffic congestion gets reduced, and average travel time of bus travelers gets reduced by 6–11%. Total money saving of Rs. 38.5 per hour per km was obtained. Study results reveals that more travelers will shift to bus, and a money value savings of about Rs. 11 per hour per km was obtained for the buses. This results shows that when a public transportation mode offers efficient, easy, less expensive, and safe travel, more travelers get attracted to use public transport like buses instead of private modes.

Further studies can develop mode shift model for calculating the shift percentage of passengers from private mode to bus. Also, land use models can be incorporated with the study to bring in more clarity on the effectiveness of exclusive bus lane. With the help of government and traffic enforcement, we can give a life to this proposal in an effective manner. Although initial implementation cost will be higher, but, once it done, it will develop our road transportation system to a better level and will lead to increase the economy of the nation.

References

1. Arasan VT, Vedagiri P (2009) Simulating heterogeneous traffic flow on roads with and without bus lanes. J Infrastruct Syst 15(4):305–312
2. Arasan VT, Vedagiri P (2010) Micro simulation study of the effect of exclusive bus lanes on heterogeneous traffic flow. J Urban Plann Dev 136(1):50–58
3. Cervero R (2013) Bus rapid transit (BRT): an efficient and competitive mode of public transport
4. Siddharth SP, Ramadurai G (2013) Calibration of VISSIM for Indian heterogeneous traffic conditions. Proc Soc Behav Sci 104:380–389
5. Chen Y, Chen G, Wu K (2016) Evaluation of performance of bus lanes on urban expressway using paramics micro-simulation model. Proc Eng 137:523–530
6. Syed U, Ghodmare SD, Khode BV (2016) Performance impact of various bus priority strategies using vissim 7.0. Int Res J Eng Technol 3(4):2479–2483
7. Abdelfatah A, Abdulwahid AR (2017) Impact of exclusive bus lanes on traffic performance in urban areas. In: Proceedings of the 2nd world congress on civil, structural and environmental engineering, pp 1–10. Spain
8. Indian highway capacity manual-draft report, CSIR-CRRI, New Delhi (2017)
9. Anand A, Ramadurai G, Vanajakshi L (2014) Data fusion-based traffic density estimation and prediction. J Intell Transp Syst 18(4):367–378

Assessment of Emission Implications of Introducing Electric Vehicles in India on the Non-renewable Energy Sector

Vikas Nimesh, Ranjana Kumari, V. Mahendra Reddy, and Arkopal K. Goswami

1 Introduction

The urban centers in India are witnessing growing rates of urban sprawl and motor vehicle usage, resulting in greater energy demand and environmental degradation. Air pollution is a major problem faced by several cities worldwide as a significant threat to the environment. The electrification of the vehicular fleet is being seen as a viable alternative to minimize vehicular emissions [1–3]. There is a significant push by various governments toward adoption of alternate fuel vehicles in order to reduce emissions and energy use [4]. The Government of India, in 2013, launched the National Electric Mobility Mission Plan (NEMMP) 2020, to encourage adoption of electric vehicles (EVs) in the country. The mission has a goal to sell more than 6 million hybrid and electric vehicles annually [4]. However, increased EV sales may lead to greater electricity demand. India's dependence on coal, which is of poorer quality as compared to other nations, may lead to increased emission of carbon dioxide (CO_2) per unit production of electric power. According to estimates, India produces 926 $_g$ of CO_2 per kilowatt of electricity generated, which was approximately 75% higher than the global average in 2012 [5].

The urban areas in India often fail to meet WHO guidelines for safe levels of air quality. According to WHO, the capital city New Delhi is one of the most polluted

V. Nimesh · R. Kumari · A. K. Goswami (✉)
Ranbir and Chitra Gupta School of Infrastructure Design and Management, Indian Institute of Technology Kharagpur, Kharagpur, West Bengal 721302, India
e-mail: akgoswami@infra.iitkgp.ac.in

V. Nimesh
e-mail: vikasnimeshdtu@iitkgp.ac.in

V. Mahendra Reddy
Department of Mechanical Engineering, Indian Institute of Technology Kharagpur, Kharagpur, West Bengal 721302, India
e-mail: mahendra@iitkgp.ac.in

M. Parida et al. (eds.), *Proceedings of the Fifth International Conference of Transportation Research Group of India* , Lecture Notes in Civil Engineering 220,
https://doi.org/10.1007/978-981-16-9925-2_15

cities in the world. Thirteen cities of India fall under the world's twenty most polluted cities. Cities of India have one of the highest levels of PM_{10} and $PM_{2.5}$ [6]. At more than 150 μg, the capital city New Delhi has the highest level of airborne particulate matter $PM_{2.5}$. These numbers are six times more than the WHO "safe" limit of 25 μg, and the rate of growth in vehicular ownership is likely to be the principal reason [3].

Some studies like Buekers et al. [7] have suggested that the electric vehicle introduction could help in reducing CO_2 emissions and the dependency on petroleum products. Both China and the USA are the two crucial EV market, and each has more than 10% of global electric vehicle sales. As they are also heavily dependent on coal for their power generation (76% and 38% in 2012, respectively) [8], they are the primary GHG contributors globally, with the transport and power sectors being the major contributors to air pollution. Studies conducted worldwide show that electric vehicles' fuel-cycle emissions vary based on the carbon intensity and cleanliness of the electricity mix. A low share of coal-based electricity (e.g., California, USA), would result in electric vehicles significantly reducing GHG and other emissions. Alternately in other regions, electric vehicles may increase the total emissions [8]. As the share of coal in India's energy mix is higher than other developed or developing countries [9], it is vital to assess the impact that electric vehicles would have on total emissions, when compared to today's internal combustion engine vehicles (ICEV).

The objective of the study is to compare an ICEV to an EV when powered with coal thermal power plant and estimate the total emissions. In the present study, the power generation scenario of India has been studied to analyze the share of renewable and non-renewable sources in the total power generation mix. The forecasted power load, as reported by Coal India Limited (CIL), has been considered to determine the future expected growth pattern of emissions from coal. The emissions and losses occurring during the electricity production process are identified. Finally, an emission index (EI) for ICEVs and EVs is developed for the Indian scenario.

2 Literature Review

2.1 Power Generation Scenario in India

The power generation sources in India range from natural gas to nuclear power to solar and finally to agricultural waste. The demand for electricity is expected to rise, which will necessitate capacity augmentation. In the year 2017–2018, the average electricity consumption was 1149 kWh per capita, whereas in 2018, the total installed capacity of power stations in India was 346,048 Megawatt (MW).

Figure 1 shows the energy production scenario in India for the year 2018. The total installed coal capacity in India was approximately 197,000 MW, whereas the capacity of hydropower generation is approximately 45,500 MW, and the natural gas capacity was only about 25,000 MW [10]. Lastly, India has about 69,000 MW of renewable energy capacity.

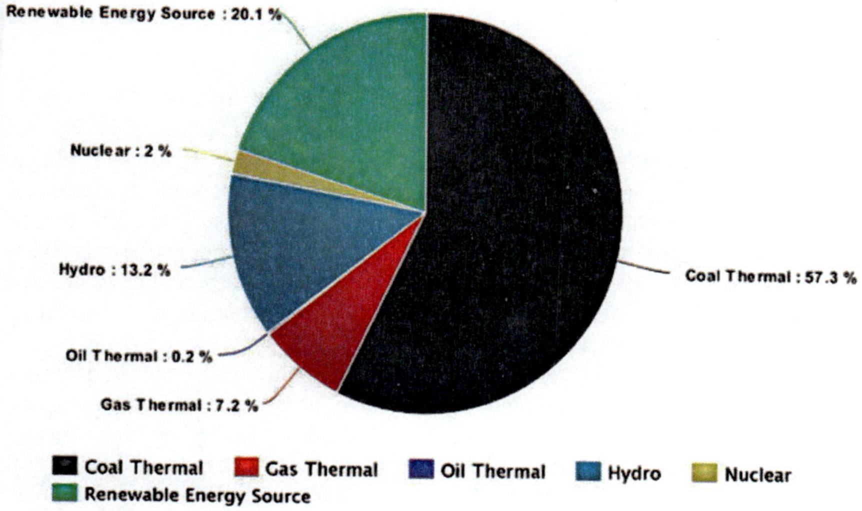

Fig. 1 Energy production in India [10]

As the number of inhabitants in India is rising, the demand for electricity is also increasing at an equitable rate. The compound annual growth rate of electricity production in India is 5.69%. There was an increase in the installed capacity of the thermal coal plant, i.e., from 53% in 2015 to 57.3% in FY2018 [10]. The power generation pattern is described in Fig. 2.

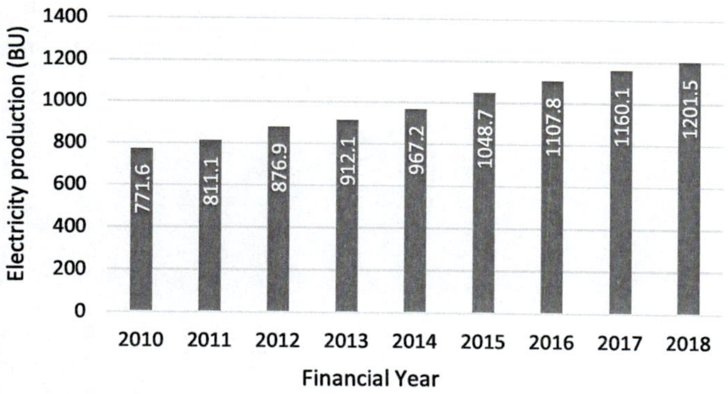

Fig. 2 Electricity production in India (BU) [10]

2.2 Power Generation

Approximately 60% of India's energy demand is fulfilled by coal [10]. India is now third in the list of coal-producing nations worldwide [11]. There are various related environmental impacts on the native ecosystem due to coal power plants. Many pollutants are released during the combustion of coal such as oxides of nitrogen and sulfur and particulate matter. Various greenhouse gases like carbon dioxide and methane are also emitted by them which are known to contribute to climate change and global warming [12]. Power generation by coal has increased in India at a very high rate. In each financial year, it has doubled from its preceding year [10]. The growth pattern can be visualized in Fig. 3.

In the present and coming years as well, India will go through a major transformation in the electricity sector. The electricity demand is increased at an accelerated pace and also expected to rise further rapidly [13]. India added power generation capacity at a very demand satisfying rate. In the future, more power generation capacity will require to fulfill consumer demand. In the next few years, electric mobility, new gadgets, and appliances will be the major part of per capita electricity consumption in India. According to Coal India report, the penetration of renewable will increase, but coal power production will also increase at a significant rate. In Coal India report, three scenarios have been generated corresponding to GDP growth rate, i.e., 8% GDP, 7% GDP, and energy-efficient scenario [13]. For this study, 8% GDP scenario has been considered and further used in estimating the future coal emissions from coal power plants (Fig. 4).

Losses in Coal power generation The process of electricity generation starts by the combustion of coal in boiler that produce's steam, which flows into the turbine section, rotating the generator to create electricity. The steam is then condensed back into water and pumped to the boiler at high pressure [14]. Burning coal breaks down

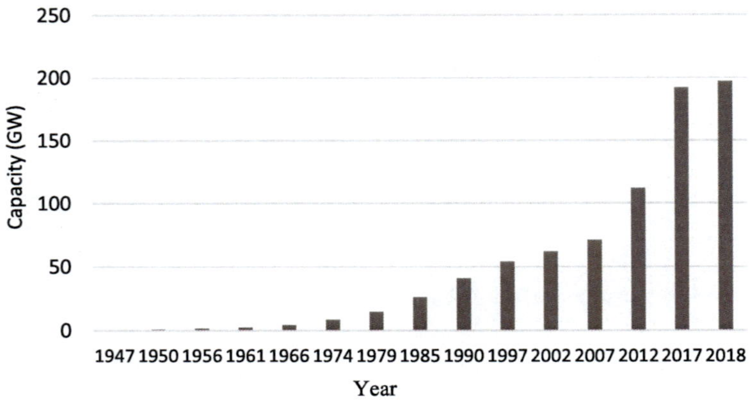

Fig. 3 Coal-capacity (MW) in India from 1948 to 2018 [10]

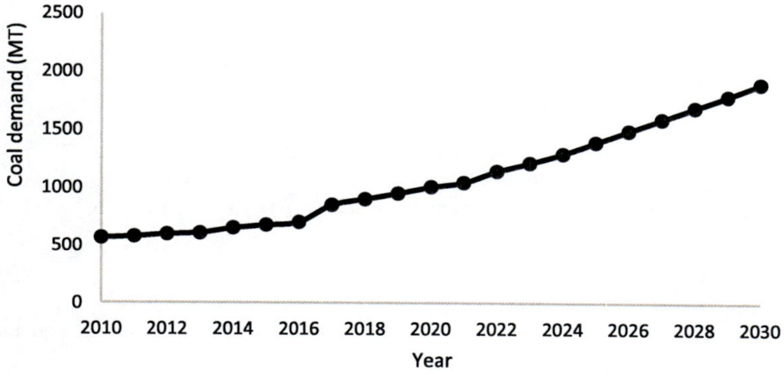

Fig. 4 Future coal demand [13]

the chemical bonds holding its carbon atoms in place, releasing energy. Additionally, airborne pollutants and heavy metals are also emitted into the environment as a result of other chemical reactions [5]. Coal combustion majorly releases SO_X, NO_X, CO_2, CO, and particulate matter (PM) in a specific ratio with the particular content of coal combustion [15]. The losses that take place during the process of power generation from coal is depicted in Fig. 5. There are losses and emissions during the excavation process as well [16].

2.3 *Conventional Vehicles*

Conventional internal combustion engine vehicles (ICEV) run on liquid or gaseous fuels like petrol, diesel, or CNG. India has the world's fourth-largest automotive industry, where the nation is the fourth-largest manufacturer of cars, and seventh-largest manufacturer of commercial vehicles [18]. Figure 6 shows the growth in the number of vehicles in India from 1981 to 2016 [19].

Emissions in petroleum production Crude oil is a raw natural resource consisting of various hydrocarbon deposits and organic products. After removing from the ground, crude oil is transferred to the refinery where they are converted into useful products like gasoline, distillate, jet fuel, etc., through fractional distillation and extraction. Subsequently, it is transported to the fuel stations to bring it into use [20]. Energy is expended, and GHG emissions are produced in each stage, as given in Table 1 [21].

Considering all the stages in Table 1, the total energy consumed and pollutant emitted, per 1000 L of fuel produced, is indicated in Table 2.

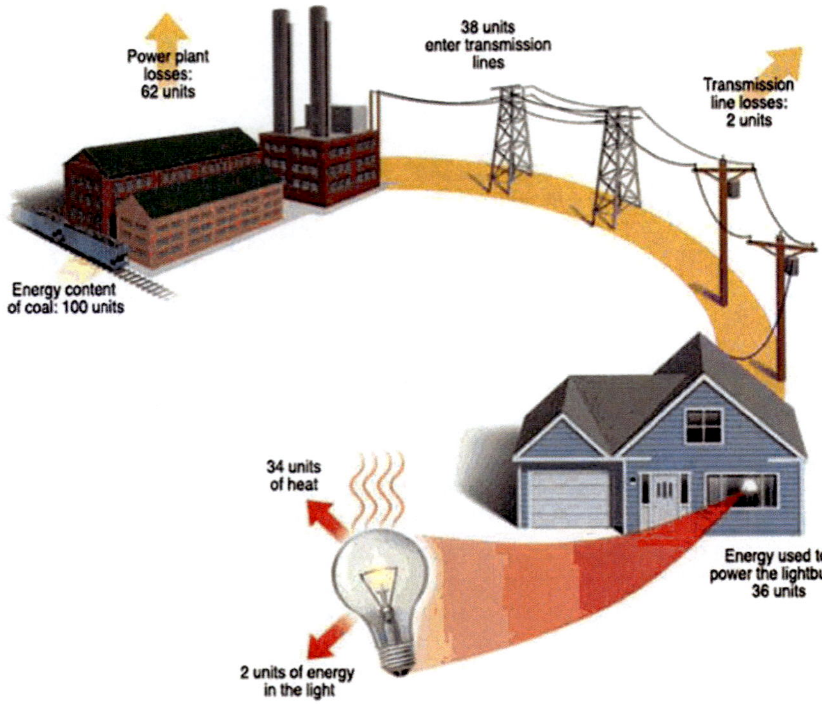

Fig. 5 Losses in coal power generation [17]

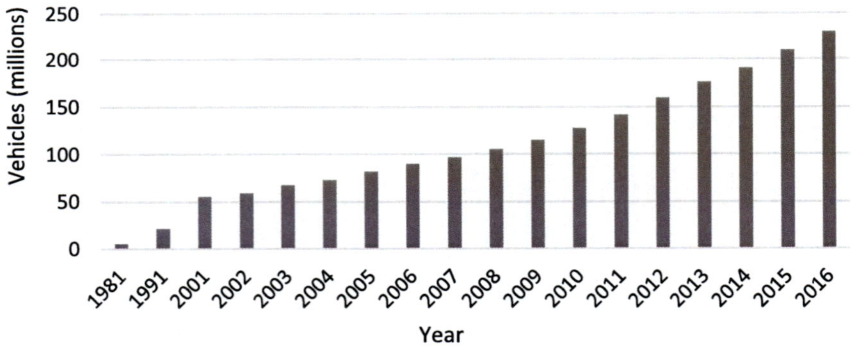

Fig. 6 Vehicular population in India (in millions) [19]

Table 1 CO_2 equivalent emissions in petroleum production [21]

Stages	Emissions during gasoline production (gm per liter)	Emissions during diesel production (gm per liter)
Crude oil extraction	41.1	58.3
Overseas transportation	32.0	35.5
Petroleum refining	299	102.7
Domestic transportation	14.3	14.3
Total	386.4	210.8

Table 2 Energy consumption and pollutant emissions [12]

	Gasoline	Diesel
Energy consumption (MJ)	3030	1960
Carbon monoxide (Kg)	0.2	0.02
Oxides of nitrogen (Kg)	0.5	0.6
Sulfur dioxide (Kg)	0.2	0.3
Volatile organic compound (Kg)	6.3	2.3

Table 3 Fuel consumption standards of passenger vehicle (Petrol equivalent), L/100 km [22]

Year	Test	Fuel consumption
FY 2016–2017 to 2020–2021	NEDC	$0.0024 \times (W{-}1037) + 5.4922$
FY 2021–2022 onward	NEDC	$0.002(W{-}1145) + 4.7694$

Consumption per unit The fuel economy standards, based on a Corporate Average Fuel Consumption (CAFC) system, given in Table 3, are adopted from [22]. The standards depend on the weighted average kerb weight (W) of all vehicles.

3 Study Methodology

The study estimates the change in emissions at the coal power generation sources as a result of the shift from ICEVs to EVs in India. In order to do so, firstly, power generation scenario of India has been studied to estimate the amount of power generated from different sources. Secondly, the scenario of the coal power plant has been studied concerning the type of coal being used in India, losses from coal mines to the end-user which has been identified with power unit losses and the emissions from the coal power plant. Thirdly, forecasted power load on coal power plant has been considered to determine the future expected growth pattern of emissions from coal. The emission factors for coal power generation, and the vehicle is identified from the literature study. Finally, a comparative analysis of EV and ICEV has been done to determine the difference in emission (Fig. 7).

To carry out this study, data is assimilated from secondary sources. The data which is used in this study is as follows: current power generation, expected coal consumption, emission factors for thermal power plant, emission factors of coal for coal mines to thermal power plant, I.C. engine vehicle emission factor, well-to-tank emission factor for petroleum products etc. as mentioned in Table 4.

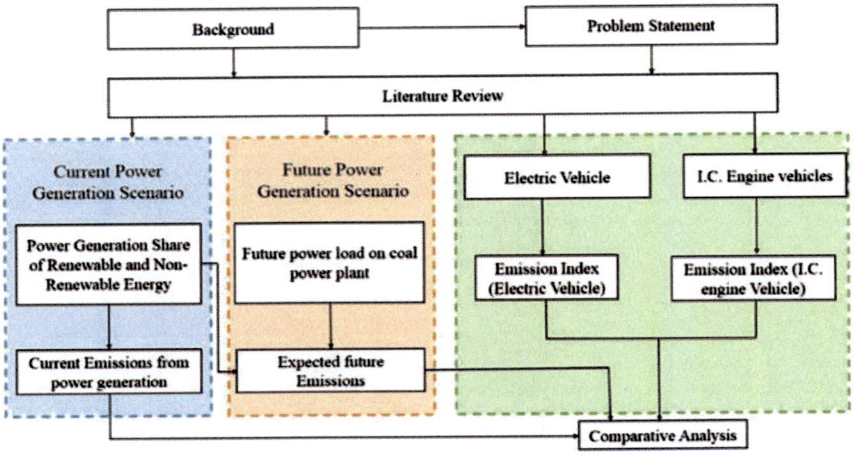

Fig. 7 Methodology of the study

Table 4 Sources and type of data used

Data type	Source and year
Current power generation	Ministry of statistics and programme implementation [10]
Expected coal consumption	Coal India Limited 2017 [13]
Emission factors for thermal power plant	Gurjar et al. 2004 [26], Chakraborty et al. 2008 [24]
Emission coefficient for coal (coal mines to thermal power plant)	Mishra [23]
I.C. engine vehicle emission factor	ARAI 2017 [22]
Well-to-tank emission factor for petroleum products	Furuholt 1995 [21], Eriksson and Ahlgren 2013 [28]

3.1 Emissions from Coal

Indian coal, with low calorific value and high ash content, coupled with the use of inefficient combustion technologies, leads to higher GHG emissions and other pollutants. Primary emissions from coal thermal power plants are CO_2, SO_x, CO, NO_x, fly ash, suspended particulate matter (SPM), carbonaceous material (soot), and other trace gases [5, 12]. The composition of Indian coal is described in Table 5.

Table 6 presents the emission coefficients as reported by earlier studies, viz., Chakraborty et al. [24], OSC [25], and Gurjar et al. [26]. The emissions depend upon the quality of coal, and plant efficiency, which further varies year to year because of maintenance.

Quantification of emission

$$E_x = M_{\text{coal}} \times E_{\text{coefficient}} \tag{1}$$

where

E_x	Emission of compound from coal.
M_{coal}	Mass of coal consumed in power generation.
$E_{\text{coefficient}}$	Emission coefficient of coal emissions.

Table 5 Percentage range for components in coal of India [23]

Constituent	Percent range
Carbon (C)	38–60
Aluminum oxide (Al_2O_3)	15–36
Volatile matter	1–36
Iron oxide (Fe_2O_3)	2–20
Water (H_2O)	3–43
Silicon oxide (SiO_2)	45–63
Ash	3–60
Sulfur (S)	0.3–8.3
Phosphorus (P)	> 0.5

Table 6 Comparison of emission coefficients

Type of gas	Emission coefficient range (g kWh^{-1})		Emission coefficient (g Kg^{-1} of coal)	
	Study by OSC (1997–1998)	Chakraborty et al. (2008)	Gurjar et al. (2004)	Chakraborty et al. (2008)
CO_2	800–1800	776–1490	1739	1639
SO_2	4–18	5.210–15.99	14.767	14.031
NO	6–13.1	1.540–3.263	0.824	4.018
CO	Not available	0.055–24.49	0.253	5.153

Table 7 Emission factors for four-wheeler (g/Km)

Emissions	Passenger 4-wheeler
CO_2	140
CO	1.2
NO_x	1.2
SO_2	0.053

3.2 *Emissions from ICEVs*

Vehicles emit pollutants directly in to the atmosphere, which could be considered as primary pollutants. In addition, secondary pollution may be generated as a result of chemical reactions between pollutants in the atmosphere [6, 27]. Burning of 1 L of gasoline emits 2.31 kg of CO_2 [28]. The emission factor for vehicle emission is given in Table 7.

Quantification of emission The on-road emissions are estimated on the basis of number of vehicles and the distance traveled by each vehicle type in one year, which is given by

$$\varepsilon_a = \sum_{i=1}^{n}(V_b \times D_b) \times E_{ab} \tag{2}$$

where

ε_a Emission of pollutant (a).
V_b Vehicle population per type (b).
L_j Distance covered by vehicle type (b) in one year.
E_{ab} Pollutant emission (a) from vehicle type (b)/km.

3.3 *Emission Index*

Emission index (E) is the ratio of mass of pollutants (M) emitted per unit of energy (kWh).

$$E = \frac{M}{\text{kWh}}. \tag{3}$$

ICEV Emission Index The conventional I.C. engine vehicles operate on the fuels like petrol, diesel, or CNG. The operation of vehicle involves power consumption at various steps. The calculation of ICEV emission index (E_{ICEV}). considers the power consumption and emission involved in each step of vehicle operation. Total ICEV emissions $\left(\acute{e}_{\text{ICEV}}\right)$. depend on the well-to-tank emissions $\left(\acute{e}_{\text{wtt}}\right)$, vehicle's engine efficiency (η_E), and drivetrain or transmission efficiency of vehicle (η_T). Furthermore,

$(\acute{e}_{\text{wtt}})$ depends on per unit transport emissions $(\acute{e}_t)$, excavation $(\acute{e}_{\text{ex}})$, and refining $(\acute{e}_{\text{ref}})$. of fuel. A study by Furuholt indicates that 0.39–0.80 kg/l of CO_2 equivalent emission is generated during the production and transportation of gasoline [21]. There are 68–72% engine losses in ICEV. The engine losses occur due to exhaust heat, pumping of fuel, friction. etc. Thereafter, drivetrain or transmission system leads to 5–6% energy loss [29]. The following expressions are used to develop the EI for I.C. engine vehicles [16, 30]:

$$E_{\text{ICEV}} = f\left(\sum \acute{e}_{\text{ICEV}}\right) \tag{4}$$

$$\acute{e}_{\text{ICEV}} = f\left(\acute{e}_{\text{wtt}}, \eta_E, \eta_T\right) \tag{5}$$

$$\acute{e}_{\text{wtt}} = f\left(\acute{e}_t, \acute{e}_{\text{ex}}, \acute{e}_{\text{ref}}\right) \tag{6}$$

EV Emission Index The electric vehicle consumes electricity which comes from various sources depending upon the power generation source. The coal power energy dominates the area of electricity production in India. Therefore, in this study, coal thermal power is a source for an electric vehicle. A few past studies have indicated that almost 93% of carbon dioxide discharged during the power plant operation and rest of the emissions occurs during the production and transportation of coal [31]. Although electric vehicles are considered to be zero tail-pipe emission vehicles, the emission index of EV, when charged with coal, (E_{EV}) are indirectly associated to the power generation source, i.e., coal power plant. $\acute{e}_{\text{EV}}$ depends on the per unit coal extraction emissions $(\acute{e}_{\text{CEE}})$, transportation $(\acute{e}_{\text{CTE}})$, power plant operations $(\acute{e}_{\text{PE}})$, and also the efficiency of electric vehicle (η_{EV}). The per unit emissions from coal thermal power plant $(\acute{e}_{\text{PE}})$ further depends on the efficiency of combustion (η_{comb}), turbine (η_{turb}), boiler (η_{Boil}), and other losses $(E_{l(o)})$ occurred during the process of power generation. Further, the mathematical relation is explained in following equations [16, 30].

$$E_{\text{EV}} = f\left(\sum \acute{e}_{\text{EV}}\right) \tag{7}$$

$$\acute{e}_{\text{EV}} = f\left(\acute{e}_{\text{CEE}}, \acute{e}_{\text{CTE}}, \acute{e}_{\text{PE}}, \eta_{\text{EV}}\right) \tag{8}$$

$$\acute{e}_{\text{PEE}} = f\left(\eta_{\text{comb}}, \eta_{\text{turb}}, \eta_{\text{Boil}}, \acute{e}_{l(o)}\right) \tag{9}$$

$$\eta_{\text{EV}} = f(\eta_b, \eta_m, \eta_t) \tag{10}$$

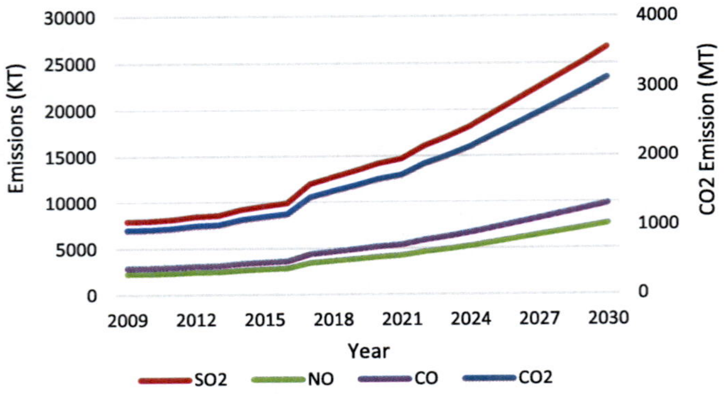

Fig. 8 Coal power plant emission estimation (SO_2, NO_x, CO, and CO_2)

4 Results

In the present study, the power generation scenario in India is presented, depicting the share of renewable and non-renewable source of energy in total power generation grid. The coal consumption for power generation in previous years and as well as expected consumption for the future is gathered from various secondary sources. This expected future coal consumption is further used to estimate future emissions from coal thermal power plant in India. The emission indices for ICEV and EV are determined from the developed formulations. The emission index measures the emissions generated per unit of energy from well to wheel activities for ICEV, and coal mine to wheel activities, for EV.

4.1 Future Emission Estimation from Coal Power Generation

Figure 8 shows the growing trend in emissions from thermal power plants as estimated by the study. The results indicate that CO_2 emissions from coal increased by 15% between the years 2009_2015, and the expected growth of CO_2 emissions in coming ten years is almost 100%, if the dependency on coal remains same.

4.2 Comparison of Emission Indices: ICEV Versus EV

The emission index is calculated as per the formulation developed in this study. It indicates the emissions generated per unit energy. The estimation of fuel consumption for ICEV has been done according to NEDC test. Emissions estimation for I.C. engine vehicle is done based on emission factors and emission standards of India.

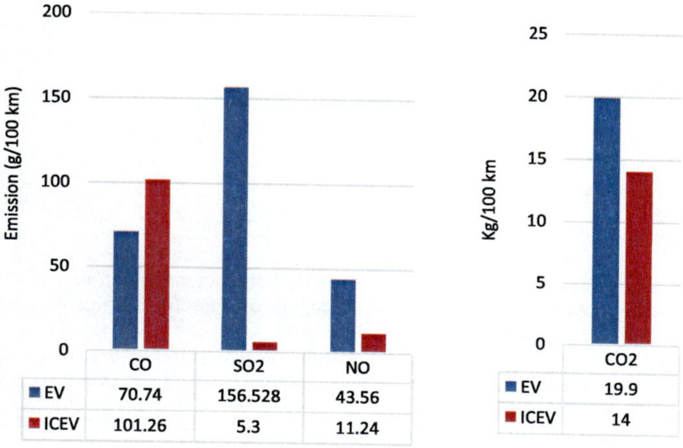

Fig. 9 Emission index comparison of EV and I.C. engine vehicle

An electric car has been assumed to get powered with the coal power plant. Therefore, the emissions estimations of electric vehicles are based on the emissions coefficients of coal power plants in India including all the activity involved from coal mines to wheel. Emissions involved in coal mines to wheel is taken from the various literature study. The results in Fig. 9 indicates that electric vehicles which solely depend on coal would generate more CO_2, SO_2, and NO as compared to ICEVs. The EVs are likely to have approximately 40% higher CO_2 emissions than ICEVs. However, EVs would have lesser CO emissions when compared to ICEVs.

5 Conclusion and Discussion

Coal is the primary source of power generation in India. Currently, 57% of total power is generated from coal power plants. As India gradually shifts from ICEVs to EVs, the load on power generation sources is likely to increase. If EVs are charged using electricity generated from coal thermal power plants, then the CO_2, SO_2, and NO emissions at the power plants would be higher when compared to ICEVs. However, CO emissions are considerably less in comparison with ICEVs. In this study, the indirect emission due to the rise of electric vehicles, which will primarily take place at coal power plants, are discussed. The emissions from other source of energy are very less and they are not considered in this study but kept as a future scope.

Nations like Sweden, Germany, China, and Scotland are moving toward renewable energy by investing in solar, wind, and other renewable energy sources to achieve clean transport and healthy environment. India's investment in electric vehicles will be further substantiated if the country makes similar significant investments in renewable sources or cleaner coal.

Acknowledgements The authors acknowledge the Indian Council of Social Science Research (ICSSR), Ministry of Education, GoI, for funding the research under their Impactful Policy Research in Social Science (IMPRESS) scheme.

References

1. Kancharla SR, Ramadurai G (2018) An adaptive large neighborhood search approach for electric vehicle routing with load-dependent energy consumption. Transp Dev Econ 4:1–9. https://doi.org/10.1007/s40890-018-0063-3
2. Vajjarapu H, Verma A, Gulzar S (2019) Adaptation policy framework for climate change impacts on transportation sector in developing countries. Transp Dev Econ 5:1–16. https://doi.org/10.1007/s40890-019-0071-y
3. WRI. Urban transport and environment n.d.
4. Hawkins TR, Gausen OM, Strømman AH (2012) Environmental impacts of hybrid and electric vehicles—a review. Int J Life Cycle Assess 17:997–1014. https://doi.org/10.1007/s11367-012-0440-9
5. Shearer C, Fofrich R, Davis SJ (2017) Future CO_2 emissions and electricity generation from proposed coal-fired power plants in India. Earth's Future 5:408–416. https://doi.org/10.1002/eft2.201
6. Zhang K, Batterman S (2013) Air pollution and health risks due to vehicle traffic. Sci Total Environ 451:307–316. https://doi.org/10.1016/j.scitotenv.2013.01.074
7. Buekers J, Van HM, Bierkens J, Int L (2014) Health and environmental benefits related to electric vehicle introduction in EU countries. Transp Res Part D 33:26–38. https://doi.org/10.1016/j.trd.2014.09.002
8. Huo H, Cai H, Zhang Q, Liu F, He K (2015) Life-cycle assessment of greenhouse gas and air emissions of electric vehicles: a comparison between China and the US. Atmos Environ 108:107–116
9. Nagamani C, Saravana Ilango G, Reddy MJB, Rani MAA, Lakaparampil ZV (2015) Renewable power generation Indian scenario: a review. Electr Power Compon Syst 43:1205–1213. https://doi.org/10.1080/15325008.2015.1013164
10. Central Statistics Office. Energy Statistics (2018). vol 25, New Delhi
11. Blondeel M, Van de Graaf T (2018) Toward a global coal mining moratorium? A comparative analysis of coal mining policies in the USA, China, India and Australia. Clim Change 150:89–101. https://doi.org/10.1007/s10584-017-2135-5
12. Mittal ML (2010) Estimates of emissions from coal fired thermal power plants in India. In: Proceedings of the international emission inventory conference, vol 39. Florida, USA, p 1–22
13. Coal India Limited. Coal vision 2030. 2018
14. Hanak DP, Biliyok C, Manovic V (2015) Efficiency improvements for the coal-fired power plant retrofit with CO_2 capture plant using chilled ammonia process. Appl Energy 151:258–272
15. McBride JP, Moore RE, Witherspoon JP, Blanco RE (1978) Radiological impact of airborne effluents of coal and nuclear plants. Science 202(4372):1045–1050. https://doi.org/10.1126/science.202.4372.1045
16. Nimesh V, Sharma D, Reddy VM, Goswami AK (2020) Implication viability assessment of shift to electric vehicles for present power generation scenario of India. Energy 195. https://doi.org/10.1016/j.energy.2020.116976
17. National Research Council 2008 (2013) What you need to know about energy 48. https://doi.org/10.5860/choice.48-1502
18. Ramachandra TV, Shwetmala (2009) Emissions from India's transport sector: statewise synthesis. Atmos Environ 43:5510–5517. https://doi.org/10.1016/j.atmosenv.2009.07.015

19. Sharma RD, Jain S, Singh K (2011) Growth rate of motor vehicles in India-impact of demographic and economic development. J Econ Soc Stud 1:137
20. Tehrani Nejad MA (2007) Allocation of CO_2 emissions in petroleum refineries to petroleum joint products: a linear programming model for practical application. Energy Econ 29:974–997. https://doi.org/10.1016/j.eneco.2006.11.005
21. Furuholt E (1995) Life cycle assessment of gasoline and diesel. Resour Conserv Recycl 3449:251–263
22. The Automotive Research Association of India (ARAI) (2018) Indian emissions regulations. Pune, India
23. Mishra UCÃ (2004) Environmental impact of coal industry and thermal power plants in India. J Environ Radioact 72:35–40. https://doi.org/10.1016/S0265-931X(03)00183-8
24. Chakraborty N, Mukherjee I, Santra AK, Chowdhury S, Chakraborty S (2008) Measurement of CO_2, CO, SO_2, and NO emissions from coal-based thermal power plants in India. Atmos Environ 42:1073–1082. https://doi.org/10.1016/j.atmosenv.2007.10.074
25. Mittal ML, Sharma C (2003) Anthropogenic emissions from energy activities in India: generation and source characterization (Part I: emissions from thermal power generation in India). http://www.osc.edu/research/pcrm/emissions/India.pdf
26. Gurjar BR, Van AJA, Lelieveld J, Mohan M (2004) Emission estimates and trends (1990–2000) for megacity Delhi and implications. Atmos Environ 38:5663–5681. https://doi.org/10.1016/j.atmosenv.2004.05.057
27. Hooftman N, Oliveira L, Messagie M, Coosemans T, Mierlo J (2016) Environmental analysis of petrol, diesel and electric passenger cars in a Belgian urban setting. Energies 3. https://doi.org/10.3390/en9020084
28. Eriksson M, Ahlgren S (2013) LCAs of petrol and diesel: a literature review. Swedish University of Agricultural Sciences, Department of Energy and Technology
29. Baglione M, Duty M, Pannone G (2007) Vehicle system energy analysis methodology and tool for determining vehicle subsystem energy supply and demand
30. Nimesh V, Kumari R, Soni N, Goswami AK, Mahendra Reddy V (2021) Implication viability assessment of electric vehicles for different regions: an approach of life cycle assessment considering energy analysis and battery degradation. Energy Convers Manage 237:114104. https://doi.org/10.1016/j.enconman.2021.114104
31. Agrawal K, Jain S, Jain AK, Dahiya S (2014) Assessment of greenhouse gas emissions from coal and natural gas thermal power plants using life cycle approach. Int J Environ Sci Technol. https://doi.org/10.1007/s13762-013-0420-z

Analysis of Drivers' Speed Behavior Along Horizontal Curves of Two-Lane Rural Highways Using Driving Simulator

Tushar Choudhari and Avijit Maji

1 Introduction

The three critical interacting elements in driving are the driver, vehicle, and infrastructure. Any improper communication among these elements increases the crash probability. In India, MoRTH's (2015) report showcased that at least 50% of the total road-related deaths occurred along non-urban highways [1]. Moreover, overspeeding and driver's error were the predominant causes of crashes [1–3]. Based on previous studies along the midblock section of rural highways, the horizontal curves were more critical than other locations. They had a significant influence on its preceding and succeeding road sections. Therefore, researchers studied the speed of drivers along rural highways to understand their basic driving behavior [4–7]. Lamm et al. [8] observed a significant relationship of speed differential with the rate of crashes. The number of crashes along horizontal curves was found higher with the high-speed differential values. Further, some developed countries started using such models during the design stage of highways to consider their driving patterns. Thus, the speed differential variable provided a suitable measure for assessing and increasing safety.

The geometric design of highways is often checked for design consistency to evaluate road safety. It showed a significant correlation with the crash risk, which depends on the geometric parameters of curves such as curvature, radius, curve length, sight distance, vertical curve, gradient, preceding tangent length, and cross-sectional elements [7, 9, 10]. The driver performance along these elements was conventionally measured using operating speed, driver workload, lateral acceleration, vehicle stability, and alignment indices [9, 11–13]. Out of these, speed (or speed reduction) showed a better relationship with accident frequency [14]. Many studies are available for high-income countries with a strong lane disciplined and homogeneous traffic conditions [15]; however, very few are available for weak lane

T. Choudhari (✉) · A. Maji
Department of Civil Engineering, IIT Bombay, Mumbai, India

M. Parida et al. (eds.), *Proceedings of the Fifth International Conference of Transportation Research Group of India* , Lecture Notes in Civil Engineering 220,
https://doi.org/10.1007/978-981-16-9925-2_16

disciplined and heterogeneous traffic environments, which is a characteristic of low- and middle-income countries [15, 16]. Hence, the speed performance of passenger car drivers accustomed to such driving environments are investigated in this study. The goal of this study is to investigate the effect of curves on the speed differential performance to study road safety. For this purpose, a fixed-base driving simulator was used having a virtual scenario of two-lane rural Indian highway.

2 Literature Review

The design consistency is often referred as designing road geometry which would satisfy the driver's expectancy of the road ahead. The risk of crashes increases if the driver's expectations are not met. Therefore, several studies developed operating speed prediction models along the horizontal curves to understand driver's expectations. Most studies observed influence of geometric parameters on the operating speed [10, 17–20]. Despite the usefulness of the developed models in evaluating design consistency, these models varied as per the data collection methodology, highway type, driver's characteristics, and traffic volume [7]. It introduced a regional bias in the developed model resulting in non-acceptance at a global level.

The operating speed differential is identified as a relevant parameter for safety evaluation. The 85th percentile of sample speed data was used for this purpose. It is characterized as a speed below which 85% of the vehicle operate below this value. It is considered as the operating speed of the vehicle. Hence, for speed prediction models, the 85th percentile value is used at each curve.

The speed differential values vary with the dataset availability [21, 22]. Lamm et al. [8] developed first-speed differential parameter, ΔV_{85}. It is the difference of 85th percentile values of the individual speed distributions each observed at preceding tangent and successive curve (midpoint). It was further used in highway safety evaluation that classified a horizontal curve into good, fair, or poor. Since the 85th percentile values of each distribution were used for speed differential calculations, it could not track the individual speed differential experienced by the drivers. Hence, Misaghi and Hassan [7] modified the speed differential calculations to consider vehicle tracking. They calculated the speed difference experienced by the drivers and used 85th percentile value of its distribution, i.e., $\Delta_{85}V$. They compared $\Delta_{85}V$ and ΔV_{85}, and concluded that $\Delta_{85}V$ was significantly greater than ΔV_{85}. Hence, Misaghi and Hassan [7] suggested $\Delta_{85}V$, which accounts for individual driver speed differential experience, as a better estimator of speed differential compared to ΔV_{85}.

Further, these two parameters consider speed data at fixed locations. Therefore, further studies obtained speed profile data to examine the actual speed differential felt by the drivers [4, 6, 16, 23]. These studies used the speed profile of a vehicle and calculated maximum speed reduction (MSR). It is the difference between maximum speed at preceding tangent and minimum speed at curve section. The 85th percentile of MSR distribution, i.e., 85MSR, was calculated at each curve. Bella [4] compared the 85MSR and $\Delta_{85}V$ and concluded that $\Delta_{85}V$ significantly underestimated the actual speed reduction experienced by the drivers since it was considerably lesser than 85MSR. Bella [4] also pointed out that these two parameters were not conceptually different since both of them required vehicle tracking. They are different because of the detailed data availability for 85MSR and only fixed location data points for the calculation of $\Delta_{85}V$.

Overall, these studies suggested that 85MSR $\geq \Delta_{85}V \geq \Delta V_{85}$. The parameter $\Delta_{85}V$ developed by Misaghi and Hassan [7] was found to be 7.55 km/h higher than the ΔV_{85}. Further, the parameter 85MSR developed by Mcfadden and Elefteriadou [23] was found approximately twice of ΔV_{85}. The mean difference between 85MSR and ΔV_{85} was observed as 9.81 kmph. Park and Saccomanno [24] observed 85MSR is on average about two times than ΔV_{85}. Further, a comparative study by Bella [4] showed the mean difference of 2.17 kmph between 85MSR and $\Delta_{85}V$. The regression equation observed the mean difference of more than 6.35 kmph between these parameters. Overall, these studies indicated that 85MSR provided better information for speed reduction compared to ΔV_{85} and $\Delta_{85}V$.

However, 85MSR only considers the maximum and minimum speed for a curve; however, the driver may not effectively reduce the speed between these two speeds. The actual deceleration observed between these speeds varies with the distance between these maximum and minimum speed locations (i.e., extreme speed locations). A reasonable distance required for low values of speed reduction is better than constrained distance having high values. Therefore, it warrants further study to find effective speed differential parameter for the two-lane rural highways.

Llopis-Castelló et al. [25] introduced the short-term memory (STM) concept for a driver based on driver's working memory. It suggested that at an instantaneous location, the driver could remember up to past 15 s of road driven. With this small period of localized driving experience, the driver builds an expectancy of the road ahead, and improvises his driving strategies accordingly. Using this STM concept, Llopis-Castelló et al. [25] calculated the inertial speed profile considering the weighted moving average of the speed profile data through this period. They defined drivers' expectancies and related them with the rural road geometry. They stated that the significant speed difference between the instantaneous speed profile and the inertial speed profile defined unexpected road stretches for a driver. They successfully validated this concept with the previous crash data. Further, they developed consistency models providing critical locations along the road. However, they considered the homogeneous road stretches and not the individual curves. Moreover, the methodology in their study used the previously developed speed models, without using the actual driving data. As a result, further study is needed to examine the localized consistency of the horizontal curves using actual driving behavior at horizontal curves.

In summary, the speed reduction parameter needs further improvement to meet the driver's expectations about the horizontal curves. As mentioned earlier, studies showed 85MSR as a better estimator of speed differential for checking the consistency of horizontal curves; however, it does not represent the effective speed reduction. Hence, this study aims to develop an improved speed reduction parameter using the driver's speed profile data. For this purpose, the STM concept was explored as an effective speed reduction variable. The required data was obtained from the driving simulator. Further, this study is extended to obtain the zone of speed reduction at the configurations.

3 Methodology

The speed profile data provided a better indication of speed reduction than the spot speed profile data for the evaluation of horizontal curves. However, speed reduction only does not consider the rate of speed differential experienced by the driver. The driver's comfort level decreased with an increased rate of speed reduction. Therefore, this study had examined a new speed reduction parameter based on the driver's expectancies about the tangent-curve configuration. Based on the study by Llopis-Castelló et al. [25], the inertial operating speed profile of a driver was obtained. The difference between the actual operating speed and the inertial operating speed is calculated for each tangent-curve configuration. The rate of change of speed reduction with respect to time was obtained to find the locations where the driver significantly started and stopped the speed reduction. The speed at such locations was extracted to calculate the speed reduction values. The 85th percentile of these values at each curve was used for model development.

3.1 Basic Assumptions

The following assumptions were adopted while using the methodology mentioned in this study.

- Drivers driving slower than their desirable speed would more likely increase their speed in the curve section. Such drivers are assumed to be comfortable along curves; hence, the speed gain observed are not considered in the study.
- Previous studies had shown that the operating speed of the drivers was not constant throughout the curve configuration [26, 27]. Therefore, the speed reduction is assumed between the operating speed and inertial speed. However, obtaining a significant speed differential using only speed data was a major task. Moreover, the drivers driving at low speed would not reduce the speed significantly along curves. Therefore, the speed reduction of more than 5 kmph was assumed to be a significant value in this study.
- As mentioned earlier, the STM rate was assumed to be linear. Hence, the moving average with linear weightage was used for the inertial speed profile calculation.

3.2 *Inertial Speed Profile*

It was calculated using the STM technique. As mentioned earlier, the driver acquires suitable driving strategies as per the distance traveled during the last STM time. The optimal STM time of 15 s obtained by Llopis-Castelló et al. [25] was used for the calculation. The moving average speed profile was obtained for the time of the last 15 s. Since the STM declines over time, the information is assumed as lost after 15 s. Therefore, the moving average with linear weightage was calculated. Such a speed profile was calculated as shown in Eq. (1).

$$V_{m,n} = \frac{\sum w_p \cdot V_{i,n}}{\sum w_p} \tag{1}$$

where,

$V_{m,n}$ is the inertial operating speed (kmph) at station *n*,
$V_{i,n}$ is the actual operating speed at station *n*,
w_p is the weight factor at point *p*.

The speed profiles obtained for drivers at each curve were used to locate the speed at which the driver started reducing the speed and reach a comfortable speed. For example, Fig. 1A presents the inertial speed profile and operating speed profile throughout the tangent-curve-tangent configuration. Figure 1B shows the speed reduction between these profiles. The rate of change of speed reduction was used to identify the significant speed reduction zone. This zone is shown in Fig. 1B indicating a significant change in the rate of speed reduction. The actual operating speed values at the start and end of the speed reduction zone were used for calculating significant speed reduction calculations. These values were subsequently used for model development.

4 Experiment Setup

Total 26 curve configurations were designed in AutoCAD Civil 3D® software which was exported to the driving simulator for scenario development. The descriptive statistics of the curve parameters are presented in Table 1. These curve configurations were isolated curves, i.e., significant tangent length was available before the actual study curve. The speed differential parameters were calculated using the speed profile data observed during the data collection process.

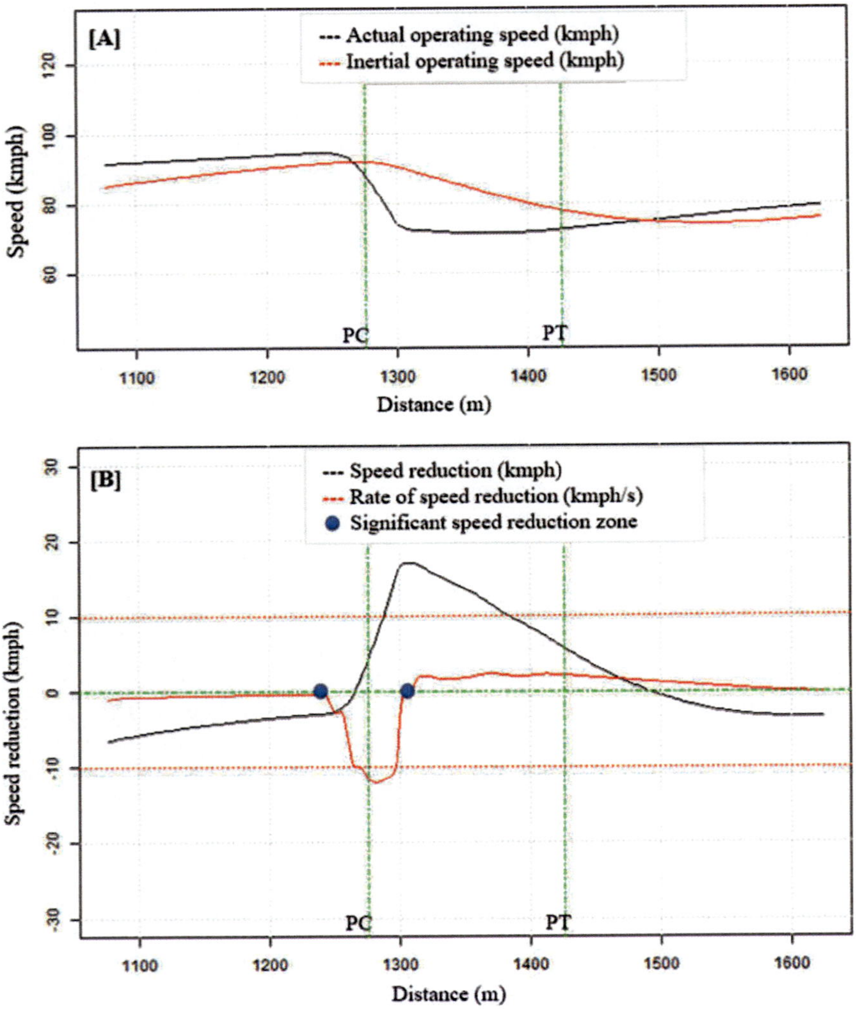

Fig. 1 Example of speed profiles obtained for a driver at a curve: [**A**] operating speed and inertial speed profiles, [**B**] speed reduction and rate of speed reduction profiles speed reduction and location

4.1 Participants

Participants with a driving license (of light motor vehicle, LMV [28]) and driving experience of 2 years were recruited for data collection. These participants were the staffs and students working in the Indian Institute of Technology Bombay and the Mumbai region's professional drivers. Overall, total 77 drivers (9 female drivers) volunteered in this study with ages between 19 and 57 years (mean 32 years).

Table 1 Descriptive statistics of curve parameters

Curve parameters	Abbreviation	Min	Max	Mean	SD
Preceding tangent length (m)	PTL	110	502	241.0	90.5
Radius (m)	*R*	50	300	165.3	88.0
Horizontal curve length (m)	CL	81	496	255.0	134.9
Gradient (%)	*G*	− 2.04	2.03	0.01	1.35
Shoulder width (m)	SW	1	2.5	1.84	0.6
Extra-widening (m)	EW	0.6	1.2	0.76	0.22
Super elevation (%)	*e*	7.00% at all study curves			

4.2 Experiment Procedure

The data collection started with a questionnaire session, followed by a test drive and a final data acquisition drive. The volunteers were guided about their role in this study. The questionnaires covered the driver demographic and driving experience information. In the test drive, all drivers initially drove through a trial section so as to get familiar with the driving simulator. It generally took 5–10 min to complete a test drive. During this process, the drivers with simulator sickness were discarded for the further experiment, and only comfortable drivers were allowed to drive the further study section. There was no ambient vehicle along the study curves to maintain free-flow conditions. However, a few vehicles were provided at the road section other than study curves to provide a scenario similar to a reasonably practical driving environment. We observed six drivers getting symptoms of simulator sickness and hence stopped the experiment for those participants. Finally, 71 drivers successfully completed the experiment process.

5 Results and Discussion

The inertial operating speed for a complete 28 km road stretch was obtained. Each study curve configuration consists of a horizontal curve and two tangents each prior to and after the curve. The observations for individual curves are noted form 200 m prior to the point of curvature (PC) of the respective curve until the 200 m after the point of tangency (PT) of the same curve. Using the methodology in this study, a total of 1846 values of speed reduction were obtained for 71 drivers driving at 26 different curves. Out of these values, only 967 (52.4%) speed reduction values were more than 5 kmph. Based on the assumption in this study, only 52.4% times significant speed reduction was observed.

5.1 Comparison Between Speed Reduction Factors

The preliminary variables presented in the literature review section for analyzing speed reduction are 85MSR, ΔV_{85}, and $\Delta_{85} V$. These variables were calculated based on the data observed along several configurations. The relation between these variables is presented in Fig. 2. It suggests that the 85MSR considers higher speed reduction values compared to ΔV_{85} and $\Delta_{85} V$. This indicates that ΔV_{85} and $\Delta_{85} V$ underestimate the actual speed reduction represented by 85MSR variable. Similar observations were available in the previous studies [4, 23]. The mean difference observed for 85MSR with $\Delta_{85} V$ is 31.2 kmph and with ΔV_{85} as 38.8 kmph.

As indicated by Bella [4], an 85MSR factor provided a better estimation of speed reduction for the evaluation of horizontal curves. It was compared with the speed reduction ΔV_{STM} factor explained in the methodology section. To compare the means of the speed reduction dataset for all study curves, the bilateral t-test for the paired sample was used.

The outcome of the test showed that the null hypothesis was rejected at the significance of 0.05 (mean of paired difference between [85MSR $-$ ΔV_{STM}] was 28.1; $t = 12.5; p < 0.05$). Consequently, 85MSR was found to overestimate the speed difference in comparison with ΔV_{STM}. Earlier, it has been mentioned that 85MSR provided more information about speed differential; however, it had some limitations. For example, only difference between maximum and minimum value of speed reduction may not reflect the effective speed differential experienced by the driver. However, this fact is considered while calculating ΔV_{STM}; hence, in this study, ΔV_{STM} is considered a better speed differential variable. It can be said that the 85MSR overestimates the speed differential values compared to ΔV_{STM}. Figure 3 showed similar observations. It also indicated the overestimation of 85MSR compared to ΔV_{STM} experienced by

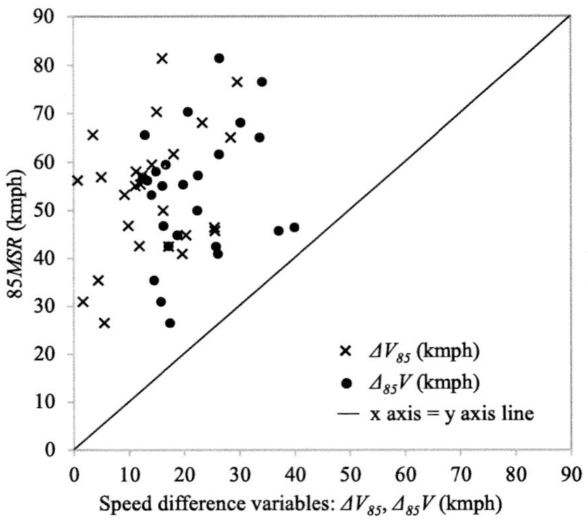

Fig. 2 Plot of 85MSR *Vs* ΔV_{85} and ΔV_{85}

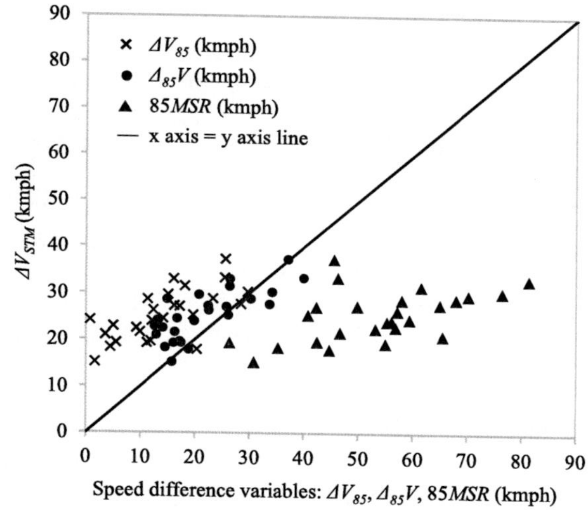

Fig. 3 Plot of ΔV_{STM} *Vs* ΔV_{85}, Δ_{85}V, and 85MSR

the driver. These results pointed out that the 85MSR cannot assure a suitable indicator of speed differential factor.

Figure 3 also presented the relation between ΔV_{STM} and previous speed differential variables such as ΔV_{85} and $\Delta_{85}V$. The paired *t*-test indicated a significant difference ($p < 0.05$) between ΔV_{STM} and these variables. The mean difference between ΔV_{STM} and ΔV_{85} is 10.7 kmph, with $\Delta_{85}V$ it is 3.2 kmph. Considering these mean differences, it can be said that the ΔV_{85} and $\Delta_{85}V$ underestimate the actual speed differential values.

5.2 *Model Development*

The multiple linear regression analysis was used for model development between the speed reduction as a dependent variable and geometric parameters as independent variables (shown in Table 1). Along with geometric parameters, the speed values observed at PC and 100 m, 200 m prior to PC were considered to develop the model. The backward elimination method is used for this purpose. With all the correlation checks, the model developed with this method is shown in Eq. 2.

$$\Delta V_{STM} = -1.01 + \frac{944.81}{R} + 0.185 \times V_{100} \quad R^2_{adj} = 0.71 \tag{2}$$

where

ΔV_{STM} is the 85th percentile speed reduction using STM concept,
R is the radius of the curve, and.
V_{100} is speed observed at 100 m prior to the PC.

The p-value for R was less than 0.001 and for V_{100} was 0.055, respectively, indicating their significance in the developed model. Previous researches also indicated curve radius as crucial parameter for calculating the speed reduction [4, 16, 29]. Moreover, the speed observed at tangent section was also an important parameter in speed reduction [4]. The speed reduction methodology provided in this study provided similar observations.

As mentioned earlier, the higher speed differential indicates a high possibility of risk of crashes along curves. Therefore, to avoid such values, the effective geometric parameters could be reiterated while designing the highways, to reduce the predicted speed differential. The model developed in Eq. 2 presented ΔV_{STM} as speed differential variable and radius of curve as an effective geometric parameter. Thus, with the increase in radius in Eq. 2, the value of ΔV_{STM} reduces. This would lead to decrease in risk of crashes along curves. Currently, the methodology presented in this paper is limited to the development of ΔV_{STM} model. Future scope arises in identifying the thresholds for ΔV_{STM} to reiterate the values of such geometric parameters.

For an application point of view, the developed model using the STM concept can be used during the design stage of road. This model can be obtained considering road geometric parameter as independent variables and speed profile data obtained from sample drivers. As mentioned earlier, the higher speed differential indicates high possibility of risk of crashes along curves. Therefore, to avoid such values, the geometric parameters could be improved while designing the highways, to reduce the predicted speed differential. The nomograms considering effective geometric parameter and driver speed differential measure can provide suitable information to improve highway design prior to its construction. Thus, the developed models could be used for safety along horizontal curves.

In this study, only significant speed reduction values were used. Hence, the percentage of drivers with significant speed reduction values were calculated according to the study curves, since the developed model indicated a significant influence of the curve sharpness on the speed reduction. These percentages are shown in Fig. 4. Further, Fig. 4 also showed the mean and standard deviation of speed reduction values for all these curve radii. It indicated a higher percentage of significant speed reduction for sharper curves compared to milder curves. Moreover, the mean speed reduction experienced by the driver also increases with the decrease in the curve radius.

The increased value of speed reduction caused a significant increase in the number of crashes. The driver's expectancies should be met with the road geometry so that the significant speed differential could be reduced. The speed calming measures such as innovative traffic signs, road markings, and speed humps could be incorporated for the sharp curves. For this purpose, it was important to find the speed reduction zone for implementing such measures. Therefore, the speed reduction zones were analyzed at each configuration of different radii. For this purpose, the percentile of drivers using the distance required for speed reduction is shown are analyzed.

Figure 5 shows the percentile of drivers using the zone required for the speed reduction. The PC point is the point of curvature for all the curve radii. For the study curves, it was observed that 85th percentile drivers perceived the curve and

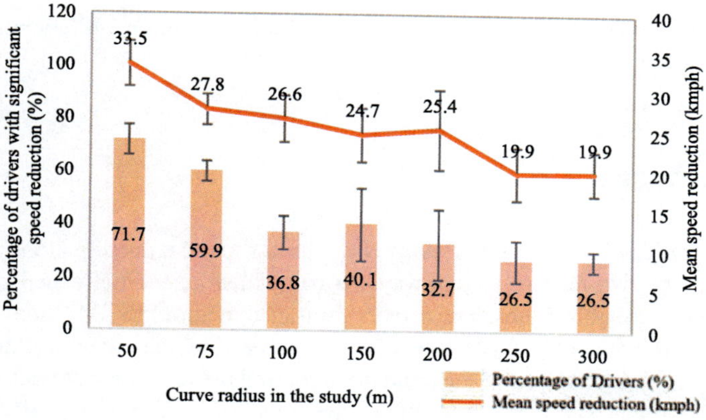

Fig. 4 Effect of curve radius on drivers and their speed reduction

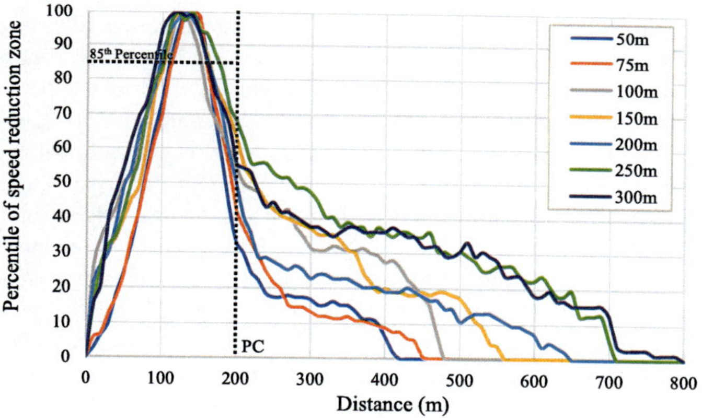

Fig. 5 Distance for speed reduction along different radii curves

started reducing speed approximately 50–120 m prior to the PC (see 85th percentile line in Fig. 5). Considering the PC location with the trends of different curve radii, approximately 30th percentile drivers continued the speed reduction in the curve of 50 m radius. This means 70th percentile drivers already reduced to their comfortable speed prior to entering this curve. This percentile decreased with the increase in curve radius, e.g., for a 75 m radius, it was 60th percentile and for 250 m curve radius, it was approximately 30th percentile. This concluded that with the increase in curve radius, the driver preferred continuing the speed reduction for milder curves. Further, beyond PC, the speed reduction zone increases with the increase in curve radius. Overall, as shown in Fig. 5, the speed calming measures can be adopted at 120 m, prior to the PC of the curve, so that more than 85 percentiles of driver start reducing speed prior to entering a curve. This would avoid a significant speed reduction while maneuvering the tangent-curve configuration.

6 Conclusion

This study included a new methodology for defining speed reduction along the horizontal curve of two-lane rural highways. It calculated an inertial speed profile to obtain speed reduction values. The drivers shot-term memory (STM) time was used for obtaining this speed profile. Drivers adjust the speed of vehicle while negotiating a curve as per his perception of the curve features and the driving strategies acquired during STM time. The regression model developed also indicated that the curve radius and speed at 100 m prior to the point of curvature (PC) as the most influential parameters among all other geometric features. The comparison with previous speed reduction factors indicated that 85MSR overestimates the speed reduction values compared to ΔV_{STM}. Also, the ΔV_{STM} has least difference with $\Delta_{85} V$ when compared with ΔV_{85} and 85MSR. The analysis of the speed reduction zone indicated that more than 85 percentiles of driver started reacting to the curve at a distance of 80 to 120 m in the tangent section prior to the curve. When compared to the developed model, this shows the speed at 100 m prior to PC as a good indicator for calculating speed reduction. Further, the significant speed reduction of the driver was observed prior to the start of sharp curves (R < 150 m); however, the drivers preferred to continue decreasing the speed in the curve section with the increase in curve radius.

The results from this study can be used in practice with the future validation in actual field condition. Currently, the methodology presented in this paper is limited to comparison of ΔV_{STM} with other speed differential variables, and development of ΔV_{STM} model. Future scope arises with identifying the thresholds for ΔV_{STM} to evaluate safety along horizontal curves. Also, the present study was restricted to cars only. Hence, the methodology could be further explored for other vehicle types.

References

1. MORTH: Road Accidents in India 2015 (2016) Transport research wing, ministry of road transport and highways, Government of India, New Delhi
2. Urie Y, Velaga NR, Maji A (2016) Cross-sectional study of road accidents and related law enforcement efficiency for 10 countries: a gap coherence analysis. Traffic Inj Prev 17(7):686–691
3. Maji A, Velaga NR, Urie Y (2018) Hierarchical clustering analysis framework of mutually exclusive crash causation parameters for regional road safety strategies. Int J Inj Contr Saf Promot 25(3):257–271
4. Bella F (2007) Parameters for evaluation of speed differential: contribution using driving simulator. Transp Res Rec 2023(1):37–43
5. Lamm R, Psarianos B, Choueiri ME, Soilemezoglou G (1995) A practical safety approach to highway geometric design international case studies: Germany, Greece, Lebanon, and The

United States. In: The international symposium on highway geometric design practices. Boston, Massachusetts

6. Choudhari T, Maji A (2019) Effect of horizontal curve geometry on the maximum speed reduction: a driving simulator-based study. Transp Dev Econ 5(2):14
7. Misaghi P, Hassan Y (2005) Modeling operating speed and speed differential on two-lane rural roads. J Transp Eng 131(6):408–418
8. Lamm R, Guenther AK, Choueiri EM (1995) Safety module for highway geometric design. Transp Res Rec J Transp Res Board 1512(9):7–15
9. Fitzpatrick K, Wooldridge MD, Tsimhoni O, Collins JM, Green P, Bauer KM, Parma KD, Koppa R, Harwood DW, Anderson IB, Krammes RA, Poggioli B (2000) Alternative design consistency rating methods for two-lane rural highways
10. Jacob A, Anjaneyulu MVLR (2013) Operating speed of different classes of vehicles at horizontal curves on two-lane rural highways. J Transp Eng 139(3):287–294
11. Heger R, Watters P, O Mahony M, Wong JT, Huang SH, Green P, Hagendoorn RGC (1998) Driving behavior and driver mental workload as criteria of highway geometric design quality, No. E-C003
12. Gibreel GM (1999) State of the art of highway geometric design consistency. J Transp Eng Am Soc Civ Eng 305–313
13. Choudhari T, Maji A (2019) Socio-demographic and experience factors affecting drivers' runoff risk along horizontal curves of two-lane rural highway. J Safety Res 71:1–11
14. Fitzpatrick K, Anderson IB, Bauer KM, Collins JM, Elefteriadou L, Green P, Harwood DW, Irizarry N, Koppa R, Krammes RA, McFadden J, Parma KD, Passetti K, Poggioli B, Tsimhoni O, Wooldridge MD (2000) Evaluation of design consistency methods for two-lane rural highways, executive summary
15. Nama S, Maurya AK, Maji A, Edara P, Sahu PK (2016) Vehicle speed characteristics and alignment design consistency for mountainous roads. Transp Dev Econ 2(2):23
16. Choudhari T, Maji A (2017) Miscellaneous study on speed reduction along horizontal alignments of rural highways: a driving simulator based approach for developing countries like India. In: Annual meeting at transportation research board, pp 17–06699
17. Camacho-Torregrosa FJ, Pérez-Zuriaga AM, Campoy-Ungría JM, García-García A (2013) New geometric design consistency model based on operating speed profiles for road safety evaluation. Accid Anal Prev 61:33–42
18. Wu X, Wang X, Lin H, He Y, Yang L (2013) Evaluating alignment consistency for mountainous expressway in design stage : a driving simulator-based approach. In: Annual meeting at transportation research board, pp 13–2309. Washington
19. Maji A, Tyagi A (2018) Speed prediction models for car and sports utility vehicle at locations along four-lane median divided horizontal curves. J Mod Transp 1–8
20. Maji A, Sil G, Tyagi A (2018) 85th and 98th percentile speed prediction models of car, light, and heavy commercial vehicles for four-lane divided rural highways. J Transp Eng Part A Syst 144(5):1–8
21. Lamm R, Psarianos B, Cafiso S (2002) Safety evaluation process for two-lane rural roads—a 10-year review. Transp Res Rec 1796(2):51–59
22. Bella F, D'Agostini G (2010) Combined effect of traffic and geometrics on rear-end collision risk. Transp Res Rec J Transp Res Board 2165:96–103
23. Mcfadden J, Elefteriadou L (2000) Evaluating horizontal alignment design consistency of two-lane rural highways development of new procedure. Transp Res Rec 1737:9–17
24. Park YJ, Saccomanno FF (2006) Evaluating speed consistency between successive elements of a two-lane rural highway. Transp Res Part A Policy Pract 40(5):375–385
25. Llopis-Castelló D, Camacho-Torregrosa FJ, García A (2018) Development of a global inertial consistency model to assess road safety on Spanish two-lane rural roads. Accid Anal Prev 119:138–148
26. Pérez-Zuriaga AM, Camacho-Torregrosa FJ, García García A (2013) Tangent-to-curve transition on two-lane rural roads based on continuous speed profiles. J Transp Eng 139(11):1048–1057

27. Montella A, Galante F, Mauriello F, Aria M (2015) Continuous speed profiles to investigate drivers' behavior on two-lane rural highways. Transp Res Rec J Transp Res Board 2521:3–11
28. MVA (1988) The motor vehicles act. ministry of road transport and highways, India
29. Choudhari T, Maji A (2018) Assessment of driver performance based on driver demography and road geometry. In: Annual meeting at transportation research board, pp 18–06523

Evaluating Locations of Future Intercity Passenger Rail Maintenance

Shailesh Chandra, Ravi Mazin, Mehran Rahmani, and R. Thirumaleswara Naik

1 Introduction

Rail transportation is vital for economic growth of a country—as it facilitates movement of people on a mass scale [1]. Maintenance and layover facilities for rail fleet are required for timely operations of the transit service [2]. The facilities need to be optimally decided and located to provide optimum service standards. There are various techniques adopted in research in determining the optimal location across a given set of variables and criteria. For example, multi-criteria decision making (MCDM) is one such technique used by decision makers in determining the best alternative among a list of alternatives. A MCDM problem deals with the evaluation of a set of alternatives (as choices) in terms of a set of decision criteria for the decision maker [3].

Similar to MCDM is the multi-objective decision-making (MODM) problem in which the goal is to identify the optimum solution to a set of quantifiable but conflicting objectives, with a set of well-defined constraints [2]. The efficient solution (also known as non-inferior solution) to MODM is the one which is the preferred value for all the conflicting factors [4].

There are also approaches that convert multi-objectives into one to try identifying the optimal location. One of the most commonly used approaches is genetic algorithm—such as multi-objective genetic algorithm [5] and non-dominate sorting genetic algorithm [6].

There are various factors and constraints that need to be satisfied for optimal location determination. However, determining an optimal location for a rail maintenance

S. Chandra (✉) · R. Mazin · M. Rahmani
Department of Civil Engineering and Construction Engineering Management, California State University Long Beach, Long Beach, CA, USA
e-mail: shailesh.chandra@csulb.edu

R. T. Naik
Department of Mechanical Engineering, Indian Institute of Science, Bengaluru, Karnataka, India

M. Parida et al. (eds.), *Proceedings of the Fifth International Conference of Transportation Research Group of India* , Lecture Notes in Civil Engineering 220,
https://doi.org/10.1007/978-981-16-9925-2_17

or a layover facility in a region is often a challenging task due to several constraints that drive the choice, including considerations for soil conditions, costs, seismicity, fire, noise, etc.

Optimal location of a future maintenance and operations facility should adhere to standards and design as per the fire code standards of the region—acknowledging that the building and fire codes vary depending on the location. Potential location choice of a facility should be governed based on the proper identification of flood hazard and regulations. There are important considerations for designing and determining locations of a facility that could be flood prone—such as potential of flood depth, velocity, and duration. Tsunami hazard should be considered for sites close to the coast.

Sustainable construction practices also govern the location considerations. For example, use of "just-in-time" transportation and construction methods—for precast concrete members, use of minimal equipment, lower traffic levels, minimize air pollution due to dust, and have feasibility and use of zero emission multiple unit trains (DMUs). Other considerations for an optimal location could be provision of easy elimination and reuse of waste on construction projects, use of methods to share equipment such as crane and lifts with other nearby construction projects, and having opportunities to use renewable energy sources—especially incorporate renewable energy such as solar devices.

Other factors that are necessary for identifying location and building an operations and maintenance facility (OMF) could include structure's dimensional needs, capacity constraints for individual elements—such as locomotives, cars, and storage.

This research will provide guidance in identifying optimal location of a rail maintenance and layover facility based on some location-specific key factors. This is demonstrated using a case study example of some select California's intercity passenger rail yards upgraded to a future rail maintenance and layover facility.

Studies on determining optimal location of a rail maintenance and layover facility have been limited to just only four in number [8]. A recent research by Tönissen et al. solve the location problem of maintenance facilities taking into consideration fleet planning, line planning, and routing [8]. A previous study analyzed an optimal facility location with respect to costs and capacity restraints [9]. Most researches do not consider site-specific limitations which are local constraints such as seismic hazards and resilience if the facility is shutdown in emergency situations. Therefore, this paper takes into account these factors that are often important for practitioners and facility managers. This research can also serve as reference for stakeholders outside the United States of America, particularly in India which has recently proposed to mechanize and automate 40 of the 100 yards [10]. Rail maintenance and layover facilities play a critical role in improving operational efficiencies of the rail cars and locomotives [11]. Strategic-level decisions are necessary to provide tactical and operational-level decisions for efficient railway management, and the location of rail yards play a crucial role in ensuring efficiency of rail operations in India [12].

2 Methodology

A modeling framework is constructed for multiple factors with constraints to identify an optimal location of a future maintenance and/or layover facility. Factors in determining an optimal location consist of the following:

- Risks associated due to seismicity and other natural hazard (S)
- Distance from the existing facilities (D) (Note: This factor ensures that if the new facility is shut down due to any emergency situation, the closest existing facility can take over and play the role of a stand-by. This supports the resilience of the new facility. With this intention, a minimum distance from an existing facility to the new facility is desirable.)
- Average time/ distance traveled to the nearest rail transit station (T), and
- Connectivity to nearest rail line (shortest unobstructed distance to a rail line, R)

The multi-objective optimization of the weighted linear summation of factors above can be represented using M and expressed as:

$$M = \alpha S + \beta D + \lambda T + \sigma R \tag{1}$$

where

α weight for risks associated due to seismicity and other natural hazards,
β weight for distance from the existing facilities,
λ weight for distance traveled to the nearest station, and
σ weight for distance to the closest point on a rail line.

Thus, the goal is to obtain a minimum value for M. The weights α, β, λ, and σ will vary from one site location to another site location. However, ensuring a minimum for all the factors factor simultaneously (expressed using Eq. 1) might not be feasible. While one factor is minimized, the other factor(s) may not achieve the minimum, and hence, can be considered as constraints. Other constraints for identifying location and building a maintenance facility could include structures dimensional needs, capacity constraints for individual elements—such as locomotives, cars, and storage. Since construction of a new maintenance facility involves massive investments, one or more candidate facility locations should be known beforehand to simplify the optimal site selected—and hence, the four factors mentioned above can also serve as mutual constraints that can be quantified and model. The constraints are better represented when ranked qualitatively and/or given quantitative values—such as high (denoted by 100), moderate denoted by 50 and low denoted by 0).

Pareto-optimal fronts (which is widely used in engineering and economics) were constructed and used to identify an optimal location for the objectives. The factor varied for each set of factors considered critical for any candidate location. The methodology for multi-criteria decision making (MCDM) was used to identify an optimal location. The locations are spatially coded using their latitude and longitude values.

3 Application Example

The decision-making framework developed under the methodology section was tested with data collected through interviews and applied to the intercity passenger rail network of California. In several ways, the passenger transport in urban cities of California is similar to India—such as the rail service provided to bay area commuters from adjacent cities is similar to services that New Delhi commuters have from National Capital Region (NCR) cities and towns.

3.1 Data Collection

Interviews and site visits were conducted at two passenger rail maintenance and layover facility in California during February 2019. The first was at the Oakland Maintenance Facility (OMF) and the second at the Stockton Altamont Corridor Express (ACE) Maintenance and Layover Facility. Interviews were carried out among managers and crew working at the facilities during the two site visits. There was only one specific question in which the interview participants were asked to rank the importance of factors listed in Eq. (1). Both the facilities are located in California—as shown in the map of Fig. 1. Both the facilities are located on an existing line of the intercity passenger rail network.

The Oakland and Stockton ACE facilities visited are very well connected to transit stations. Car dispatch can be carried out after service maintenance for operations. However, in case of overload or space limitations, cars are often dispatched to nearby layover facilities from Oakland OMF. This indicates proximity to nearby stations, and layover or maintenance facility can also be considered critical for future facility selection. Future intercity rail maintenance and layover facility should be able to serve multimodal needs of Californians through an integrated, state-of-the-art rail system as laid out in the 2018 California Transportation Rail Plan for future [7]. This requires that the future facility be located at an optimal location to stations, railway corridor, and nearby existing layover facilities.

There were five candidate site locations that were identified for analysis purposes. The spatial representation of the five locations is shown in the map of Fig. 2 for California. All of the five site locations are closest to either Amtrak, BNSF, or the Union Pacific freight rail that are often shared by passenger rail.

The methodology laid out in this paper is very generic and can be replicated for any other region. Thus, the five candidate sites selected for analysis are for illustration purposes only. Based on the information gathered from the data collection and interviews with Oakland and Stockton ACE facilities, quantification of the importance of factors were carried out for the five candidate site locations. These have been presented in Table 1. The numerical quantification of the importance of each factor based on the interview findings is as follows: Low $= 0$, Medium $= 5$ and High $= 10$.

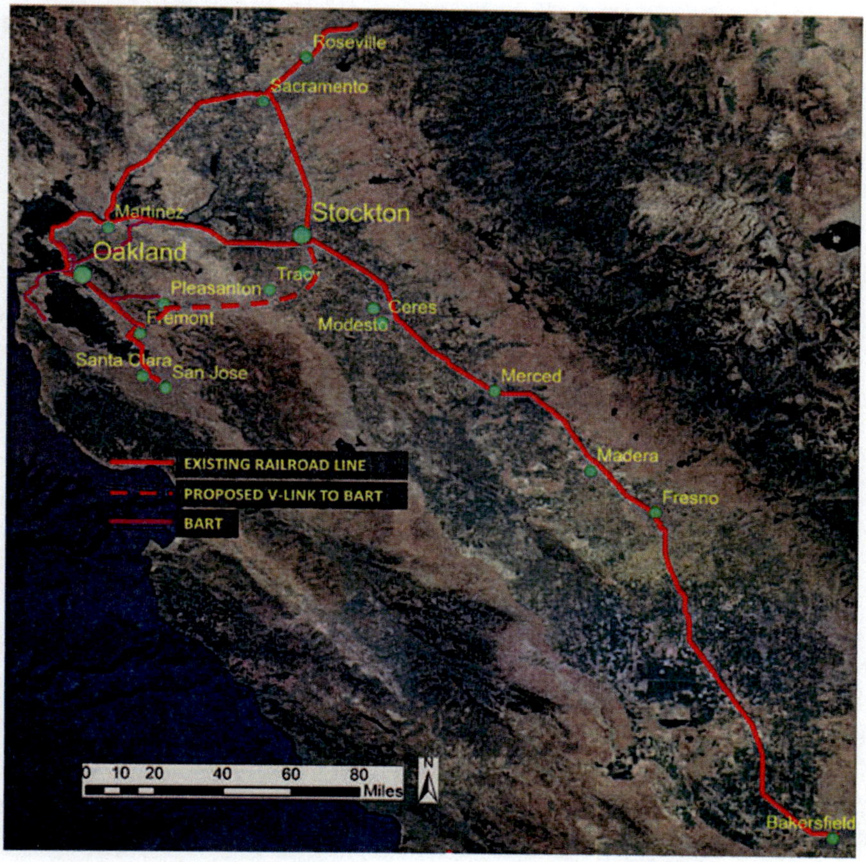

Fig. 1 Geographical location of Oakland and Stockton facilities visited

The determination of the qualitative information presented in Table 1 is based on threshold level for each intensity of the factor as presented in Table 2.

4 Results

4.1 Sensitivity Analysis

Based on the variation of the weights for the parameters, α, β, λ, and σ will vary from one site location to another site location. Since obtaining a unique site location might not be possible due to multiple factors being considered simultaneously, sensitivity

Fig. 2 Candidate site locations for analysis from California

Table 1 Compilation of information on factors based on interviews

	Seismicity risk, natural hazards, etc.	Distance from another existing facility	Distance from the nearest rail station	Connectivity to an existing rail line
Site location 1	Medium	Low	High	Medium
Site location 2	Medium	High	High	Low
Site location 3	High	Medium	Medium	High
Site location 4	High	Low	High	High
Site location 5	High	Low	Low	Medium

Table 2 Classification of qualitative assessment of factors based on thresholds

Factors considered	Qualitative classification
	Low
Seismicity risk, natural hazards, etc.	Not vulnerable from natural hazards (**assigned value = 0**)
Distance from another existing facility	Closest facility within a mile from the site location (**assigned value = 0**)
Distance from the nearest rail station	Closest station within a mile from the site location (**assigned value = 0**)
Connectivity to an existing rail line	Shortest distance to a rail line within a mile from the site location (**assigned value = 0**)
Factors considered	Qualitative classification
	Medium
Seismicity risk, natural hazards, etc.	Average vulnerability from natural hazards (**assigned value = 50**)
Distance from another existing facility	Closest facility within one and two miles from the site location (**assigned value = 50**)
Distance from the nearest rail station	Closest station within one and two miles from the site location (**assigned value = 50**)
Connectivity to an existing rail line	Shortest distance to a rail line within one and two miles from the site location (**assigned value = 50**)
Factors considered	Qualitative classification
	High
Seismicity risk, natural hazards, etc.	Highly vulnerable from natural hazards (**assigned value = 100**)
Distance from another existing facility	Closest facility beyond two miles from the site location (**assigned value = 100**)
Distance from the nearest rail station	Closest station beyond two miles from the site location (**assigned value = 100**)
Connectivity to an existing rail line	Shortest distance to a rail line beyond two miles from the site location (**assigned value = 100**)

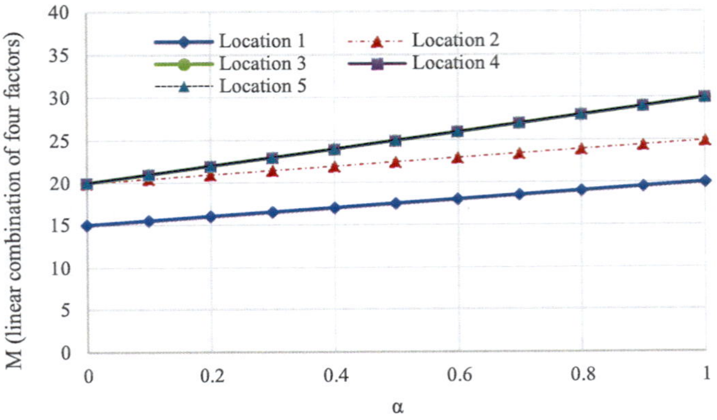

Fig. 3 Variation of weighted linear combination of factors versus the weight associated due to risks from seismicity and other natural hazards

analysis was performed to identify the optimal site for variation in the values for weights α, β, λ, and σ. The range for these weights is variable from 0 to 1.

The chart in Fig. 3 shows the variation in weighted linear summation of the four factors—risks associated due to seismicity and other natural hazards, distance from the existing facilities, average time/distance traveled to the nearest station, and distance to the nearest rail line—expressed as M with respect to α (i.e., weight associated with risks due to seismicity and other potential natural hazards of the site location in California) while keeping rest of the weights β, λ, and σ equal to 1. The chart in Fig. 3 suggests that Site Location 1 in Fig. 3 is the most preferred location for a future rail maintenance facility if the goal is to select the safest location from any seismic or natural hazard danger. It is also noted from the chart of Fig. 3 that Site Locations 2–5 are the next preferred locations if the influence of seismicity and other natural hazards is ignored—evident for $\alpha = 0$.

The chart in Fig. 4 shows the variation in M with respect to weight of the distance from the nearest another rail maintenance facility, while keeping rest of the weights α, λ, and σ are each equal to 1. It is observed that Site Location 2 should be the most preferred location for a rail maintenance facility for the value of weight β up to 0.5. However, with further increase in the value of β, Site Location 1 becomes the most preferred location for a future rail maintenance facility in California. The chart in Fig. 4 also shows that the weighted linear combination of the four factors for site location does not change with any variation in the value of weight β.

Based on the chart of Fig. 5 (in which weights α, β, and σ are each equal to 1), it is evident that the Site Location 1 should be the most preferred location for a future maintenance facility in California irrespective of the influence of λ—the weight associated due to distance from the closest rail transit station. The weighted linear combination of the four factors for Site Location 1 is found to be minimum among all other site locations with increase in λ which is the weight associated for

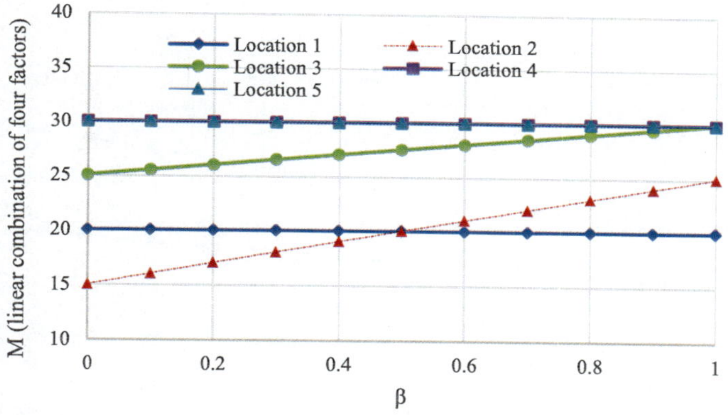

Fig. 4 Variation of weighted linear combination of factors versus the weight associated due to distance from the closest another rail maintenance facility

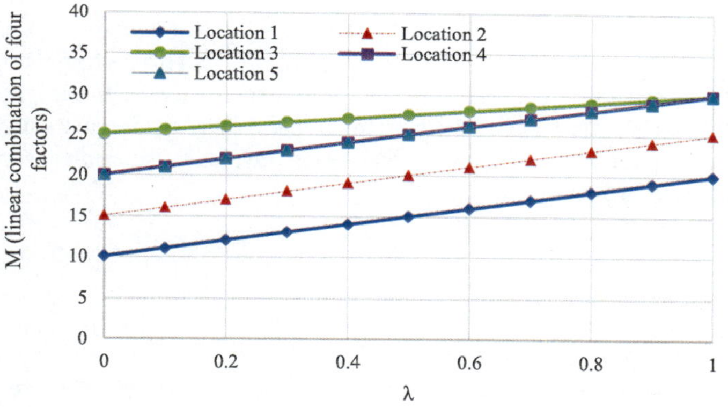

Fig. 5 Variation of weighted linear combination of factors versus the weight associated due to distance from the closest rail transit station

the distance of the site location from the nearest rail transit station. If the influence of distance of the site location to a rail line is ignored (i.e., $\lambda = 0$), the second preferred location is Site Location 2, third and fourth are both the Site Locations 4 and 5, and Site Location 3 is the last.

The variation in M with respect to the weight associated with the inverse of the distance of the site location to a rail line as shown in the chart of Fig. 6 (in which weights α, β, and λ are each equal to 1) illustrates that Site Location 1 should be the most preferred location for a future rail maintenance facility in California. Ignoring the influence of the inverse of the distance to the closest rail line, Site Locations 3, 4, and 5 are the second, third, as well as the fourth preferred site locations. Site Location 2 is the fifth or the last preferred location up to $\sigma = 0.5$.

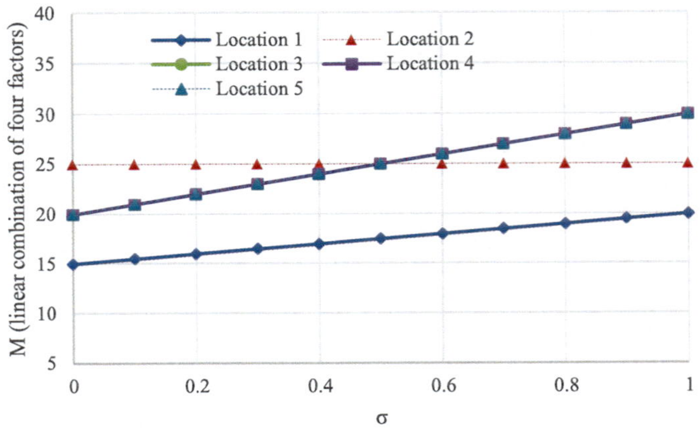

Fig. 6 Variation of weighted linear combination of factors versus the weight associated due to distance from the nearest rail line

Based on the findings from the charts of Figs. 3, 4, 5, and 6, it is found that overall Site Location 1 turns out to be the most preferred location across the weighted linear combination of risks associated due to seismicity and other natural hazards, distance from the existing facilities, average time/distance traveled to the nearest station, and shortest distance to a nearest rail line.

5 Conclusions

Public rail transit is an important mode for transportation of people. An intercity passenger rail is essential for the economic growth of a nation. Rail transit is also quite popular due to its right-of-way track, higher operating speeds, and reliability in transportation. In countries, like India, Indian Railways provides daily transportation services to several millions of passengers. However, with increase in service capacities of these rail, there is also need for regular maintenance of the cars of the transit. Maintaining these rail cars and various other components require timely adherence to regulations and policies—as well as strategic location needs of the maintenance and layover facilities.

This research develops a methodology of locating and identifying the most preferred site for future location of a passenger rail maintenance facility in any given region. Often there is an opportunity to model components of a site that can be evaluated before identifying such as location. The methodology consists of identifying factors that are considered important/essential to be included in the goal of location identification—and those that can be easily quantified. A linear mathematical optimization framework is present, and these factors (which also are treated as part of the factors) consist of risk level associated due to seismicity and other natural hazards, distance from the existing facilities, distance traveled to the nearest station, and distance to the closest point on a rail line. The mathematical optimization model setup did not consider space as a constraint as it was assumed that a maintenance facility would be having enough space needed along a rail right-of-way.

An application example is presented with five future potential site locations in California. These sites are tentative based on interviews conducted with the engineers and staff at two existing rail maintenance and layover facilities in California—the Oakland Operations and Maintenance Facility (OMF) and the Stockton Altamont Corridor Express (ACE) Maintenance and Layover Facility. Based on the methodology outlined in this research, it was found that one of the sites located close to the Union Pacific rail line could be the most preferred site for a future location of a maintenance and a layover facility. This site satisfied all factors considered in this research, namely minimum risk level associated due to seismicity and other natural hazards, minimum distance from the existing facilities, minimum distance traveled to the nearest station, and minimum distance to the closest point on a rail line.

Future research consists of building upon the methodology proposed in this research to incorporate other factors that might be considered important for regions similar to California for setting up a future intercity rail maintenance and layover facility.

References

1. Chandra S, Vadali S (2014) Evaluating accessibility impacts of the proposed America 2050 high-speed rail corridor for the Appalachian Region. J Transp Geogr 37:28–46
2. Farahani RZ, SteadieSeifi M, Asgari N (2010) Multiple criteria facility location problems: a survey. Appl Math Model 34(7):1689–1709
3. Triantaphyllou E, Shu B, Sanchez SN, Ray T (1998) Multi-criteria decision making: an operations research approach. Encycl Electr Electron Eng 15:175–186 (1998)
4. Ulungu EL, Teghem J (1994) Multi-objective combinatorial optimization problems: a survey. J Multi-Criteria Decis Anal 3(2):83–104
5. Fonseca CM, Fleming PJ (1993) Multiobjective genetic algorithms. IEE Colloq Genet Algor Control 6/1–6/5
6. Srinivas N, Deb K (1994) Multiobjective optimization using nondominated sorting in genetic algorithms. Evol Comput 2(3):221–248
7. 2018 California State Rail Plan—Caltrans, Accessed on May 22, 2019
8. Tönissen DD, Arts JJ, Shen ZJ (2019) Maintenance location routing for rolling stock under line and fleet planning uncertainty. Transp Sci 53(5)
9. Xie W, Ouyang Y, Somani K (2016) Optimizing location and capacity for multiple types of locomotive maintenance shops. Comput-Aided Civil Infrast Eng 31(3):163–175

10. Press Releases, Department of Industrial Policy and Promotion, Press information Bureau, Media reports, Railways Budget 2016–17
11. Lin B, Liu S, Lin R, Wang J, Sun M, Wang X, Liu C, Wu J, Xiao J (2019) The location-allocation model for multi-classification-yard location problem. Transp Res Part E Logistics Transp Rev 122:283–308
12. Premkumar P, Kumar PR (2019) Literature Review of locomotive assignment problem from service operations perspective: the case of Indian railways. IIM Kozhikode Soc Manage Rev 8(1):74–86

Evaluation of Appropriate Performance Grade Bitumen for Road Construction According to Indian Climate Change Scenario

Siksha Swaroopa Kar, Ambika Behl, and M. N. Nagabhushana

1 Introduction

Roads in India are, in majority, flexible pavement type wherein the paved surfaces mostly have bituminous layers. India has presently a road network of over 5.9 million km according to sources, the second largest road network. Qualitatively Indian roads are mix of modern highways and narrow unpaved roads; only 62% of the road network is paved [1]. The fast growing Indian economy demand for a high quality pavement structure as it will have to carry very heavy traffic both in terms of number and axle loading. So proper material selection along with better design, construction and maintenance techniques are required.

Bitumen is a visco-elastic material and its stiffness is temperature dependent. In India bitumen specifications were first brought out in the year 1950 (IS: 73–1950) and were classified based on penetration test, which were subsequently revised in 1962 and 1992. To improve bitumen grading IS: 73–1992 was revised to include the viscosity grade and was published in July 2006. The viscosity at 60 °C was considered and four viscosity grade binders were identified VG 10, VG 20, VG 30 and VG 40. Indian Oil commenced marketing of Bitumen as per Viscosity Grade specifications conforming to IS: 73–1992 from all its refineries from Aug 2009. Therefore, the Penetration grades have been replaced by Viscosity grade Bitumen. Tests are conducted at 60 and 135 °C, which represent the temperature of road surface during summer (hot climate, similar to northern parts of India) and mixing temperature, respectively. As per Viscosity grading higher the grade stiffer will be the Bitumen.

Being a visco-elastic material; behaviour of bitumen depends on both temperature and rate of loading. Increased traffic loads and change in climatic condition led to the development of Superpave Performance Grade (PG) binder for a better performing

S. S. Kar (✉) · A. Behl · M. N. Nagabhushana
CSIR-Central Road Research Institute, Mathura Road, New Delhi 110025, India

M. Parida et al. (eds.), *Proceedings of the Fifth International Conference of Transportation Research Group of India* , Lecture Notes in Civil Engineering 220,
https://doi.org/10.1007/978-981-16-9925-2_18

and durable roads in USA. The relationship between binder properties and conditions of use are more complete and precise in Superpave PG system. However this concept is yet to be adopted in Indian applications. A Superpave Performance Grading mapping with commercially available binders is expected to streamline the usage for best performance of pavements. However, this needs a vigorous and rigorous exercise of characterizing the various bitumens from different sources at different grades.

In the study, it was envisaged to collect bitumen of different grades from various sources and evaluate for basic and also visco-elastic properties using these and then map on an Indian PG grade chart. As a first step towards this direction, Indian binders need to be classified in to performance grading system. The present study has attempted to classify commercially available and CSIR-CRRI developed Viscosity Grade binders for performance through a well-conceived mapping system.

2 Literature Review

The goal of the Superpave system was to improve asphalt pavement performance. Superpave was admittedly designed with an emphasis on high traffic levels and critical roadways. However, the system was also designed with an eye towards lower traffic levels and less demanding applications. The ultimate goal is to select, specify, design and produce the right mixture for the particular application. Development of PG binders were the major output of the study by SHRP.

The Superpave Performance Graded Binder Specification (AASHTO MP-1) is based on providing a binder that is resistant to rutting, fatigue cracking and low temperature cracking at specific pavement temperatures. The binder temperature ranges in the specification are based on the high and low temperatures at which a binder reaches critical values of distress-predicting properties. Binder grade is selected based on design high and low pavement temperatures expected at the construction site and on desired reliability. The most common method of selecting a binder grade is to determine the design air temperature range for the specific project and then to establish the corresponding design pavement temperatures.

Performance Graded (PG) System is the most advanced asphalt binder grading system, which is result from the $50 million Strategic Highway Research Programme (SHRP) and has been adopted by all of the other 49 states in the U.S. The system includes new test specification that use new binder physical properties [2].

In Asphalt Institute (1995) various test processes are given [3]. The Superpave performance graded asphalt specification (AASHTO Designation M320-03) uses grading designations which correspond to the maximum and minimum pavement temperature.

As a part of Superpave research new test and specifications for binders were developed that fully define its use in Hot Mix Asphalt (HMA) pavements. The tests and specifications are specifically to address pavement performance parameters such as rutting, fatigue cracking and thermal cracking. Binder Grading in Strategic

Highway Research Programme (SHRP) is designated as Performance Grade (PG) and is classified based on traffic and temperature.

Researchers in India have tried to compare the design of bituminous concrete by the Superpave and Marshall methods of mix design [4, 5]. Detailed laboratory studies have been reported using different stone aggregates and bitumen from the known sources. Even, laboratory evaluation of modified bitumen using Superpave methodology under Indian conditions has been reported [6, 7]. However, there are still more need to be done. Specifications governing Superpave PG binder and mix analysis are the need of the day. There is also a need of mapping the available bitumen on a PG measure. To meet the above requirements, the present study has been initiated.

3 Materials and Methods

VG 40 two different sources, VG 30 from eight different sources and VG 10 from two different sources were evaluated in this study. Also VG 20 and VG 50 were formulated in CSIR-CRRI laboratory using refinery pitch product. Bitumen characterization was done by performing the elementary test usually conducted according to the IS and ASTM specifications for all the original binders. The various test carried out was listed below:

- Penetration Test at 25 °C, IS:1203–1978
- Softening Point Test, IS:1205–1978
- Ductility Test at 27 °C, IS:1208–1978
- Viscosity test @60 °C, ASTM D2171-10
- Viscosity Test @ 135 °C, ASTM D4402-13
- Specific Gravity Test, IS:1202–1978
- Dynamic Shear Rheometer (DSR) Test

Tests were carried out on fresh samples as well as aged binders through rotating thin-film oven test (RTFOT) according to AASHTO T240.

The test observations of different viscosity grade bitumen binders are summarized in Tables 1, 2 and 3.

4 Results and Discussion

4.1 Determination of Tentative Specifications for Indian Performance Grade (IPG) Binder

It is well known that bituminous binders play an important role in the performance of the pavements and therefore selection of quality and quantity of binder is crucial. For

Table 1 Properties of VG-10 and VG-20 grade bitumen

Properties	VG20		VG10	
	D1-20	D2-20	S1-10	S2-10
Penetration (1/10th mm)	62	59	95.6	97
Softening point (°C)	50.3	46.4	46.7	43
Viscosity at 60 °C (Poise)	2270	1788	1250	1417
Viscosity at 135 °C (cP)	480	–	400	350
Dynamic Shear, SHRP B-003, $G^*/\sin\delta$, Minimum, 1.00 kPa Test Temp @ 10 rad/s, °C for unaged binder	62	63	58	58
Dynamic Shear, SHRP B-003, $G^*/\sin\delta$, Minimum, 2.2 kPa Test Temp @ 10 rad/s, ° for RTFOT aged binder	72.7	71.1	68	69
Dynamic Shear, SHRP B-003, $G^*\sin\delta$, Maximum, 5 MPa Test Temp @ 10 rad/s, °C for PAV aged binders	10.1	12	7.6	6.5
BBR, Creep Stiffness, SHRP B-002 S, Maximum, 300 MPa, *m*—value, Minimum, 0.300 Test Temp @ 60 s, °C	−17	−17	−18	−17.5

Table 2 Properties of VG-30 grade bitumen

Properties	VG30							
	S1-30	S2-30	S3-30	S4-30	S5-30	S6-30	S7-30	S8-30
Penetration (1/10th mm)	67	79	65	63	64	52.7	61	47.3
Softening point (°C)	52	49.6	51.3	54.3	52.2	53.1	53.2	50.2
Viscosity at 60 °C (Poise)	2260	2140	2590	3510	3360	2550	3150	3230
Viscosity at 135 °C (cP)	410	390	490	650	520	545	600	525
Dynamic Shear, SHRP B-003, G*/sinδ, Minimum, 1.00 kPa Test Temp @ 10 rad/s, °C for unaged binder	69	69	68.8	75.8	73.6	69	76	69
Dynamic Shear, SHRP B-003, G*/sinδ, Minimum, 2.2 kPa Test Temp @ 10 rad/s, ° for RTFOT aged binder	74	74.2	72.2	78	77.1	73.4	79.2	73
Dynamic Shear, SHRP B-003, G*sinδ, Maximum, 5 MPa Test Temp @ 10 rad/s, °C for PAV aged binders	21.5	23.2	14.4	14.4	27	21.8	22	18.9
BBR, Creep Stiffness, SHRP B-002 S, Maximum, 300 MPa, *m*—value, Minimum, 0.300 Test Temp @ 60 s, °C	−15	−15	−15	−17	−16	−15	−15	−15

Table 3 Properties of VG-40 and VG-50 grade bitumen

Properties	VG 40		VG50 D1-50
	S1-40	S2-40	
Penetration (1/10th mm)	44	42	25
Softening point (°C)	54	51.7	55
Viscosity at 60 °C (Poise)	4325	3275	5027
Viscosity at 135 °C (cP)	625	710	645
Dynamic Shear, SHRP B-003, G*/sinδ, Minimum, 1.00 kPa Test Temp @ 10 rad/s, °C for unaged binder	73	75.9	85
Dynamic Shear, SHRP B-003, G*/sinδ, Minimum, 2.2 kPa Test Temp @ 10 rad/s, ° for RTFOT aged binder	74	77	85
Dynamic Shear, SHRP B-003, G*sinδ, Maximum, 5 MPa Test Temp @ 10 rad/s, °C for PAV aged binders	18	19	22.3
BBR, Creep Stiffness, SHRP B-002 S, Maximum, 300 MPa, *m*—value, Minimum, 0.300 Test Temp @ 60 s, °C	−13	−13	−12

any location, pavement temperature and expected traffic are the general parameters considered for choosing type of binder and mix design for amount of binder. Neat normal, polymer, crumb rubber modified binders are being currently used in India to build high volume roads and these binders are classified based on the parameters such as (i) viscosity at 60 °C (ii) penetration value and (iii) softening point value of the binders, respectively.

Bituminous binders are classified by viscosity grading in India and a number of empirical tests are specified as per Indian standards. No proper guidelines are available for selection of binder for pavement construction at a given location. Based on extensive analysis and inferences drawn, 'Tentative Specifications for Indian Bitumens (VG) for Performance Grading' were drawn and have been shown in Fig. 1,

Indian performance Grade (IPG)	IPG-58				IPG-64				IPG-70				IPG-76			
Average 7-day pavement Design Temperature, °C, Max	58	58	58	58	64	64	64	64	70	70	70	70	76	76	76	76
Minimum Pavement Design Temperature °C	-16	-22	-28	-34	-16	-22	-28	-34	-10	-16	-22	-28	-10	-16	-22	-28
	Original Binder															
Flash Point Temperature IS-1209 min, °C	230															
Viscosity, IS-1206. Max-3 Pas. Test Temp. °C	135															
Dynamic Shear, SHRP B-003: G*/Sin(δ), Min, 1.0 kPa Test Temperature @10rad/s, °C	58	58	58	58	64	64	64	64	70	70	70	70	76	76	76	76
	Rolling Thin Film Over Test (AASHTO T240, ASTM D2872)															
Mass Loss, Max, Percent	1.0	1.0	1.0	1.0	1.0	1.0	1.0	1.0	1.0	1.0	1.0	1.0	1.0	1.0	1.0	1.0
Dynamic Shear, SHRP B-003: G*/Sin(δ), Min, 2.2 kPa Test Temperature @10rad/s, RTFO Aging Temp, °C	58	58	58	58	64	64	64	64	70	70	70	70	76	76	76	76
	Pressure Aging Vessel Residue (SHRP B-005)															
PAV Aging Temp, °C	100	100	100	100	100	100	100	100	100	100	100	100	100	100	100	100
Dynamic Shear, SHRP B-003 : G*Sin(δ), Max, 5000 kPa Test Temperature @10rad/s, °C	25	22	19	16	25	22	19	16	28	25	22	19	28	25	22	19
Creep Stiffness, SHRP B-002: S,Max, 300,000 kPa, M-value, Min, 0.30, Test Temp@60sec, °C	-6	-12	-18	-24	-6	-12	-18	-24	0	-6	-12	-18	0	-6	-12	-18
Direct Tension, SHRP B-006: Failure Strain, Min, 1.0% Test Temp @ 1.0mm/min, °C	-6	-12	-18	-24	-6	-12	-18	-24	0	-6	-12	-18	0	-6	-12	-18
Average Air Temp, °C, max	40	40	40	40	45	45	45	45	50	50	50	50	55	55	55	55

Fig. 1 Tentative specifications for Indian performance grade (IPG) binder [8]

Table 4 Conformity of available VG grade bitumen with Indian tentative specifications

S. No	Grade	Source	Conformity with Indian tentative specifications
1	VG30	S1	PG64-15
2		S2	PG64-13
3		S3	PG64-15
4		S4	PG70-16
5		S5	PG70-16
6		S6	PG64-15
7		S7	PG 70-15
8		S8	PG 64-15
9	VG10	S1	PG 58-18
10		S2	PG 58-18
11	VG20	D1	PG58-15
12	VG40	S1	PG70-13
13		S2	PG82-13
14	VG50	D1	PG 76-12

wherein the PG temperatures have been marked based on the observations made through performance tests [8, 9].

From Table 2, it can be seen that S2-30 and S1-30 does not comply the specification criteria. Hence, performance grading should be considered. In India, major reason for failure of pavement is high temperature and high loading. In India, minimum temperature will not go in general much below to consider as a significant contributor. Hence, the minimum temperature for performance grading was considered same as the result obtained from BBR testing in Table 4.

The test observations of different viscosity grade bitumen binders are summarized in Tables 1, 2 and 3 and based on these, the conformity of these binders is given in Table 4.

4.2 Determination of Guidelines on the Use of an Appropriate Performance Grade Bitumen for Indian Scenario

An attempt has been made to arrive at the possible pavement temperatures using the atmospheric temperature data (Source: Indian Meteorological Agency) of some states and places. Available relations (Witzack and Brutton) have been used for the purpose as no Indian recommended work is available. Further, the pavement temperature-based recommendations have been made for use of an appropriate viscosity grade bitumen, as per the established tentative 'Indian Performance Grade (IPG)' Specifications above (Fig. 1), since no IPG designations are available in Indian market. Based on altitude and air temperature of different regions of India, expected pavement temperatures were estimated, which were employed as indicators for the appropriate binder selection. For estimation of pavement temperatures, the relation suggested by Witczack, as below, was used.

$$\text{AMPT}(^{\circ}\text{C}) = (1.05 \times \text{AMAT}) + 5.0 \tag{1}$$

where AMPT is annual maximum pavement temperature and AMAT is annual maximum air temperature.

Since the low temperature is not conspicuously dictating the performance of Indian bituminous roads, the resulting minimum pavement temperatures were directly incorporated. However, the high pavement temperatures are observed to be at least an additional ten degrees above the high air temperatures, the relative pavement temperatures are worked out accordingly (i.e. by adding ten degrees to the high air temperatures*). Finally, by the established PG selection criteria, the suggested performance grade of bitumen for Indian places wherever air temperature data is available, has been provided in the Table 5 and based on this appropriate grade of binder to be used was suggested for better performance. The nearly equivalent viscosity grade bitumen is to be then chosen by the respective user.

5 Study Recommendations

The study envisages to identify, characterize and categorize, the available bitumen from various sources for the requirements of a 'Performance Grade' concept, more specifically for the Indian conditions.

This study also recommends performance-based grading of binder with number of performance tests on binder to control different pavement distresses and uses pavement temperature and traffic parameters for selecting binder for a region. Though maximum efforts were made to procure bitumen samples, the market is too vast to cover all the supplies and hence a typical scenario has been encompassed. Through the present study, it can be concluded that the present Indian gradation system for bitumen is viscosity based, but not all the four designated grades are readily available. Also, based on Indian temperature region, harder grade bitumen needs to be developed for better performance.

Also, relationship between air and pavement temperature according to Indian scenario is also developed based on existing Witzack equation. However, a comprehensive study is required to systematically collect the air and pavement temperature data over longer regimes to develop a temperature database which can give more conclusive directions for conversion of air temperature to pavement temperature to enable better selection of bitumen.

Table 5 Temperature-based suggested performance grade binders for Indian regions

State/UT	City	Temperature, °C						Suggested performance grading bitumen
		Air		Pavement				
				After Witzack		Indian*		
		Max	Min	Max	Min	Max	Min	
Himachal Pradesh	Dharamshala	43	9	50	14	53	14	PG64-13
	Shimla	32	5	39	10	42	10	PG58-15
Rajasthan	Udaipur	43	13	50	19	53	19	PG64-13
	Barmer	44	16	51	22	54	22	PG64-13
	Kota	45	15	52	21	55	21	PG 70-15
	Bikaner	44	15	51	21	54	21	PG64-13
	Jodhpur	43	15	50	21	53	21	PG64-13
	Ajmer	42	15	49	21	52	21	PG64-13
	Jaipur/Sanganer	44	14	51	20	54	20	PG64-13
	Churu	44	14	51	20	54	20	PG64-13
	Ganganagar	46	12	53	18	56	18	PG 70-15
	Bikaner	44	15	51	21	54	21	PG64-13
Assam	Guwahati/Bhorjor	36	12	43	18	46	18	PG58-15
	Silchar/Kumbhigr	37	12	44	18	47	18	PG58-15
	Tezpur	37	12	44	18	47	18	PG58-15
	North Lakhimpur	37	13	44	19	47	19	PG58-15
	Dibrugarh/Mohanb	37	11	44	17	47	17	PG58-15
West Bengal	Kolkata/Dum Dum	40	20	47	26	50	26	PG64-13
	Kolkata/Diamond harbour	40	11	47	17	50	17	PG64-13
	Asansol/Panagarh	43	17	50	23	53	23	PG64-13
	Shanti-Niketan	44	18	51	24	54	24	PG64-13
Andaman Nicobar	Port Blair	35	26	42	32	45	32	PG58-15
Maharashtra	Nagpur/Sonegaon	45	17	52	23	55	23	PG 70-15
	Ozar	43	14	50	20	53	20	PG64-13
	Akola	43	17	50	23	53	23	PG64-13
	Dahanu	36	20	43	26	46	26	PG58-15
	Bombay/Santa Cruz	40	22	47	28	50	28	PG64-13
	Aurangabad/Chik	41	18	48	24	51	24	PG64-13
	Parbhani	42	16	49	22	52	22	PG64-13
	Bombay	40	22	47	28	50	28	PG64-13
	Alibagh	39	22	46	28	49	28	PG64-13

(continued)

Table 5 (continued)

State/UT	City	Temperature, °C						Suggested performance grading bitumen
		Air		Pavement				
				After Witzack		Indian*		
		Max	Min	Max	Min	Max	Min	
	Ratnagiri	36	23	43	29	46	29	PG58-15
	Mahabaleshwar	40	15	47	21	50	21	PG64-13
	Kolhapur	41	17	48	23	51	23	PG64-13
	Sholapur	41	19	48	25	51	25	PG64-13
	Pune	42	16	49	22	52	22	PG64-13
Tamil Nadu	Kodaikanal	37	16	44	22	47	22	PG58-15
	Tiruchirapalli	41	21	48	27	51	27	PG64-13
	Nagapattinam	37	23	44	29	47	29	PG58-15
	Adirampattinam	38	22	45	28	48	28	PG58-15
	Madurai	40	23	47	29	50	29	PG64-13
	Coimbatore	40	16	47	22	50	22	PG64-13
	Cuddalore	38	23	45	29	48	29	PG58-15
	Tondi	39	22	46	28	49	28	PG64-13
	Kanniyakumari	37	23	44	29	47	29	PG58-15
	Madras/Minambakk	37	24	44	30	47	30	PG58-15
	Tuticorin	37	24	44	30	47	30	PG58-15
	Pamban	41	18	48	24	51	24	PG64-13

References

1. Basic Road Statistics of India 2015–16, Government of India Ministry of Road Transport and Highways Transport Research Wing
2. Kennedy TW, Huber GA, Harrigan ET, Cominsky RJ, Hughes CS, Von Quintus H, Moulthrop JS (1994) Superior performing asphalt pavements (Superpave): the product of the SHRP asphalt research program.
3. Asphalt Institute (1995) SUPERPAVE. Superpave series no 2 (SP2), Asphalt Institute, Lexington
4. Roy N, Veeraragavan A, Krishnan JM (2013) Influence of air voids of hot mix asphalt on rutting within the framework of mechanistic-empirical pavement design. Procedia Soc Behav Sci 104:99–108
5. Chakroborty P, Das A, Ghosh P (2009) Determining reliability of an asphalt mix design: case of marshall method. J Transp Eng 136(1):31–37
6. Chattaraj R (2011) Bitumen Grading system–from penetration grading to viscosity grading–a step towards better quality control. Indian Highways
7. Palit SK, Reddy KS, Pandey BB (2004) Laboratory evaluation of crumb rubber modified asphalt mixes. J Mater Civ Eng 16(1):45–53

8. Development of technology for superior performing bituminous pavements (SUPERBIT-PAVE)—WP 12—A 12th five year plan project by CRRI on development and application of technologies for sustainable transportation (SUSTRANS) in joint collaboration with NAL, Bangalore; CSIO, Chandigarh and CLRI, Chennai
9. Kar SS, Arora K, Mani C, Jain PK (2016) Characterization of bituminous mixes containing harder grade bitumen. Transp Res Procedia 17:349–358

Accident Severity on National Highways in the Presence of Liquor Shop: A Case Study of National Highway 5, India

Satya Lakshmi Aparna Noojilla and Saurabh Dandapat

1 Introduction

The risk of road traffic accidents is very acute as road accidents claim more than 1.2 million lives each year. The road accident leaves a great impact not only the victims' health but also on the country's development. World Health Organization (WHO) estimated that, by the year 2020, road traffic accidents might be the third-highest threat to public health, outranking other serious killer health problems. [1]. In India, during the year 2017, over 4, 64,910 road crashes have claimed 1, 47,913 lives and caused a different degree of injury to more than 4, 70,975 persons [2]. In fact, those figures are probably an underestimate, as all injuries/accidents are not reported to the police [1]. National highways (NH), which comprises only 1.8% of total road lengths, accounted for 30.4% of total road accidents and 36.0% of the total deaths due to road accidents. During the period of 2005 and 2017, accident severity (number of persons killed per 100 accidents) increased from 21.6 to 31.8 indicating a drastic upsurge in risk associated with the road accidents [3]. With the major risk involved in the accidents on national highways, it is important to understand about the causative factors, crash frequency, and the severity of accident for developing effective road safety policy in order to minimize the loss of lives and reduce the injury severity.

Numerous studies have been devoted to investigate the influence of factors associated with the vehicle, driver, road characteristics, crash characteristics, environment, and traffic characteristics on the crash frequency and severity of crashes. The accident location (intersection, non-intersection) and cause of the accident (speed, running a

S. L. Aparna Noojilla (✉)
Doctoral Research Scholar, Department of Civil Engineering, Indian Institute of Technology, Kharagpur, India
e-mail: aparnanoojilla@iitkgp.ac.in

S. Dandapat
RCG School of Infrastructure Design and Management, Indian Institute of Technology, Kharagpur, India

M. Parida et al. (eds.), *Proceedings of the Fifth International Conference of Transportation Research Group of India* , Lecture Notes in Civil Engineering 220,
https://doi.org/10.1007/978-981-16-9925-2_19

red light, wrong way violation, etc.) found to influence severity [4]. Ulfarsson and Mannering (2004) [5] identified a significant difference in injury severity between male and female drivers. Younger and older drivers are found more prone to involve in fatal crashes, and thus, they have critical highway safety needs [6]. Dissanayake et al. [7] showed that beside gender, vehicle type also influences the probability of fatal crashes. Several studies identified the vehicle type as an influencing variable for determining the injury severity [8–11]. Crash characteristics such as type of crash (head-on, rare end, side swipe, right turn collusion, etc.,), environmental factors such as lighting condition, weather (clear sky, cloudy, rainy, etc.), traffic characteristics such composition of traffic, roadway characteristics such as width of median, and presence of curve have been found to influence the crash frequency and severity [12–14]. Several studies have also been carried out to identify the influence of different factors in the Indian context [1, 15–17]. These studies have identified various factors influencing accident frequency and severity. Drunken driving was found to be one of the important factors [18, 19]. In the past few years, the issue of drunken driving has become a major point of debate in India. In 2016, the Supreme Court of India pronounced verdict to restrict any liquor shop within 500 m from the boundaries of highways [20]. However, the judgment is yet to be implemented in most of the states. Although the influence of drunken driving on road crash has been reported in several studies, there is no study, at least to our knowledge, has been reported the influence of the presence of liquor shop, which is one of the primary source of liquor around the national highways and involve substantial activities during night time, on the accident frequency and injury severity. The present study is possible one of the first in this direction.

With this background, the present paper aims to carry out a study to understand the influence of liquor shop and several other factors on the accident frequency and severity on a national highway, namely NH5 (now renamed as NH16). The reminder of the paper is organized in six sections. Section 2 includes a review of literature in the area of accident frequency and severity modeling and factors. Section 3 discusses the database used in the present study. An analysis of the factors influencing the crash frequency is presented in Sect. 4. Similarly, the crash severity analysis is presented in Sect. 5 which is followed by conclusions derived from the study.

2 Literature Review

This section presents a review of literature pertaining to road accidents and accident severity. The review is presented in two sections. First, a review is presented with respect to the identification of the factors influencing the crash frequency and injury severity, which is followed by a discussion on the modeling techniques used by the researchers for understanding the relation between various contributing factor and the level of severity.

2.1 Factor Influencing Accident Frequency and Severity

Numerous studies have been reported about identifying the factors influencing the occurrence of accident and severity level. Traffic Safety Facts 2015 [6] highlighted various factors related to time, location, circumstances, driver and driving behavior, vehicle type, etc., which are influencing the accident and injury severity to the drivers, passengers, and other victims of accidents. Abdel-Aty [21] highlighted driver's injury severity on basic roadway sections, signalized intersections, and toll plazas in Central Florida, USA, and identified that driver's age, gender, seat belt use, point of impact, speed, and vehicle type have a significant impact on the injury severity level. Gray et al. [9] investigated specific crash characteristics that increase the likelihood of higher orders of injury severity. Bedard et al. [22] established that propensity to fatal accident gets influenced with individual characteristics such as age, alcohol concentration in blood, gender, crash characteristics (i.e., type of crash), and vehicle characteristics such as vehicle type and speed of vehicle, and argued that increasing seatbelt use, reducing speed, and reducing the number and engineering measures to reduce driver-side impacts may prevent fatalities. Khattak et al. [23] indicated in a study which analyzed the accident data of the State of Iowa, USA, that older drivers (⩾65 years) becomes more prone to severe injury when they consume alcohol and the severity becomes higher in the cases of farm vehicle as compared to other vehicles. Another study carried out considering the crash data from a two-lane rural highway in the USA developed crash prediction models in order to investigate the effect of geometric features of the road such curve, grade, and deflection angle of the horizontal curve [24]. Khorashadi et al. [25] asserted that severe and fatal injured crashes are less likely to occur during morning time (5:31–8:00) in both urban and rural areas. Naik et al. [26] investigated the relationship between single-vehicle truck crash injury severity and detailed weather conditions and subsequently claimed that the wind speed, rain, humidity, and air temperature were related with higher orders of injury severity. Wang and Zhang [27] studied the traffic crash severity under the combination of the different roadway and environmental conditions and asserted that the factors such as road type, crash location, light condition, road alignment, road surface condition, and speed limit have significant impacts on crash severity. The significant impact of the built up area was observed by Shrestha and Shrestha [28]. This study is significant as in the case of highways as a significant part of the highway pass through both rural and urban landscape. In addition, several researchers have studied the effect of alcohol (drunken driving) on the crash frequency and crash severity [29–33]. Thomas et al. [34] indicated that the number of injuries per crash increases at an average of 0.71 when the at-fault driver is drunk. Schneider IV and Savolainen [35] concluded that the impact of relevant crash factors such as crash type and location on the occurrence and severity of crashes proliferates when high speed or alcohol is involved. Overall, a wide range of factors that are influencing the road accident frequency and severity have been identified by various researchers. However, the factors are found to vary widely in different context based on the type of roads and vehicles. Similarly, the causes of accidents also vary. Accordingly, the countermeasures for reducing the accident risk and severity levels demand context-specific considerations for becoming effective.

2.2 Modeling of Accident Frequency and Severity

Accident modeling has been a popular tool among the researchers for developing a greater insight about the influence of different factors on accident risk and severity. Accident modeling can be broadly divided into two categories, namely accident frequency modeling and accident severity modeling. While the accident frequency modeling helps to understand the influence of different factors on the occurrence of the accident or the accident risk, the accident severity modeling is useful to understand the influence of the factors on accident severity level. The insight developed from these models altogether helps to frame broad policy and suggest suitable countermeasures to reduce the risk of accident and also to minimize the injury when accidents occur.

Accident frequency is generally modeled using Poisson and Binomial modeling [36–42] where accident counts (as dependent variable) are predicted as a function of different factors (independent variables). On the other hand, crash severity modeling aims to establish a relationship between the severity levels of individual accident (as dependent variable) and different contribution factors. Multivariate analyses and discrete-response models have been used extensively for modeling accident severity [23, 43–45]. As the crash severity is monotonously increasing function in terms of different levels of injury (low to high), thus, order response models such as ordered logit and ordered probit [9, 10, 46–48] have been a popular choice for severity modeling. These models are believed to better recognize the indexed nature of various response variables compared to the multivariate models such as multinomial logit model [5, 25, 37, 49, 50]. However, the multinomial logit models have been proved to be useful in several severities modeling [5, 37, 41, 51, 52]. The strength of such models (i.e., multinomial logit model) lies in their flexibly to allow greater variation in the parameters associated with the independent factors as such models relax the restrictions imposed by the ordered models [35, 53–56].

The results from the ordered probit and ordered logit models are found to be fairly similar, and thus, both type of models may be employed to model accident severity. However, studies suggested that ordered logit models have limitations as they do not represent the random test variation and exhibiting restrictive substitution patterns due to the independence of irrelevant alternatives (IIA) property [57–60]. On the other hand, probit models can handle random test variation and also allow any pattern of substitution [57]. Therefore, the ordered probit models have been extensively used by the researchers [21, 44, 55] for crash severity modeling on different facilities such as different roadways including highways, signalized and un-signalized intersections, and toll plazas. As the present study focuses on understanding the influence of different factors including the location of the liquor shops on the accident severity, ordered probit model is used in the present study for examining the relationship between different factors and crash security levels. Maximum likelihood technique was adopted for the estimation of probit model [23, 47].

3 Database Formation

3.1 Crash Data

The database was developed based on the crash data record of National Highway section. The study corridor is identified as Rajahmundry to Tuni section (chainage 799.998–901.753) which is a four lane divided carriageway with a raised median. In India, National Highway Authority maintains the accident database which is generally developed based on the police reports. The present database includes 1536 accident records reported from January 2013 to April 2017. The format of the database contains various details on each of the accident includes information related to time of accident (clock time, date, month, and year), location of the accident (urban, rural), type accident (overturning, head-on collision, rear-end collision, collision brush or side wipe, left turn collision, skidding, right turn collision, and others), counts of affected individuals for different severity levels (fatal, grievous injury, minor injury, and non-injury), information related to road alignment and geometry at the accident location (straight road, slight curve, sharp curve, flat road, gentle incline, steep incline, hump, and dip), weather condition at the time of accident (fine, mist/fog, cloudy, light rain, heavy rain, hail/sleet, snow, strong wind, dust storm, very hot, very cold, and other extraordinary condition). The cause of accident is also recorded with a five-level classification system (drunken, over-speeding, vehicle out of control, fault of the driver of "motor vehicle/driver of other vehicle/cyclist/pedestrian/passenger," and defect in mechanical condition of motor vehicle/road condition). Although several factors are described in terms of a number of levels, in most of the cases, the occurrences are limited to a few levels or sometimes the records are missing. For example, the factor "road condition" is described by eight levels, but only two to three levels are generally found in the record. The missing information may primarily be attributed to the lack of proper training of accident data record to the police personnel who are responsible for recording the accident data. The missing data in accident record is a common issue in many of the countries, especially in the lower income and middle income countries [1].

3.2 Data Related to the Location of Liquor Shops

The data related to the location (i.e., chainage) of the liquor shops along the study section were also collected from the Highway Authority of India (NHAI). This data is not a part of the accident data record, but the information is generally available with the NHAI or local police. The liquor shops adjoin to the highway or located within 100 m from the edge of the shoulder which were considered in this study. A total 18 such shops were found along the study corridor of length 101.755 km indicating a density of one shop per 5.5 km segment.

4 Analysis of Crash Frequency

The data was analyzed to develop an insight into the crash frequency on the study corridor and understand the influence of different factors on the crash frequency. For the descriptive factors, the crash frequency was estimated across different levels of the factors. On the other hand, for linear factor such as the distance of the crash location from the nearest liquor shop, the crash frequency was analyzed by examining the distribution of crashes across different distances. For analyzing the distribution, a bin size of 0.5 km was found suitable for the present context. It may be noted that the distances between the crash locations and the nearest liquor shops were estimated using the chainage of the accident location and the chainage of the nearest liquor shop. The findings from this analysis are discussed in the following sections.

4.1 Descriptive Factors

The frequency and composition of the occurrence of accidents across various levels of each of the factors are presented in Table 1. It may be observed from Table 1 that in some cases, a significant number of data are not known related to several factors such as accident location type (35.29%) and road condition (18.23). Thus, these factors are not used in this analysis. It may be observed that among the levels of the lighting condition, the duration from 12:00 a.m. to 6:00 a.m. in the morning observed the highest number of accidents. Overturning crashes dominate (45.38%) among all types of crashes. The reason behind such crashes may be attributed to a number of possible causes such as turning maneuver with higher speed and heavy load, sudden breaking, and instant maneuvering of the vehicle due to the appearance of unexpected obstacle on the road [61, 62]. It is interesting to note that the frequency of overturning accidents is significantly high closer to the liquor shops (Fig. 1). Over 40% of total overturning crashes occurred within 1.5 km of the liquor shops, and then, crash frequency reduces as the distance from the liquor shop increases. Such higher concentration of accident around the liquor shops is observed possibly due to sudden braking and appearance of unexpected obstacles caused due to various activities around the locations of liquor shops including peoples' entry to the highway under the influence of alcohol.

The frequency of accidents involving various severity levels shows a considerably high rate of severe crashes at this section of the highway. The fatal and grievous injury crashes together constitutes over 42% of the total crashes. It may be noted

Table 1 Crash frequency across different levels of the factors

Factors variable	Levels	Frequency (Share)
Accident location	Rural	878(57.16%)
	Urban	116 (7.55%)
	Unknown	542(35.29%)
Accident month	February–May	542(35.29%)
	June–September	485(31.58%)
	October–January	460(29.95%)
Accident day	Week end	241(15.69%)
	Week day	1295(84.31%)
Lighting condition	(12 AM–6 AM)	825(53.71%)
	(6 AM–12 PM)	39(2.54%)
	(12 PM–6 PM)	355(23.11%)
	(6 PM–12 AM)	317(20.64%)
Type of accident	Overturning	697(45.38%)
	Head on	250(16.28%)
	Rear-end	345(22.46%)
	brush/Side wipe	108(7.03%)
	Left turn	7(0.46%)
	Skidding	81(5.27%)
	Right turn	8(0.52%)
	Others	40(2.60%)
Severity	Fatal	223(14.52%)
	Grievous	425(27.67%)
	Minor	482(31.38%)
	No injury	406(26.43%)
Causes	Drunken	63(4.1%)
	Over-speeding	905(58.92%)
	Vehicle out of control	318(20.7%)
	Fault of driver /Cyclist/Pedestrian/Passenger	182(11.85%)
	Unknown	68(4.43%)
Road condition	Straight road	654(42.58%)
	Curve	561(36.52%)
	Unknown	321(20.9%)
Weather Condition	Fine	1062(69.14%)
	Adverse	194(12.63%
	Unknown	280(18.23%)
Help	Ambulance	646(42.06%)
	Patrol vehicle	890(57.94%)

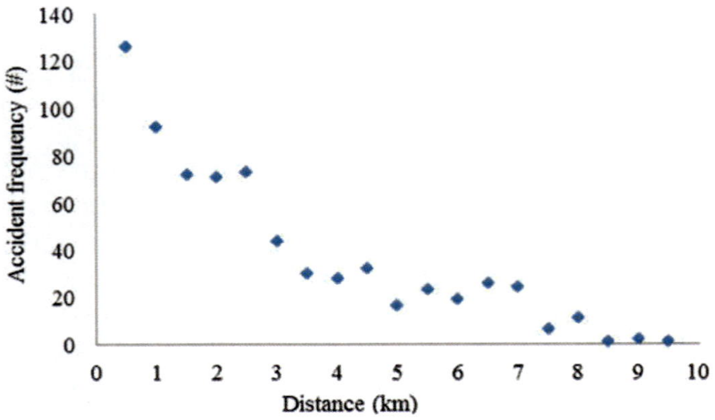

Fig. 1 Distribution of overturning crashes based on the distance of from the nearest liquor shop

that the reliability of accident data is a major concern all over the world, especially in developing countries such as India [63–65]. Although the data related to severe crashes such as fatal and grievous injury crashes are more or less reliable, the minor injury and no-injury crashes are generally under reported in India [1]. However, the total number of fatal and grievous crashes at the study corridor is considerably high which demand proactive safety measure to enhance the road safety standard.

Among the reported cause of accidents, over-speeding dominates with a share of nearly 60% of total crashes, while the cause "drunken-driving" constitutes only 4% of the total crashes. The general tendency of reporting the cause of the accident as over-speeding dominates in most of the developing countries [63]. The actual speed of the vehicle at the time of crash generally remains unknown as there is no system in place to monitor the speed of vehicles in Indian highways. The police personnel reports the cause primarily based on the feedback from eye witnesses. On the other hand, drunken driving as a cause of accidents is sometimes under reported primarily to avoid complexity in insurance claims for the victims [18, 66, 67].

It may also be observed from Table 1 that altogether only 12% of the total crashes were reported to occur under adverse weather condition such as cloudy, rainy, and foggy. Although for nearly 20% cases, the weather conditions during the accident were not recorded, thus, remained unknown, the majority of these are likely to be under fine weather condition.

4.2 *Effect of Liquor Shop*

The crash frequency distribution with respect to the distance of accident location from the nearest liquor shop is shown in Fig. 2. It is interesting to observe from Fig. 2 that the highest frequency of crash is recorded near the liquor shops. Over

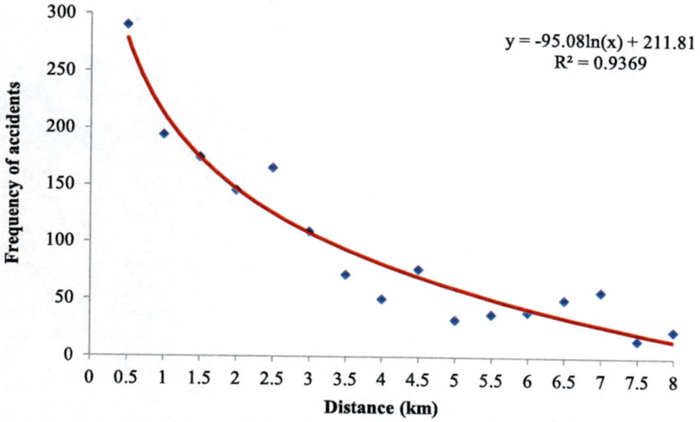

Fig. 2 Accident frequency distribution over the distance to the accident spot from the nearest liquor shop

42% of the total crashes were recorded occurred within the one kilometer distance from the liquor shops. The reason behind such observation may be attributed to the presence of drunken peoples' activity around the shops, unsafe entry of vehicles to the highways, and an unexpected sudden hindrance to the main stream traffic flow. The crash frequency reduces with the increase of the distance from the liquor shop increases. However, the rate of reduction of crash frequency diminishes gradually with the increases of distance. It appears that a logarithmic function may be well suited to explain this variation. The findings also justify the recent judgment of the Supreme Court of India [20] regarding the ban on the liquor shops within 500 m from the national highways as a measure of enhancing road safety. However, unfortunately, the judgment has largely been ignored by the implementing authorities as the liquor shops are generally seen as a decent source of revenue.

5 Accident Severities

The accident severity was first analyzed by classifying the crashes based on various factors and their levels. The effects of different factors were then analyzed by developing an ordered probit model.

5.1 Descriptive Analysis

The crashes were classified in four levels based on the severity of the injury. The severity levels were defined as fatal, grievous, minor injury, and no injury. Fatal

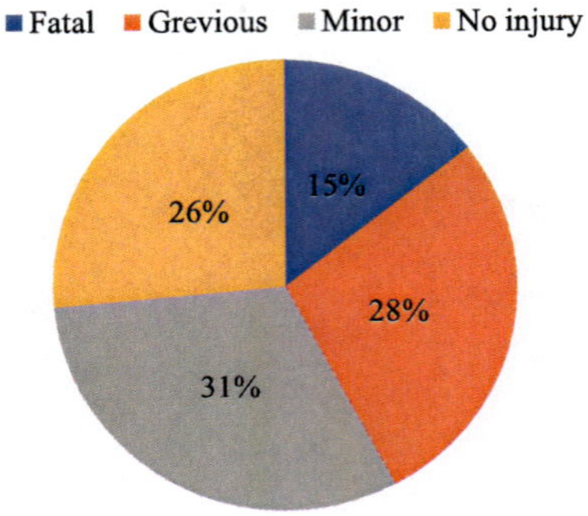

Fig. 3 Distribution of severity of accidents

injury is defined as the accident which involves at least one death of road user [68]. It was found that out of the total crashes, no-injury crashes, minor injury crashes, and grievous crashes comprised around 26, 31, and 28%, whereas fatal crashes comprised only 15% of the total recorded crashes (Fig. 3). The fatal and grievous injury crashes together comprise over 42% of the total crashes which is considerably high and a matter of grave concern in the view point of road safety. However, it may be noted that the reflection of such high share of fatal/grievous may have been an over estimate, as like many other developing countries, in India also the minor and no-injury crashes are often not reported and thus the record of such crashes are not available [69].

Table 2 shows the trend of accidents with different levels of severity occurred from January 2013 to April 2017. It may be observed from the table that although the crash rate (number of crash/year/km) has dropped drastically to 2.71 crashes/year/km recorded during 2016 from a rate of 3.8 crashes/year/km recorded during 2013, the rate of fatal injury remained more or less unchanged. The year 2015 observed the highest rate of crashes showing a rate of 4.27 crash/year/km. Overall, these estimates of crashes over only a few years (i.e., year January 2013 to April 2017) may not be sufficient to infer on the trend of crashes over the time period and further study may

Table 2 Average accident rates for different severity levels during 2013–2017

Year	Number of accidents					
	Fatal injury	Grievous injury	Minor injury	No injury/property damage	Total accidents	Accident/km/year
2013	49	114	119	99	381	3.8
2014	52	83	133	102	370	3.7
2015	56	143	117	111	427	4.27
2016	55	70	72	74	271	2.71
2017 (Till April)	10	16	40	20	86	–

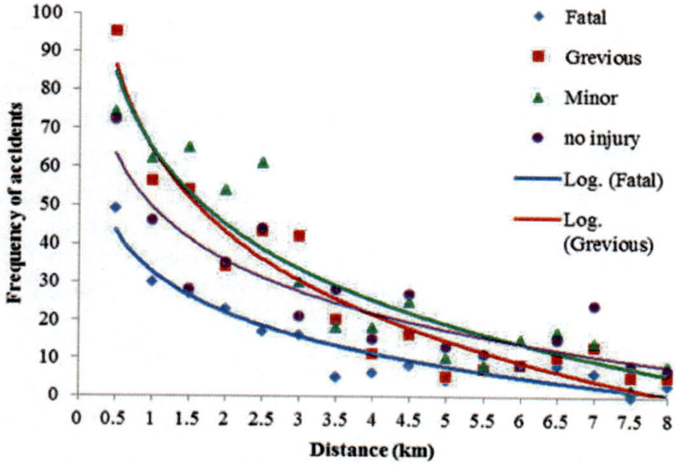

Fig. 4 Accident frequency distribution for different levels of severity

be necessary including data for a longer period of time in order to make a meaningful inference in this regard.

5.2 *Effect of Liquor Shops*

The accident frequency for each level of severity is found to reduce with the increase of the distance from the liquor shops (Fig. 4). It may be observed from Fig. 4 that like other crashes, higher severity crashes such as fatal and grievous injury crashes are also found to be more frequent near the liquor shops. However, as the distance from the liquor shops increases the crash frequency with different severity levels tend to converge.

5.3 *Order Probit Model*

Ordered probit model was developed to analyze the effect of different factors related to accident on the severity level of accident. The details on the modeling and results are discussed in the following sections.

Database framing. Accident data analysis and the severity modeling is a challenging task as the unbalanced nature of crash data is a very common issue which the analysts face. The representation of fatal crashes generally remains low as compared to the related to other severity levels [28, 35]. However, this generally is not a severe problem in the case of logistic models [70]. In order to avoid the representation bias, the severity levels fatal and grievous are clubbed together to indicate the high level of severity and coded as 2. Other levels of severity were coded as 1 for minor injury and 0 for non-injury. Different levels of the independent variables were coded using dummy coding scheme [61]. The coding scheme used for each level of the independent variable is summarized in Table 3.

Model description. The study aims to investigate the influence of different factors related to accident on the severity level. A review of different modeling techniques used by the researchers has been presented in Sect. 2.2. However, a brief discussion is presented in this section with a focus on the ordered probit model.

Table 3 Ordered probit model estimates for injury severity

Factors	Description		Coefficient	Standard error	*t*-stat	*P* value
Constant			0.233	0.172	1.35	0.178
Time (T)	(12 AM–6 AM) –1, 0–otherwise		−0.405***	0.059	−6.83	0.000
Type of accident	Overturning (OT)	1—OT, 0—Otherwise	0.534***	0.103	5.19	0.000
	Head-on collision (HO)	1—HO, 0—Otherwise	0.559***	0.118	4.72	0.000
	Rear-end collision (RE)	1—RE, 0—Otherwise	0.547***	0.113	4.86	0.000
	Skidding (SK)	1—SK, 0—Otherwise	0.661***	0.159	4.16	0.000
	Right turn collision (RT)	1—RT, 0—Otherwise	1.009**	0.453	2.23	0.025
Cause of accident	Drunken (DK)	1—DK, 0—Otherwise	0.720***	0.207	3.49	0.000
	Over-speeding(OS)	1—OS, 0—Otherwise	0.265*	0.145	1.83	0.067
	Vehicle out of control (VO)	1—VO, 0—Otherwise	0.330**	0.153	2.15	0.031
	Fault of driver/pedestrian, etc. (FD)	1—FD, 0—Otherwise	0.405**	0.164	2.47	0.013
Weather	Adverse weather condition (AW)	1—AW, 0—Otherwise	−0.226**	0.088	−2.56	0.010
Liquor shop	Distance to liquor shop	Scaled variable	−0.054***	0.014	−3.90	0.000
	Mu(01)		0.840***	0.034	24.61	0.000
	Adjusted Rho-squared		0.0345			
	Log likelihood		−1597.24			

***, **, * —Significance at 1, 5, 10% level

Most popularly used modeling techniques for accident severity model are ordered logit and ordered probit models [10, 71–74]. The difference between these two models lies in the assumption of the distribution of the random error. In the ordered logit model, it is assumed that the errors are independently identically distributed with a logistic distribution, whereas the ordered probit model assumes the distribution of errors as a multivariate normal. However, several researchers indicated that the results from ordered probit and ordered logit models are fairly similar [71, 75]. Ordered probit model is a class of probit models extensively used in accident severity modeling because of the ordered nature of the response variable, i.e., severity. The general formation of the ordered probit model with reference to the present context is given below.

The predictor variables, x_i, for observation i are expressed in terms of unobserved latent variable Y_i^*, as shown in Eq. (1).

$$Y_i^* = \beta x_i + \varepsilon_i \tag{1}$$

where β represents the estimated coefficients and ε_i is a normally distributed error term. The predicted crash severity can be found using the measurement equation (Eq. 2)

$$Y_i^* = \begin{cases} 0 - \text{no injury}, & \text{if } Y_i < 0 \\ 1 - \text{minor injury} & \text{if } 0 < Y_i < \mu_1 \\ 2 - \text{fatal or grievous injury} & \text{if } Y_i > \mu_1 \end{cases} \tag{2}$$

The maximum likelihood (ML) estimation technique is used to estimate the coefficients of the regression of Y_i^*. The threshold values, μ_n, for the predicted values of Y_i^* are estimated during the estimation of the parameter vector β.

During the model development, the method of elimination was adopted for identifying the factors influencing the crash severity. Initially, all the factors and their levels were considered. The factors which have fewer representations and the factors/levels of factors with insignificant coefficient estimates were then removed one by one. Following this approach, the final model was developed with the factors which give significant coefficient estimates. The results of the final model are presented below.

Results. The result of the order probit model is presented in Table 3. It may be observed from Table 3 that the all the coefficient estimates are statistically significant at 5% level of significance, except the coefficient with respect to over-speeding which is significant at 10% level of significance. The adjusted rho-square value of the model, though found low, is comparable to several other studies reported in the literature [10, 74–77]. In fact, several researchers have argued that the adjusted rho-square value should not be considered as a measure of goodness of feet for such models [78].

It may be observed from Table 3 that the factor distance to the accident location from the nearest liquor shop is negatively correlated with the fatal/grievous crashes, which is in agreement with the observation made in the general analysis in the previous section. This observation eventually indicates that the crashes occurring closer to the liquor shop are like to be more severe as compared to the other crashes.

The negative sign to the coefficient of time period, mid-night to morning (12:00 a. m. to 6:00 a.m.), indicates that this time period is more critical to the occurrence of higher severe crashes. During this period, the traffic volume on the highways generally remains lower and remains almost free from external interferences, thus drivers tend to drive at a higher speed which increases the probability of the crashes to be more severe. This is in agreement with the researchers [22, 23, 27] reporting about the higher odds ratio of the accidents during night time. Drunken driving and unavailability of immediate assistance during this period may also be other reasons behind increased severity level in these crashes.

Among different accident types, head-on collision, rare end collision, overturning, skidding, and right turn showed higher propensity to more severity to the accident. According to the researchers [1, 19], in India, head-on collisions comprise significant share of the total crashes on four-lane divided highways. Though divided highways are justified on the basis that these would eliminate the occurrence of head-on collisions, many vehicles are going the wrong way on divided highways. This is probably because the vehicles travel the wrong way when they exit from roadside businesses and the cut in the median is too far away [19]. They also reported that the severity levels of rear-end collisions are high on highways which shows that availability of more space on wider roads is leading to higher frequency and severity of rear-end crashes. This is probably due to poor visibility of vehicles rather than road design itself. Similarly, the cause such as drunken driving, over-speeding, and vehicle out of control showed a propensity to higher severity.

6 Conclusions

The study presents an investigation to identify the factors influencing the crash frequency and injury severity with special focus on liquor shops in the close proximity to the national highways. Present work brings out new finding related to the influence of liquor shop on the crash frequency and injury severity. The frequency and severity of accident found to increase in the presence of liquor shop and the crash rate across all severity levels reduces as the distance from the liquor shop increases. Therefore, it is important to implement stringent regulation to restrict liquor shops closer to the national highways in order to improve the safety standard in the highways. The execution of the judgment of the Supreme Court of India, which was passed during 2016, may be instrumental to enforce the regulation and restrict the liquor shops within 500 m of the highways. The study also identified several other factors related to the time of accidents, type of accident, cause of the accident, and weather which influence the accident frequency and severity. It is also extremely important to intensify the strategic campaign to make citizen aware of the danger of drunken driving and over-speeding on the highways.

Furthermore, as the present study is limited to only one National Highway, it is important to carry out additional studies in the similar direction considering other national highways to develop more insight about the influence of the liquor shops on accident and its implications on the road safety policy. It could also be interesting to investigate the combined effect of different factors on the accident frequency and severity which was not addressed in present study. This would certainly be helpful in developing more rational understanding about the accident occurrence. Also, the present data did not provide any information about the responsible road user, which, if available, it could bring in more comprehensive understanding about the impact of drunken driving by identifying the responsible road users, say drunken people on road or drunken drivers. Thus, there is an immediate need to address the issue of under—reporting of the accident data in India as the reporting of the accident data in a systematic manner will improve the quality of conclusions based on which policy decisions can be made.

Acknowledgements Authors express their sincere thanks to the National Highways Authorities of India, Rajahmundry, for all the cooperation during this study and providing access to the accident records of the selected section of the National Highway 5 (NH5).

References

1. Mohan D (2014) Road Accidents in India. IATSS Res 33:75–79
2. MoSPI (2017) Motor vehicle—statistical year book India 2017. In: Minist. Stat. Program. Implement. http://mospi.nic.in/statistical-year-book-india/2017/189
3. Vaidya C (2009) Urban Issues, Reforms and Way Forward in India
4. Al-Ghamdi AS (2002) Using logistic regression to estimate the influence of accident factors on accident severity. Accid Anal Prev 34:729–741
5. Ulfarsson GF, Mannering FL (2004) Differences in male and female injury severities in sport-utility vehicle, minivan, pickup and passenger car accidents. Accid Anal Prev 36:135–147
6. NHTSA (2016) Traffic safety facts: 2015
7. Dissanayake S, Lu JJ, Chu X, Turner P (1999) Evaluation of critical highway safety needs of special population groups. ITE J (Institute Transp Eng) 69:28–30
8. Chang LY, Wang HW (2006) Analysis of traffic injury severity: an application of non-parametric classification tree techniques. Accid Anal Prev 38:1019–1027. https://doi.org/10.1016/j.aap.2006.04.009
9. Gray RC, Quddus MA, Evans A (2008) Injury severity analysis of accidents involving young male drivers in Great Britain. J Safety Res 39:483–495
10. Kockelman KM, Kweon YJ (2002) Driver injury severity: an application of ordered probit models. Accid Anal Prev 34:313–321
11. George Y, Athanasios T, George P (2017) Investigation of road accident severity per vehicle type. Transp Res Procedia 25:2076–2083
12. Zou W, Wang X, Zhang D (2017) Truck crash severity in New York city: an investigation of the spatial and the time of day effects. Accid Anal Prev 99:249–261
13. Mohamed SA, Mohamed K, Al-Harthi HA (2017) Investigating factors affecting the occurrence and severity of rear-end crashes. Transp Res Procedia 25:2098–2107
14. Shinstine DS, Wulff SS, Ksaibati K (2016) Factors associated with crash severity on rural roadways in Wyoming. J Traffic Transp Eng English Ed 3:308–323

15. Bandyopadhyaya R, Mitra S (2013) Modelling severity level in multi-vehicle collision on indian highways. Procedia Soc Behav Sci 104:1011–1019
16. Dinu RR, Veeraragavan A (2011) Random parameter models for accident prediction on two-lane undivided highways in India. J Safety Res 42:39–42
17. Balasubramanian V, Bhardwaj R (2018) Pedestrians' perception and response towards vehicles during road-crossing at nighttime. Accid Anal Prev 110:128–135
18. Arora P, Chanana A, Tejpal HR (2013) Estimation of blood alcohol concentration in deaths due to roadside accidents. J Forensic Leg Med 20:300–304
19. Dinesh Mohan, Tiwari G, Bhalla K (2015) Road safety in India: Status Report
20. Chandrachud DY (2016) Supreme court of india civil appellate jurisdiction
21. Abdel-Aty M (2003) Analysis of driver injury severity levels at multiple locations using ordered probit models. J Safety Res 34:597–603
22. Bédard M, Guyatt GH, Stones MJ, Hirdes JP (2002) The independent contribution of driver, crash, and vehicle characteristics to driver fatalities. Accid Anal Prev 34:717–727
23. Khattak AJ, Pawlovich MD, Souleyrette RR, Hallmark SL (2002) Factors related to more severe older driver traffic crash injuries. J Transp Eng 128:243–249
24. Saleem T, Persaud B (2017) Another look at the safety effects of horizontal curvature on rural two-lane highways. Accid Anal Prev 106:149–159
25. Khorashadi A, Niemeier D, Shankar V, Mannering F (2005) Differences in rural and urban driver-injury severities in accidents involving large-trucks: an exploratory analysis. Accid Anal Prev 37:910–921
26. Naik B, Tung LW, Zhao S, Khattak AJ (2016) Weather impacts on single-vehicle truck crash injury severity. J Safety Res 58:57–65
27. Wang Y, Zhang W (2017) Analysis of roadway and environmental factors affecting traffic crash severities. Transp Res Procedia 25:2119–2125
28. Shrestha PP, Shrestha KJ (2017) Factors associated with crash severities in built-up areas along rural highways of Nevada: a case study of 11 towns. J Traffic Transp Eng (English Ed) 4:96–102
29. Vaez M, Laflamme L (2005) Impaired driving and motor vehicle crashes among Swedish youth: an investigation into drivers' sociodemographic characteristics. Accid Anal Prev 37:605–611
30. Chakroborty P, Deb K, Subrahmanyam PS (1995) Optimal scheduling of urban transit systems using genetic algorithms. J Transp Eng 121:544–553
31. Zambon F, Hasselberg M (2006) Factors affecting the severity of injuries among young motorcyclists—a Swedish nationwide cohort study. Traffic Inj Prev 7:143–149
32. Papadimitriou E, Theofilatos A, Yannis G (2014) Motorcycle riding under the influence of alcohol: results from the SARTRE-4 survey. Accid Anal Prev 70:121–130
33. Christophersen AS, Gjerde H (2015) Prevalence of alcohol and drugs among motorcycle riders killed in road crashes in Norway during 2001–2010. Accid Anal Prev 80:236–242
34. Traynor TL (2005) The impact of driver alcohol use on crash severity: a crash specific analysis. Transp Res Part E Logist Transp Rev 41:421–437
35. Savolainen PT, Mannering FL, Lord D, Quddus MA (2011) The statistical analysis of highway crash-injury severities: a review and assessment of methodological alternatives. Accid Anal Prev 43:1666–1676
36. Lord D, Mannering F (2010) The statistical analysis of crash-frequency data: a review and assessment of methodological alternatives. Transp Res Part A Policy Pract 44:291–305
37. Carson J, Mannering F (2001) The effect of ice warning signs on ice-accident frequencies and severities. Accid Anal Prev 33:99–109
38. Ye X, Pendyala RM, Shankar V, Konduri KC (2013) A simultaneous equations model of crash frequency by severity level for freeway sections. Accid Anal Prev 57:140–149
39. Anderson J, Hernandez S (2017) Analytic methods in accident research roadway classifications and the accident injury severities of heavy-vehicle drivers. Anal Methods Accid Res 15:17–28
40. Mannering FL, Bhat CR (2014) Analytic methods in accident research: methodological frontier and future directions. Anal Methods Accid Res 1:1–22
41. Zong F, Zhang H, Xu H (2013) Predicting severity and duration of road traffic accident. Math Probl Eng 1–9

42. Zhu X, Srinivasan S (2011) A comprehensive analysis of factors influencing the injury severity of large-truck crashes. Accid Anal Prev 43:49–57
43. Pahukula J, Hernandez S, Unnikrishnan A (2015) A time of day analysis of crashes involving large trucks in urban areas. Accid Anal Prev 75:155–163
44. Mergia WY, Eustace D, Chimba D, Qumsiyeh M (2013) Exploring factors contributing to injury severity at freeway merging and diverging locations in Ohio. Accid Anal Prev 55:202–210
45. Lemp JD, Kockelman KM, Unnikrishnan A (2011) Analysis of large truck crash severity using heteroskedastic ordered probit models. Accid Anal Prev 43:370–380
46. Makaiwi M (2006) Modeling crash frequency and severity using historical traffic and weather data : truck involved crashes on I-80 in Iowa. Iowa State University
47. Duncan CS, Khattak AJ, Council FM (1998) Applying the ordered probit model to injury severity in truck- passenger car rear- end collisions. Transp Res Rec 63–71
48. Xie Y, Zhang Y, Liang F (2008) Crash injury severity analysis using Bayesian ordered probit models. J Transp Eng 135:18–25
49. Islam S, Mannering F (2006) Driver aging and its effect on male and female single-vehicle accident injuries: some additional evidence. J Safety Res 37:267–276
50. Schneider WH, Savolainen PT (2012) Comparison of severity of motorcyclist injury by crash types. Transp Res Rec J Transp Res Board 2265:70–80
51. Shankar V, Mannering F, Barfield W (1995) Effect of roadway geometrics and environmental factors on rural freeway accident frequencies. Accid Anal Prev 27:371–389
52. Zhang Y (2011) Exploring single vehicle crash severity on rural, two-lane highways with crash-level and occupant-level multinomial logit models. The University of Utah
53. Eluru N, Bhat CR, Hensher DA (2008) A mixed generalized ordered response model for examining pedestrian and bicyclist injury severity level in traffic crashes. Accid Anal Prev 40:1033–1054
54. Eustace D, Indupuru VK, Hovey P (2011) Identification of risk factors associated with motorcycle-related fatalities in Ohio. J Transp Eng 137:474–480
55. Abdel-Aty M, Wang X (2007) Crash estimation at signalized intersections along corridors: analyzing spatial effect and identifying significant factors. Transp Res Rec J Transp Res Board, 1953:11–98
56. Zhang H (2010) Identifying and quantifying factors affecting traffic crash severity in Louisiana. Civ Environ Eng 4:207–213
57. Zhang H (2010) Identifying and quantifying factors affecting traffic crash severity in Louisiana. Louisiana State University and Agriculture and Mechanical College
58. Benson AR, Kumar R, Tomkins A (2017) On the relevance of irrelevant alternatives. In: International world wide web conference committee, pp 963–973
59. Kroll EB, Vogt B (2012) The relevance of irrelevant alternatives. Econ Lett 115:435–437
60. Train KE (2003) Discrete choice methods with simulation. Cambridge University Press, Cambridge
61. Khattak AJ, Ph D, Schneider RJ (2003) Risk factors in large truck rollovers and injury severity : analysis of single-vehicle collisions. In: TRB 82nd annual meeting. Washington, pp 1–22
62. Bambach MR, Grzebieta RH, McIntosh AS (2013) Thoracic injuries to contained and restrained occupants in single-vehicle pure rollover crashes. Accid Anal Prev 50:115–121
63. Jadaan K, Al-Braizat E, Al-Rafayah S et al (2019) Traffic safety in developed and developing countries: a comparative analysis. J Traffic Logist Eng 6:1–5
64. Mohan D (2004) The road ahead traffic injuries and fatalities in India
65. Samuel JC, Sankhulani E, Qureshi JS, et al (2012) Under-reporting of road traffic mortality in developing countries: application of a capture-recapture statistical model to refine mortality estimates. PLoS One 7
66. Das A, Gjerde H, Gopalan SS, Normann PT (2012) Alcohol, drugs, and road traffic crashes in india: a systematic review. Traffic Inj Prev 13:544–553
67. Sivakumar T, Krishnaraj R (2012) Road traffic accidents (RTAs) due to drunken driving in India—challenges in prevention. IRACST Int J Res Manag Technol 2:401–406
68. MoRTH (2015) Basic road statistics of India 2015–16

69. Gururaj G, Aeron Thomas A (2000) RM: underreporting of road traffic injuries in Bangalore. Implications for road safety policies and programmes. In: Proceedings of the 5th world conference on injury prevention and control
70. Mussone L, Bassani M, Masci P (2017) Analysis of factors affecting the severity of crashes in urban road intersections. Accid Anal Prev 103:112–122
71. O'Donnell CJ, Connor DH (1996) Predicting the severity of motor vehicle accident injuries using models of ordered multiple choice. Accid Anal Prev 28:739–753
72. Long JS (2002) Regression models for categorical and limited dependent variables: advanced quantitative techniques in the social sciences, first. Thousand Oaks
73. Khattak AJ (2007) Injury severity in multivehicle rear-end crashes. Transp Res Rec J Transp Res Board 1746:59–68
74. Quddus MA, Noland RB, Chin HC (2002) An analysis of motorcycle injury and vehicle damage severity using ordered probit models. J Safety Res 33:445–462
75. Renski H, Hill C, Carolina N, Council FM (1998) Effect of speed limit increases on crash injury severity: analysis of single-vehicle crashes on north Carolina interstate highways. Transp Res Rec 1665:100–109
76. Tay R, Barua U, Kattan L (2009) Factors contributing to hit-and-run in fatal crashes. Accid Anal Prev 41:227–233
77. Chen H (2007) Stochastic optimization in computing multiple headways for a single bus line. J Chin Inst Ind Eng 24:316–323
78. Rifaat SM, Chin HC (2005) Accident severity analysis using ordered probit model. J Adv Transp 41:91–114

A Comprehensive Review of Motorcyclists' Riding Behaviour and Safety

Abhishek Chakraborty, Sudeshna Mitra, and Bhargab Maitra

1 Introduction

Motorcycle is a flexible mode of transportation widely used for both work and leisure trips. The number of motorcycles or motorized two-wheelers (MTWs) plying on roads across the world has been exponentially increasing. In the United States, the number has been doubled to 8.4 million in the last fifteen years [1]. In developing countries, a large percentage of trips (especially short and commuting type) are served by motorized two-wheelers [2]. Especially, in case of the metropolitan cities of such developing countries, as traffic congestion is one of the major concerns among the citizens, people prefer riding on motorized two-wheelers due to several advantages over a car: cheap price, ease in parking, lesser travel time on urban streets, more mileage, etc. For example, based on the survey conducted by Ministry of Urban Development (MoUD) in 2008, it is observed that in India (one of such developing country), on an average, motorcycles serve 16% of trips (if walking is considered as a mode of transport). Moreover, in mid-sized and large Indian cities, the mode share of motorcycles is much higher (26–29%) [3]. Since MTWs provide the flexibility of travel at a reasonable cost, they have become an integral part of the transportation system in most of the developing countries.

At the same time, accidents and fatalities of MTWs are over-represented when compared to other modes of transport and the numbers are increasing. The safety performance of MTWs has been a major concern not only in developing countries but also worldwide. From the Global Status Reports on Road Safety, 2015 and 2018 (published by WHO), has been found that the proportion of road fatalities are shared

A. Chakraborty (✉) · S. Mitra · B. Maitra
Department of Civil Engineering, Indian Institute of Technology, Kharagpur 721302, India

S. Mitra
e-mail: sudeshna@civil.iitkgp.ernet.in

B. Maitra
e-mail: bhargab@civil.iitkgp.ernet.in

M. Parida et al. (eds.), *Proceedings of the Fifth International Conference of Transportation Research Group of India* , Lecture Notes in Civil Engineering 220,
https://doi.org/10.1007/978-981-16-9925-2_20

by motorized two and three wheelers increased from 23 to 28%. In South-East Asian countries, these numbers increased from 34 to 43% [4, 5]. In case of developing countries, for example in India, the report of Ministry of Road Transport and Highway [6], documented that the MTW users share the highest percentage of road fatalities (35%). Global Burden of Disease [7] was also reported that in India, 31% of the victims of fatal road accident victims are motorcyclists. In Indian metro cities, approximately 25% of the road accident victims are motorcyclists [8]. In the case of Hyderabad, India, it was found that motorcyclists are the second most risk-prone road users after pedestrians. It was also observed that motorcyclists posed higher risks at unsignalised intersections in the city of Hyderabad [9].

Now, the safety situation can be improved either by reactive approaches or by proactive measures. Traditionally, to identify the risk factors, researchers use safety performance functions and identified different factors responsible for crashes. In the case of developing countries, non-fatal crashes are mostly under-reported and proper accident databases are not maintained. Even if an accident gets reported, due to unavailability of proper accident reporting forms, at many cases, gathering useful information related to the crash becomes impossible, which are required to develop a crash prediction model [4, 5, 10–14]. On the other hand, a proactive approach generally consists of assessing the safety situation by expert opinion through road safety audits (RSA). RSA is conducted in different project stages to identify potential crash-prone locations and the corresponding design deficiencies which have an impact on the safety of the road users. This formal procedure is carried by an expert team with the help of checklists, as mentioned in manuals or codes followed in different countries [12, 15–19]. The concept of 'safe system approach' evolved recently which takes into account all groups within the road system [20–22]. All of these approaches are dependent on expert opinion. But, in case of these proactive techniques, differences in opinions of road safety auditors may arise. However, these concepts are still new in developing countries.

Furthermore, human factors such as riding behaviours of motorcyclists have been found to be equally responsible for leading to crashes [23–26]. Unsafe riding can cause accidents even if there does not exist any infrastructural or design deficiency on the road. All the aforementioned countermeasures or proactive measures do not take into account the riding behaviours. Behaviours such as disobeying to traffic rules, aggressive riding or sensation seeking behaviour were found to be related to one's attitude or intention to perform such acts [27–29]. Studies on these behaviours have been existing in case of several developed countries [30–33], but the number of studies related to motorcyclists' riding behaviours in the context of developing countries are very few [34]. The built environment, traffic composition, driving behaviour and road users' perceptions are significantly different in a developing country compared to those of developed countries and other Motorcycle Dependent Cities (MDCs) in countries such as Vietnam, Taiwan and Indonesia.

To assess the present motorcyclists' safety situation in heterogeneous traffic flow conditions, the aim of this review study is to identify a comprehensive set of factors from past studies to capture the riding behaviour of motorcyclists and to explore the reasons behind performing unsafe riding practices of motorcyclists, which would help the stakeholders to modify motorcyclists' behaviour through training and education, infrastructural interventions and enforcement.

2 Understanding Motorcyclists' Riding Behaviour

Riding skills and experience are directly correlated to the risk of accidents [26]. For motorcyclists, chances of getting involved in an accident increases with exposure and decreases with age and experience [35, 36]. Researchers have explored the attributes related to riding behaviours such as speeding, effect of alcohol to develop the relationship with accident risks [36–38]. Till 1990s, no such proper methodology was there to assess the driving behaviours in a proper systematic way. To reduce the crash involvements of motorcyclists, researchers felt the need to explore and study the riding behaviours of motorcyclists. Most of the studies in this regard framed a questionnaire survey or used an established survey form to explore the riding behaviours across the world.

In regard to the conditions of developing countries, the risky behaviours of motorcyclists in India were assessed and the factors such as use of helmet, effective law enforcement, good vehicle condition, improvement of licensing system were found to be influential in reducing crashes and severity of crashes involving motorcyclists [39].

2.1 Capturing Aberrant Riding Behaviours

Reason et al. [40] first framed a set of driver behaviour questionnaire (DBQ) in order to assess the drivers' behaviours in United Kingdom (UK) and classified aberrant driving behaviours (errors and violations) into five classes in the questionnaire, named: 'slips', 'lapses', 'mistakes', 'unintended violations' and 'deliberate violations'. Also, all the behaviours were categorized into three risk classes, named low, intermediate and high-risk groups. According to them, errors were stated as the 'failure of planned actions to achieve their intended consequences', whereas violations were stated as 'deliberate deviations from those practices believed necessary to maintain the safe operations of a potentially hazardous system'. The researchers came up with a three factored solution named 'dangerous violations', 'dangerous errors' and 'silly errors'. These three factors could able to take into account for 33% of the total variance.

However, a few other researchers argued on this and claimed that violations were observed to be more correlated with car accidents. [41–43]. Later on, the studies have been extended by researchers and the category 'violations' were further split in two distinct scales named 'aggressive violations' and 'ordinary code violations', and the ordinary one was further split into 'fast driving' and 'maintaining progress' violations [44]. Through factor analysis, the researchers found a three factored solution, named 'errors', 'code violations' and 'aggressive violations' which could able to explain up to 49.5% of the total variance.

Studies involving motorcycles, i.e. assessment of riding behaviours of motorcyclists were first done by Elliott et al. [30] in 2007 and they developed Motorcycle Rider Behaviour Questionnaire (MRBQ) consisting of 43 items by surveying over a very large (close to 9000) UK riders. The researchers considered motorcycle crashes occur due to contribution of machine element, environmental element and human element, acting independently or in a group. Aggressive violations were not considered by them. They came up with a five-factor solution after performing principal component analysis of the data which could able to explain up to 41.2% of the total variance. These factors were 'traffic errors', 'speed violations', 'stunts', 'safety equipment' and 'control errors'. This new set of five factors of MRBQ ware found to be more effective in explaining the aberrant riding behaviours compared to the three factored solution of DBQ.

Later on, risk-taking riding behaviours, impulsive sensation seeking and aggression were included to frame a 61-item questionnaire survey to explore the relationship between the crash risk and rider behaviours and to identify the most vulnerable motorcycle rider group in Singapore [45]. Aggression and risk-taking factors were found to be influential to predict crash involvements of motorcyclists and riders with these characteristics were identified as most vulnerable ones. In Manila (Philippines), 'riding habit score' and 'aggression score' were found to be most influential on crash frequency when a survey was carried out among more than 2500 motorcyclists and other road users along five most important road corridors [46].

146 young moped riders in Netherlands participated in a survey conducted by developing moped rider questionnaire (MRQ) based on the Dutch DBQ (as proposed by Verschuur [47]) and a three factored solution came up [32]. The factors were identified as 'errors', 'lapses' and 'violations' which accounted for 34% variance of the aberrant behaviours. In the context of China, a Chinese Motorcycle Rider Driving Violation (CMRDV) scale was developed and a two-factored solution was obtained [48]. 'Driving violations' and 'aggressive violations' factors explained 60% of the total variances and the CMRDV was concluded to be valid and reliable to assess the riding behaviours of Chinese motorcyclists.

Though very few studies have been carried out in this regard in developing countries, the MRBQ was modified so that the questionnaire can suit the conditions of Nigeria [49]. The MRBQ was modified by rephrasing some of the questions and replacing some questions by new ones to assess the riding behaviours. 'Control safety', 'stunt', 'error' and 'speed hurrying' were the four factors suggested which accounted for 32.5% of the variance explained.

Özkan et al. [50] investigated the applicability of the factor structure in Turkey as proposed by Elliott et al. [30] and found similar results as it was obtained in the context of UK. To test the applicability of MRBQ survey on novice riders in Australia, Sakashita et al. [51] studied the psychometric properties of MRBQ as it was examined in UK [30] and in Turkey [50]. They further extended the study and they to determine predictive validity with respect to both historical recorded crash data and self-reported crashes and near-misses.

Surveying with the Manchester Drivers Behaviour Questionnaire (DBQ) among the taxi drivers in Iran led to conclude that young drivers commit more mistakes and offenses as they have lack of experience and skill [52]. Moreover, in prediction of self-reported accidents, intentional offences were found to be most influential variable. Later on, Stephens et al. [53] included intentional and unintentional behaviours of motorcyclists associated with crash risks and modified the MRBQ. They concluded that stunt behaviours have direct relationship with crash involvements, while violations of speed and errors significantly relates to near-crash involvements in Australian context. Behaviours such as reckless overtaking or riding on sidewalks were found to be more dominant among young riders than the habit of using cell phones in context of Vietnam [54].

The summary of the relevant studies conducted to capture the riding behaviours across world are mentioned in Table 1.

2.2 *Crash Risk and Riding Behaviours*

Researchers have developed the relationship among the aberrant riding behaviours and crashes. In most of the studies, respondents were asked to report the number of crashes experienced by them in recent past, and statistical models have been developed in various studies to predict the self-reported crashes using the factors obtained from MRBQ studies and other variables. Composite scale scores from each of the factors obtained from the MRBQ surveys were considered to be independent variables in such modellings.

Elliott et al. [30] collected the self-reported crash data from the respondents for number of 'all crashes' they experienced in last one year, as well as the number of 'blame crashes' where the respondents accepted their faults in those accidents. General linear models (GLM) were developed for both the crash types considering MRBQ composite scales along with riders' age, experience and annual mileage as independent variables. Traffic errors were found to be the significantly predicting 'all crashes', whereas in case of predicting 'blame crashes', control errors and speed violations were also found to be significant variables.

Respondents from Turkey were asked to report the number of times they were involved in active accidents (blame crashes), passive accidents (non-blame) and penalized for traffic offences [50]. Hierarchical regression models revealed that annual mileage was a significant predictor in all the cases and age was found to be significant in case of active and passive crash predictions. Among the MRBQ factors, performing stunts was found to be significant attribute to predict active crashes and offenses, whereas speed violations were significant as a predictor of offences.

In the studies conducted in Australia, zero-inflated log-link regression models were developed to predict self-reported crashes, self-reported near-miss cases and police-reported offences [51]. Results showed that errors and speed violations (MRBQ factors) along with age and exposures were the significant variables in predicting self-reported crashes and near-misses. Gender of the responded was also

Table 1 Summary of the studies conducted to capture the riding behaviours across world

Studies	Reason of study	Questionnaire used	Nos. of items	Factors obtained					Total variance explained (%)
				1	2	3	4	5	
Reason et al. [40]	United Kingdom	DBQ (Manchester)	50	Dangerous violations	Dangerous errors	Silly errors	–	–	33
Parker et al. [41]	United Kingdom	DBQ (Shortened)	24	Errors	Violations	Lapses	–	–	37.4
Lawton et al. [44]	United Kingdom	DBQ (Shortened)	16	Errors	Deliberate violations	Violations (other)	–	–	45
Elliott et al. [30]	United Kingdom	MRBQ	43	Traffic errors	Speed violations	Stunts	Safety equipment	Control errors	41.2
Steg and Brussel [32]	Netherlands	MRQ based on DBQ (Dutch)	43	Errors	Lapses	Violations	–	–	34
Cheng and Ng [48]	China	CMRDV	19	Aggressive violations	Ordinary violations	–	–	–	60
Sunday and Akintola [49]	Nigeria	Modified MRBQ	40	Control and safety	Stunt	Error	Speed hurrying	–	32.5
Özkan et al. [50]	Turkey	MRBQ	43	Speed violations	Traffic errors	Safety equipment	Stunts	Control errors	45
Sakashita et al. [51]	Australia	MRBQ	43	Errors	Speed violations	Stunts	Protective gear	–	23.6
Haghi et al. [52]	Iran	DBQ (Manchester)	50	Lapse	Errors	Ordinary violations	Aggressive violations	–	–
Stephens et al. [53]	Australia	MRBQ	46	Traffic errors	Speed violations	Stunts	Safety equipment	Control errors	54

observed as a significant variable in case of self-reported near-crashes. On the other hand, 'protective gear' and gender of the responded came out to be significant predictors in case of police-reported offences prediction model. Another study in Australia investigated the association between MRBQ factor scores and self-reported number of crashes, near-crashes and speed-related traffic violations [53]. Through logistic regression analysis, the researchers found out that the stunts (MRBQ factor) and age of the respondents to be significant in predicting crashes.

Although researchers could able to develop crash prediction models (mostly self-reported) as per the studies conducted in few countries, in case of the data from Netherlands, MRBQ factors alone or in combination with other demographic variables could not able to establish a significant relationship with crash involvements of the riders. [32]. However, the different crash prediction modelling approaches used by researchers are mentioned in Table 2.

Table 2 Modelling approaches and factors used in crash prediction models

Studies	Crash data type	Crash data duration	Modelling approach	Factors considered
Elliott et al. [30]	Self-reported	Last 12 months	General linear modelling	MRBQ factors along with age, experience and annual mileage
Özkan et al. [50]	Self-reported	Last 36 months	Hierarchical regression	MRBQ factors along with age and last year mileage
Sakashita et al. [51]	Self-reported and Police records	Last 12 months	Zero-inflated Poisson log-link regression	Age, gender, exposure along with MRBQ factors
Stephens et al. [53]	Self-reported	Last 12 and 36 months	Logistic regression	Age, performance of stunts (for past 12 months crash involvements); Riding purpose, speed violations and control errors (for past 12 months near-crash involvements) and speed violation (for past 3 years speed related traffic violations)

3 Exploration of Reasons Behind Performing Different Riding Behaviours

3.1 Psychological Models and Human Behaviours

In the context to find out the reasons or factors influencing an individual's behaviour, psychological theories and models can be used.

Health belief model (HBM) is one of the old psychological models which consists of two constructs, named 'threat perception' and 'behavioural evaluation' [55, 56]. 'Threat perception' gets controlled by two factors, 'perceived susceptibility' and 'perceived severity'; on the other hand, 'behavioural evaluation' consists of 'perceived benefits' and 'perceived barriers'. Apart from these factors, 'motivation' and 'cues to action' (i.e. social influence to perform the action) also influences the behavioural action [57].

Rotter [58] developed an internal–external (I-E) scale in 1966 to describe the 'internal beliefs' and 'external beliefs' which control an individual's behaviour. Later on, from this basic concept, several other scales were developed, and ultimately this theory was first applied in assessing driving behaviour by Montag and Comrey in 1987 [59]. Özkan and Lajunen [29] came up with the traffic locus of control (T-LOC) scale in order to assess the driving behaviours which are related to involvement in accidents.

Theory of reasoned action (TRA) is one of the old psychological theories. According to the basic theory of the reasoned action, behavioural intentions influence one's behaviour and the behavioural intentions are controlled by behavioural beliefs and normative beliefs [60–62]. Behavioural beliefs influence on an individual's attitude in doing a particular behaviour, whereas the normative beliefs effect on the subjective norms. Subjective norms are the beliefs which consider how other people thinks in performing the behaviour. Later on, the concept of theory of planned behaviour (TPB) evolved from the basic concept of TRA. Similar to the TRA, TPB consists of three main constructs, named 'attitude', 'subjective norms' and 'perceived behavioural control'. Apart from riding behaviours, TPB has also been applied to determine the factors on mode-choice behaviours in motorcycle dependent Taiwan [63].

Though the HBM and T-LOC models have been used by very few researchers in context of motorcyclists' riding behaviour, TPB has been widely used in several studies. The framework and the basic concept of the TPB is explained in the next sub-section.

The Concept of Theory of Planned Behaviour (TPB). TPB is a common psychological theory which has been applied in various fields. TPB has been used widely in the field of riding behaviour, i.e. in identification of reasons behind safe or aberrant riding behaviours performed by motorcyclists. The basic structure of the TPB [64–66], is represented schematically in Fig. 1. According to this theory, three constructs

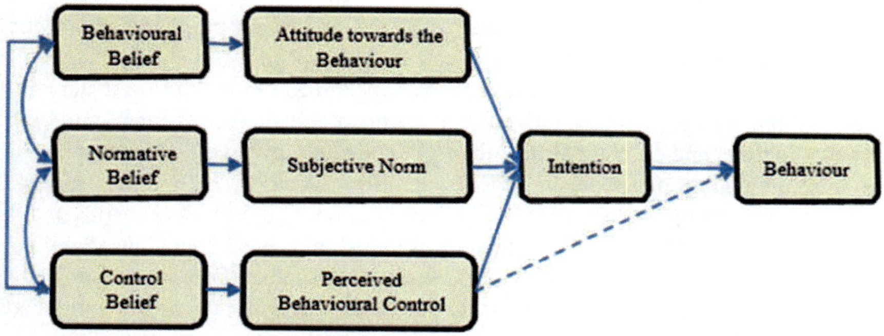

Fig. 1 The standard TPB model architecture

are generally responsible for performing any behaviour by an individual. These are attitude, subjective norms and perceived behavioural control.

Behavioural beliefs set the 'attitude' towards the behaviour of an individual. Normative beliefs control the 'subjective norm' which is basically the understanding how others or the community think on performing a particular behaviour. Lastly, control beliefs triggers 'perceived behavioural control', i.e. how an individual perceives a particular activity to perform. These beliefs help an individual decides whether performing any activity is right or wrong. All these three constructs collectively induce one's 'intention' to perform the particular behaviour and then, finally that controls whether the individual will perform that action or not (which is the observed behaviour of that person).

In several studies, researchers modified the basic structure of the TPB by introducing other psychological constructs which may impart effects on the intention to perform an action. Further, in some cases, other variables such as demographic characteristics of the respondents have found to be significant in triggering the intentions of an individual.

3.2 Psychological Models to Explain Riding Behaviours

Psychological models have been used in regards to explore the underlying reasons in performing safe or aberrant riding behaviours. Questionnaire for perception-based study on the British motorcyclists was developed to capture the riders' beliefs on safe riding with the help of two psychological model theories, named TRA and health belief model (HBM) [67]. They concluded that involvement in crashes can be predicted by an individual's belief on safe riding.

Theory of planned behaviour (TPB) was first used in the context of motorcyclists by Jamson [27] in UK. In Australia, Watson et al. [31] carried out an in depth study on identification of motorcyclists riding behaviours and used TPB to determine the underlying constructs influencing the intentions to perform some safe and unsafe riding behaviours. They found ‘attitude’ to be the significant predictor among the other TPB constructs for intentions to ride at extreme speeds.

To identify motorcyclists’ intention in over-speeding, Elliott [68] used standard ‘theory of planned behaviour’ (TPB) constructs in combination with constructs from ‘identity theory’ and ‘social identity theory’. The study involved in developing of two multiple regression models in order to identify the reasons behind intentions to ‘speeding on a 30 mph urban roads’ and ‘speeding on a 70 mph dual carriageways / motorways’. In both the cases, ‘affective attitudes’ and ‘perceived controllability’ factors were obtained to be significantly influential to predict those intentions.

Wong et al. [69] identified three basic personality traits of young riders named ‘sensation seeking’, ‘amiability’ and ‘impatience’ in Taiwan. They concluded that impatient riders who have low riding confidence, lack in traffic awareness are prone to risky riding behaviours.

Tunnicliff et al. [70] explored the psychological factors controlling motorcyclists’ intention to perform some safe and risky behaviours. Apart from the standard TPB constructs, the researchers used some additional predictors along with demographic variables to explore the reasons behind few safe and unsafe riding behaviours. Chorlton et al. [71] used TPB in association with additional constructs such as moral norms, anticipated regret, self-identity and perceived susceptibility to understand the reasons of over-speeding tendency of motorcyclists in United Kingdom. These factors could able to explain 57% of the total variance when the ‘intentions to speed up’ while riding in group was studied.

Özkan et al. [50] investigated the applicability of the factor structure in Turkey as proposed by Elliott [30] in UK. Three psychological models were developed based on ‘theory of planned behaviour (TPB)’, ‘health belief model (HBM)’ and ‘traffic locus of control (T-LOC). Five-factored HBM model consisted of perceived susceptibility, perceived severity, perceived benefit, perceived barriers, cues to action and health motivation. ‘Fate’ construct of the T-LOC model was found to be influential in predicting stunts and speeding behaviours. Simultaneously, ‘cues to action’ and ‘perceived barrier’ were found to be better predictors in case of HBM model.

Reasons behind reported repetitive violation behaviours were studied in Indonesia using TPB [34]. ‘Disregarding traffic rules’ were observed to be significantly predicted by the TPB constructs. Moreover, the researchers concluded that young riders were more prone in violating traffic rules in Indonesia.

The summary of such psychological modelling approaches is mentioned in Table 3. Apart from these modelling constructs, several behaviours such as speeding, aggressive driving, thrill-seeking driving, traffic violations, etc. were also tried to be explored by assessment of personality traits. In case of general drivers, driving behaviours have been assessed by conducting survey based on ‘aggression questionnaire’ [72], ‘sensation seeking scales’ [73], ‘aggressive driving behaviour questionnaire’ [74]. Inspired from these studies, the concepts were used by researchers in combinations with standard TPB constructs to explain the riding behaviours more rigorously [31, 70].

Table 3 Summary of the psychological modelling approaches

Studies	Region of study	Modelling approach	Target behaviour	Factors considered
Watson et al. [31]	Australia	Hierarchical regression	Intention to perform safe and unsafe riding behaviours	Standard TPB constructs, and extended constructs (group norm, sensation seeking, aggression, demographic variables etc.)
Elliott [68]	United Kingdom	Multiple regression	Speeding on urban roads and dual carriageways/motorways	TPB constructs and identity constructs
Chen and Lai [63]	Taiwan	Probit model	Mode choice behaviour	Standard TPB constructs, variables related to respondent profile, habit
Tunnicliff et al. [70]	Australia	Hierarchical regression	Safe and unsafe riding intentions	TPB constructs and extended constructs (group norm, sensation seeking, aggression etc.)
Özkan et al. [50]	Turkey	Full mediational model	Motorcycle rider behaviour (MRBQ)	Standard TPB constructs
		Hierarchical regression	Motorcycle rider behaviour (MRBQ)	Traffic locus of control variables, age, annual mileage etc
		Full saturated regression model	Motorcycle rider behaviour (MRBQ)	HBM constructs
Chorlton et al. [71]	United Kingdom	Multiple regression	Over-speeding on motorways, on rural roads and when in a group	TPB constructs with extended constructs (anticipated regret, self-identity etc.), demographic and motorcycling characteristics
Susilo et al. [34]	Indonesia	Structural equation modelling	Reported repetitive violation behavours	TPB constructs, socio-demographic characteristics, travel frequency and trip purpose

4 Discussion and Conclusions

4.1 Evidences from the Past Studies: Summary

From the reports published by different agencies, it has been clear that the motorcyclists account for the maximum fatality rate across the world, especially in the South-East Asia [4, 5]. In absence of proper accident database, the safety situation would not be possible to improve effectively by reactive measures only [4, 5, 10–14]. In the context of developing countries, there exist some practical limitations to encounter the situation proactively. Moreover, human factors such as riding behaviours have been found to be equally responsible for leading to crashes [23–26]. Studies on the behaviours have been existing in case of several developed countries [30–33], but the number of studies related to motorcyclists' riding behaviours in the context of developing countries are very few [34]. All these facts indicate the research need to assess the riding behaviours of motorcyclists in heterogeneous traffic flow conditions and to explore the reasons behind performing safe or aberrant riding behaviours.

Exploration of Perception-based Motorcycle Riding Behaviours. Reason et al. in 1990 [40] first framed a set of driver behaviour questionnaire (DBQ) in order to assess the drivers' behaviours in United Kingdom (UK) and classified aberrant driving behaviours (errors and violations) into five classes in the questionnaire, named: 'slips', 'lapses', 'mistakes', 'unintended violations' and 'deliberate violations'. Later on, for assessing motorcyclists' riding behaviour, a proper 43-item motorcycle rider behaviour questionnaire (MRBQ) was developed in UK [30]. A five-factored solution of riding behaviours were obtained, named 'traffic errors', 'speed violations', 'stunts', 'safety equipment' and 'control errors'. Özkan et al. [50] investigated the applicability of the factor structure in Turkey as proposed by Elliott et al. in 2007 [30] and came up with similar five-factored solution. Stephans et al. [53] in Australia also obtained similar factored solution.

However, such questionnaire survey led to get a four-factored solution in Nigeria [49], named 'control and safety', 'stunt', 'error' and 'speed hurrying'. Surveying based on a newly developed CMRDV in China, a two-factored solution (aggressive violations and ordinary violations) was obtained [48].

So, it can be concluded that the self-reported behavioural factors showed somewhat consistency when the MRBQ survey was carried out in developed countries. The factored solutions were obtained in completely different nature in the studies conducted in Nigeria and China. It indicates the need of a thorough study on riding behaviours of motorcyclists in other developing countries where the situations are entirely different.

Riding Behaviours and Crash Involvements. In both of studies done by Elliott et al. [30] and Özkan et al. [50] the behavioural factors were used to predict the self-reported crashes using statistical models. Though the factored solution obtained from the study in Turkey was found to be same as it was obtained in the context of UK, in English conditions, 'traffic errors' was found to be significant in prediction of self-reported crashes, whereas in case of Turkish case, it was found to be 'stunts' factor.

In case of Australian riders, it was observed that the stunt behaviours to be directly related to crash involvements, while violations of speed and errors significantly related to near-crash involvements [53]. Shakasita et al. [51] showed that errors and speed violations (MRBQ factors) along with age and exposures were the significant variables in predicting self-reported crashes and near-misses in Australia. Behaviours such as reckless overtaking or riding on sidewalks were found to be more dominant among young riders than the habit of using cell phones while riding in the context of Vietnam. In the case of Netherlands, researchers could able to develop crash prediction models (mostly self-reported). However, the MRBQ factors alone or in combination with other demographic variables could not able to establish a significant relationship with crash involvements [32].

The relationships are observed to be inconsistent when these behavioural factors were used as predictors of self-reported crashes or near-misses in different countries. Since the road environment, traffic characteristics and human behaviours are different across the world, there is a need for country-specific researches for establishing the relationship between behavioural factors and the self-reported crashes.

Identification of Reasons Underlying Aberrant Riding Behaviours. Researches have used different psychological models to identify the reasons behind safe or unsafe riding behaviours. Health belief model (HBM), traffic locus of control (T-LOC) and theory of planned behaviour (TPB) are three such theories applied in the context of motorcyclists. However, TPB is the most common and widely used theory used by several researches in this field.

In case of UK, 'affective attitudes' and 'perceived controllability' factors were obtained to be significantly influential to predict speeding intentions [68]. Three basic personality traits of young riders, named sensation seeking, amiability and impatience were identified in Taiwan and impatient riders with low riding confidence were found to be prone to risky riding behaviours [69]. Apart from the standard TPB constructs, Tunnicliff et al. [70] used some additional predictors along with demographic variables to explore the reasons behind few safe and unsafe riding behaviours in Australia]. Additional constructs (apart from standard TPB constructs) such as moral norms, anticipated regret, self-identity and perceived susceptibility were considered to understand the reasons of over-speeding tendency of motorcyclists in United Kingdom [71]. Disregarding traffic rules were observed to be significantly predicted by the TPB constructs and the young riders were found to be more prone in violating traffic rules in Indonesia [34].

Lastly, it can be concluded that riding behaviours or intentions to act such behaviours were investigated by researchers in order to identify the underlying predictors. To capture the behaviours more precisely, researchers introduced other constructs as predictors in the TPB models. Behaviours such as aggressive driving or sensation-seeking can be accounted by separate questionnaire designing [72–74].

Evidences from developing countries. If the cases of motorcycle dependent regions and the cases of developed countries are kept aside, a very few literatures exist on the topic of assessing riding behaviours of motorcyclists in developing countries. Results obtained from the MRBQ survey conducted in Nigeria and Iran [49, 52] have depicted that the factored structure of general riding behaviours to be different than that of developed countries. In both the cases, a four-factored solution was obtained by the researchers. Moreover, the questionnaire used in developed countries to capture the behaviours might not be applicable to use in other countries. Since the riding behaviours of motorcyclists in developing countries are yet to be explored in depth, a significant research gap exists in this regard.

4.2 Future Research Direction

The behaviours of motorcyclists observed in developing countries are completely different compared to that of developed countries due to several factors such as difference in road traffic conditions and environment, education and socio-economic conditions of the road users, etc. The factored solution obtained from the statistical analysis of riding behaviours from the studies conducted in United Kingdom, Australia, Turkey or Netherlands were found to be different compared to the same obtained from countries such as Nigeria, Iran or China. Since the perception of riders depends on their socio-economic background, knowledge and education and several other factors, the type of errors or violations committed by riders in developed and developing countries are highly like to be different, which is not explored in depth yet. Many of the questions which were developed to explore the riding behaviours in one country, may not be suitable to use in a different traffic conditions of another country. Moreover, the laws and policies related to traffic operations also varies across countries. Hence, the questionnaire prepared in UK or Australia to capture riding behaviours might not be applicable in other regions.

For this purpose, firstly, a set of questionnaire is required to be framed to conduct a survey to explore the behavioural factors of riders in developing countries. Analysing the data obtained from such surveys would help to identify and cluster the general riding behaviours of motorcyclists. Further, the aberrant riding behaviours observed in these regions need to be identified to improve the overall safety situation. This might be done by framing a new set of questionnaire based on some psychological models, such as TPB or HBM or T-LOC, etc., so that the reasons underlying those riding behaviours can be investigated.

The reasons behind the unsafe riding behaviours practised in developing countries are still unexplored yet and hence, the survey based on psychological models may lead to new insights in this regard. The outcome of such analysis would help the policy makers and stakeholders to consider as research evidences to think of the ways to improve the safety of motorcyclists. Based on the results, necessary training and education programmes, infrastructural interventions and enforcement related strategies may be proposed in order to change the behaviour of common riders. Lastly, assessing both the short-term and long-term effectiveness of such strategies to influence one's riding behaviour, lie in one of the sectors which are required to be explored in future.

References

1. Teoh ER (2017) Motorcycles registered in the United States, 2002–17. Insurance Institute for Highway Safety
2. Mohan D, Tsimhoni O, Sivak M, Flannagan MJ (2009) Road safety in India: challenges and opportunities
3. Ministry of Urban Development (MoUD) (2008) Traffic & transportation policies and strategies in Urban areas in India: Final Report
4. World Health Organization (2018) Global status report on road safety 2018. World Health Organization
5. World Health Organization (2015) Global status report on road safety 2015. World Health Organization
6. Transport Research Wing (2017) Road accidents in India—2016. Ministry of Road Transport & Highways, Government of India
7. Institute for Health Metrics and Evaluation (IHME) (2017) GBD compare data visualization. IHME, University of Washington
8. Kolkata Police (2011–2016) Accident review report. Lal Bazar, West Bengal, India
9. Akkelagunta P, Chatterjee S, Mitra S (2018) Identification of risk factors for fatal crashes occurring at various network locations in Hyderabad City. In: 97th Annual meeting, transportation research board. No. 18–06649
10. World Health Organization (2013) Global status report on road safety 2013. World Health Organization
11. De Ceunynck T (2017) Defining and applying surrogate safety measures and behavioural indicators through site-based observations. Hasselt University
12. Mahmud SM, Ferreira L, Hoque M, Tavassoli A (2018) Reviewing traffic conflict techniques for potential application to developing countries. J Eng Sci Technol 13(6):1869–1890
13. Hoque MM, Mahmud SS (2009) Road safety engineering challenges in Bangladesh. Bangladesh university of Engineering and Technology, Accident Research Institute
14. Dandona R, Kumar GA, Ameer MA, Reddy GB, Dandona L (2008) Under-reporting of road traffic injuries to the police: results from two data sources in urban India. Inj Prev 14(6):360–365
15. Indian Road Congress (2010) Manual on road safety audit. IRC: SP: 88–2010
16. Proctor S, Belcher M, Cook P (2001) Practical road safety auditing. Thomas Telford
17. Mahmud SS, Hoque MS, Qazi AS (2009) Road safety problems in Bangladesh: some major initiatives, constraints and requirements. Transp Commun Bull Asia Pac 79:61–89
18. Chatterjee S, Mitra S (2019) Safety assessment of two-lane highway using a combined Proactive and reactive approach: case study from indian national highways. Transportation Research Record, 0361198119846479
19. Jain SS, Singh PK, Parida M (2011) Road safety audit for four lane national highways. In: 3rd international conference on road safety and simulation
20. World Road Association, https://roadsafety.piarc.org/en/road-safety-management/safe-system-approach
21. Vardaki S, Papadimitriou F, Kopelias P (2014) Road safety audit on a major freeway: implementing safety improvements. Eur Transp Res Rev 6(4):387–395
22. Mooren L, Grzebieta R, Job S (2011) Safe system–comparisons of this approach in Australia. In: Australasian college of road safety national conference, Melbourne
23. Lewin I (1982) Driver training: a perceptual-motor skill approach. Ergonomics 25(10):917–924
24. Rumar K (1985) The role of perceptual and cognitive filters in observed behavior. In: Human behavior and traffic safety. Springer, Boston, pp 151–170

25. Elander J, West R, French D (1993) Behavioral correlates of individual differences in roadtraffic crash risk: an examination of methods and findings. Psychol Bull 113(2):279
26. Chang HL, Yeh TH (2006) Risk factors to driver fatalities in single-vehicle crashes: comparisons between non-motorcycle drivers and motorcyclists. J Transp Eng 132(3):227–236
27. Jamson SL (2004) The prevalence and antecedents of risky motorcycling behaviour. In: Behavioural research in road safety: fourteenth seminar. Department for Transport, London, pp 38–52
28. Jamson S, Chorlton K, Conner M (2005) The older motorcyclist. (Road Safety Research Report No. 55). Department for Transport, London, England
29. Özkan T, Lajunen T (2005) Multidimensional traffic locus of control scale (T-LOC): factor structure and relationship to risky driving. Pers Individ Differ 38:533–545
30. Elliott MA, Baughan CJ, Sexton BF (2007) Errors and violations in relation to motorcyclists' crash risk. Accid Anal Prev 39(3):491–499
31. Watson BTD, White K, Schonfeld C, Wishart D (2007) Psychological and social factors influencing motorcycle rider intentions and behavior. Centre for Accident Research and Road Safety (CARRS-Q). Australian Transport Safety Bureau, University of Technology, Queensland
32. Steg L, van Brussel A (2009) Accidents, aberrant behaviours, and speeding of young moped riders. Transp Res F Traffic Psychol Behav 12(6):503–511
33. Musselwhite CB, Avineri E, Susilo YO, Bhattachary D (2012) Public attitudes towards motorcyclists' safety: a qualitative study from the United Kingdom. Accid Anal Prev 49:105–113
34. Susilo YO, Joewono TB, Vandebona U (2015) Reasons underlying behaviour of motorcyclists disregarding traffic regulations in urban areas of Indonesia. Accid Anal Prev 75:272–284
35. Sexton BF, Baughan CJ, Elliott MA, Maycock G (2004) The accident risk of motorcyclists
36. Lin MR, Chang SH, Pai L, Keyl PM (2003) A longitudinal study of risk factors for motorcycle crashes among junior college students in Taiwan. Accid Anal Prev 35(2):243–252
37. Wells P (1986) Observations of motorcycle riders at junctions. (TRRL Research Report RR39). Transport and Road Research Laboratory, Crowthorne, England
38. Fell JC, Nash CE (1989) The nature of the alcohol problem in US fatal crashes. Health Educ Quart 16(3):335–343
39. Dandona R, Kumar GA, Dandona L (2006) Risky behavior of drivers of motorized two wheeled vehicles in India. J Safety Res 37(2):149–158
40. Reason J, Manstead A, Stradling S, Baxter J, Campbell K (1990) Errors and violations on the roads: a real distinction? Ergonomics 33(10–11):1315–1332
41. Parker D, Reason JT, Manstead ASR, Stradling S (1995) Driving errors, driving violations and accident involvement. Ergonomics 39:1036–1048
42. Parker D, West R, Stradling SG, Manstead ASR (1995) Behavioural traits and road traffic accident involvement. Accid Anal Prev 27:571–581
43. Quimby A, Maycock G, Palmer C, Buttress S (1999) The factors that influence a driver's choice of speed—A questionnaire study. (TRL Report 326). Transport Research Laboratory, Crowthorne, England
44. Lawton R, Parker D, Manstead ASR, Stradling S (1997) The role of affect in predicting social behaviors: the case of road traffic violations. J Appl Soc Psychol 27:1258–1276
45. Haque MM, Chin HC, Lim BC (2010) Effects of impulsive sensation seeking, aggression and risk-taking behaviours on the vulnerability of motorcyclists. Asian Transp Stud 1(2):165–180
46. Uy FAA, Regidor JRF (2011) A study on motorcycle rider characteristic and behavior in metro Manila. J Eastern Asia Soc Transp Stud 9
47. Verschuur WLG (2003) Onderzoek uitgevoerd met de 'Driver Behaviour Questionnaire'. Eindrapport. (Research based on the 'Driver Behaviour Questionnaire'. Final report). Faculteit der Sociale Wetenschappen, Rijksuniversiteit te Leiden, Werkgroep Veiligheid/R-03/68 (inDutch)
48. Cheng ASK, Ng TCK (2010) Development of a Chinese motorcycle rider driving violation questionnaire. Accid Anal Prev 42(4):1250–1256

49. Sunday OK, Akintola L (2011) The motorcycle rider behaviour questionnaire (MRBQ) and commercial motorcycle riders in Nigeria. Driver Behav Training 4:193–209
50. Özkan T, Lajunen T, Doğruyol B, Yıldırım Z, Çoymak A (2012) Motorcycle accidents, rider behaviour, and psychological models. Accid Anal Prev 49:124–132
51. Sakashita C, Senserrick T, Lo S, Boufous S, Rome LD, Ivers R (2014) The motorcycle rider behavior questionnaire: psychometric properties and application amongst novice riders in Australia. Transp Res Part F 22:126–139
52. Haghi A, Ketabi D, Ghanbari M, Rajabi H (2014) Assessment of human errors in driving accidents; analysis of the causes based on aberrant behaviors. Life Sci J 11(9):414–420
53. Stephens AN, Brwon J, Rome LD, Baldock MRJ, Fernandes R, Fitzharris M (2017) The relationship between motorcycle rider behaviour questionnaire scores and crashes for riders in Australia. Accid Anal Prev 102:202–212
54. Truong LT, Nguyen HT, De Gruyter C (2018) Correlations between mobile phone use and other risky behaviours while riding a motorcycle. Accid Anal Prev 118:125–130
55. Rosenstock IM (1966) Why people use health services. Milbank Memorial Fund Q 44(supplement):94–127
56. Janz NK, Becker MH (1984) The health belief model: A decade later. Health Educ Q 11(1):1–47
57. Sheeran P, Abraham C (1996) The health belief model. Predicting Health Behav 2:29–80
58. Rotter JB (1966) Generalized expectancies for internal versus external control of reinforcement. Psychol Monogr Gen Appl 80(1):1
59. Montag I, Comrey AL (1987) Internality and externality as correlates of involvement in fatal driving accidents. J Appl Psychol 72(3):339
60. Fishbein M, Ajzen I (1975) Belief, attitude, and behavior: aAn introduction to theory and research. Addison Wessley, Reading, Mass
61. Ajzen I, Fishbein M (1980) Understanding attitudes and predicting social behaviour
62. Madden TJ, Ellen PS, Ajzen I (1992) A comparison of the theory of planned behavior and the theory of reasoned action. Pers Soc Psychol Bull 18(1):3–9
63. Chen CF, Lai WT (2011) The effects of rational and habitual factors on mode choice behaviors in a motorcycle-dependent region: Evidence from Taiwan. Transp Policy 18(5):711–718
64. Ajzen I (1991) The theory of planned behavior. Organ Behav Hum Decis Process 50(2):179–211
65. Ajzen I (2002) Constructing a TPB questionnaire: Conceptual and methodological considerations
66. Ajzen I (20066) Constructing a theory of planned behavior questionnaire
67. Rutter DR, Quine L, Chesham DJ (1995) Predicting safe riding behaviour and accidents: demography, beliefs, and behaviour in motorcycling safety. Psychol Health 10(5):369–386
68. Elliott MA (2010) Predicting motorcyclists' intentions to speed: effects of selected cognitions from the theory of planned behaviour, self-identity and social identity. Accid Anal Prev 42(2):718–725
69. Wong JT, Chung YS, Huang SH (2010) Determinants behind young motorcyclists' risky riding behavior. Accid Anal Prev 42(1):275–281
70. Tunnicliff DJ, Watson BC, White KM, Hyde MK, Schonfeld CC, Wishart DE (2012) Understanding the factors influencing safe and unsafe motorcycle rider intentions. Accid Anal Prev 49:133–141
71. Chorlton K, Conner M, Jamson S (2012) Identifying the psychological determinants of risky riding: an application of an extended Theory of Planned Behaviour. Accid Anal Prev 49:142–153
72. Buss AH, Perry M (1992) The aggression questionnaire. J Pers Soc Psychol 63(3):452
73. Stradling S, Meadows M, Beatty S (2002) Speeding, violating, and thrill-seeking drivers
74. Čubranić-Dobrodolac M, Lipovac K, Čičević S, Antić B (2017) A model for traffic accidents prediction based on driver personality traits assessment. Promet Traffic Transp 29(6):631–642

A Study on the Factors Affecting the Commuters' Propensity to Bicycle in Srinagar City

Mahim Khan and Shafi Mir

1 Introduction

The unsustainable increase of private vehicles in urban areas of developing countries has led to severe congestion, environmental pollution, high cost of travel, high dependence on fossil fuel, ill-health, etc. Many studies in the past have shown a significant reduction in greenhouse gas emissions due to changed travel behavior. Therefore, shifting to sustainable mode of transport and curbing the use of motorized vehicles is a viable approach to address emission-related problems. Non-motorized transport (NMT) like bicycling, also known as active or human-driven transport, are ecological means of transport and have zero carbon footprint [1–4]. Yet, NMT is often underestimated and is barely considered as an essential mode of transport [5]. Also its immense health benefits are often overlooked and environmental advantages disregarded. Promoting cycling as a viable clean mode of transport can, therefore, be an effective way to reduce greenhouse gas emissions [6, 7] and combat climate change.

Over the past decades, prioritization of motor vehicles on the roads of India, have made pedestrians and cyclists more vulnerable to road accidents. A study conducted at four major cities of Mumbai, Delhi, Kota and Vadodara showed that cyclist and motorized two-wheelers accounted for 60–90% of all traffic casualties [8]. This is because cyclists are usually stereotyped as unconventional marginal people and this has given rise to safety concerns which act as a barrier to bicycle commuting. Although gender, car ownership, income, age, etc., do influence mode choice behavior of an individual, these factors however are not the only determining factors. There are other critical factors that have a major impact on the decision to bicycle. Individual's attitude and perception toward a particular mode of transport

M. Khan (✉) · S. Mir
Civil Engineering Department, N.I.T Srinagar, Srinagar, J&K 190006, India

S. Mir
e-mail: shafi@nitsri.net

M. Parida et al. (eds.), *Proceedings of the Fifth International Conference of Transportation Research Group of India* , Lecture Notes in Civil Engineering 220,
https://doi.org/10.1007/978-981-16-9925-2_21

have a great impact on the trip making behavior and modal choice [9]. Therefore, it is of immense importance to study psychological factors associated with bicycle commuting.

The current study is carried out in Srinagar Metropolitan Area (SMA), J and K, India. The city is experiencing accelerated use of private motorized vehicles. There is an insignificant number of cyclists and pedestrians. Therefore the aim of the study is to identify the critical factors associated with bicycle commuting. The study also identifies major perceived deterrents responsible for the low proportion of bicycle usage. The key motivational (positive) factors that can improve the propensity to bicycle are also detected. The difference in perception associated with bicycle mode choice was further tested on the private car and public transport commuters.

2 Literature Review

Theory of planned behavior (TPB) [10] explains that attitude, along with normative beliefs and perceived behavioral control, impacts the intention of an individual to carry out a particular behavior. Therefore, individual's attitude toward bicycle commuting is one of the pivotal factors in determining their bicycle mode choice behavior [11]. Those who are environmentally conscious, usually have a positive attitude toward cycling [9, 11, 12]. An individual's decision to commute by bike, however depends on several other factors like road infrastructure, income, gender, race, age, trip distance, trip number, marital status, safety concerns [13–15]. In some studies, lack of bicycle lanes is found to be the main deterrent and the predilection toward bicycle commuting has been associated with the availability of physical infrastructure [3, 16]. It is, however, argued that having world-class infrastructure still does not guarantee bicycle commuting if commuters are not confident enough in commuting by bicycle [17]. This is due to the perceived behavioral control, which is the ease or difficulty to perform a particular task [10]. Individuals who are unenthusiastic and in fact have negative attitude toward bicycle commuting and recognize the deterrents to be drastically strong, they have weak behavioral control in opposed to those having a positive attitude and do not perceive these barriers to be strong enough, have relatively stronger behavioral control [18]. A survey, carried out in the USA, investigated whether bicycling was considered as a non-motorized alternative for relatively shorter trips and it was found that people did not opt cycling even for shorter trips of just 2.25 kms and preferred to walk instead [6]. This was backed by many studies, which proved safety to be the main deterrent to bicycle commuting and therefore having safe bicycle lane network, prioritizing cyclists at intersections, etc., can notably change the perception of the commuters and can further lead to their behavioral change [13–15, 19–21]. The safety concerns are also caused due to the other road users; particularly motorized drivers, who often treat cyclists as marginalized commuters on the roads [9]. This further demotivates them and drives them away from commuting by bicycle. Therefore efforts need to be made to make bicycle commuting mainstream, thereby improving socio-cultural environment for the cyclists [22]. Social support and influence of peer (friends, family colleagues) is believed to have a dramatic effect on people's decision to choose a particular mode of transport and can in fact prove positive for bicycle commuting [23]. The bicycling environment can gradually change the attitude of people and can attract more people toward bicycle commuting [11]. Therefore, having an encouraging bicycle-oriented environment, with decent bicycle facilities can also have a positive impact on the perception of people. This is particularly true for workplaces, where people get inspired by their colleagues and friends [24]. Employees decision to commute by cycling however, also depends on their daily travel distance, which can also impede their bicycle mode choice [11]. Lack of proper cycling infrastructure can also act as a major obstacle against cycling [9, 18]. Providing better facilities with separate bicycle lanes, signals and parking for the better mobility of bicyclist would act as an important motivating factor and more people will adopt bicycle commuting [19]. Moreover, the perception of the quality of bike lanes influences the decision to cycle [25]. A study [26] showed that those who assumed their path with bicycle lanes separated from traffic noise, air pollution, motorized traffic and with pleasing landscape were much more willing to bicycle for commuting. It was also found that the presence of steep elevations may reduce the attractiveness of walking and cycling [23]. In addition, providing even topography results in less exhaustion to the cyclist [20]. It is further argued that just having bicycle facilities like lanes and parking spaces may not be effective in changing the attitude of commuters. If a person is found to have a negative attitude toward motorized vehicles, it could actually have a positive influence on the usage of bicycles [27]. Therefore, at workplaces levying high parking fee on cars may encourage people to cycle to work and abandon their car [11]. Apart from social and physical barriers there are few other subtler psychological factors that discourage cycle commuting. One is the effect of climatic/weather conditions [28]. In sunny days, the number of cyclists are relatively higher compared to the figure in the wind, snow and high humidity [28, 29]. A survey conducted in Canada showed that most participants (71%) stated that they would prefer bicycling at colder temperatures like −20 °C or colder, provided that the bicycle lanes are well-maintained. It was also found that in cold weather, males prefer cycling 5.5 times over to females [30]. Therefore, mode choice behavior also depends on socio-economic attributes like income, age, gender, education, etc., and need to be studied further [15, 24, 31, 32].

3 Study Methodology

Psychological traits such as habits, social norms and attitudes play a prominent role on individual's compliance to adopt a particular behavior [10]. This conceptual framework has been used to identify the factors affecting the bicycle mode choice among the commuters. A well-structured survey based on the theory of planned behavior was developed and users responses were recorded on 5 points Likert scale (1 = highly disagree to 5 = highly agree). Principal component analysis (PCA) was

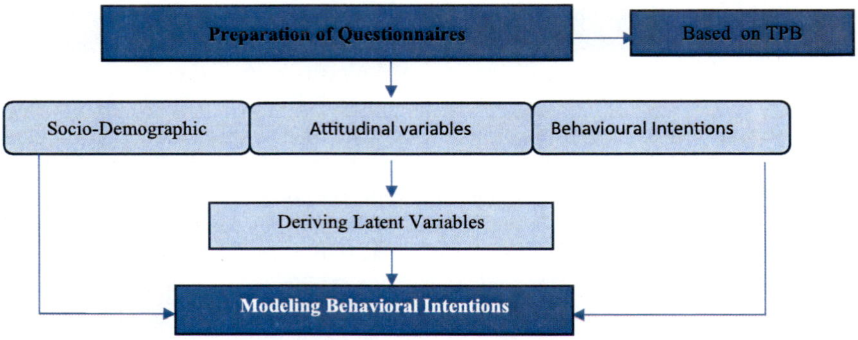

Fig. 1 Methodological framework of study

used to identify unobserved factors explaining commuters' perception toward bicycle mode choice. These latent factors are further used as predictors in regression for further analysis. The aim of the paper being, to understand the circumstances poignant for the propensity of bicycle usage for commuting—the behavioral intention was therefore taken as a dependent variable. Figure 1 demonstrates the methodological framework of the study.

Further, behavioral intention being a qualitative variable, trans-theoretical model (TTM) [33], was used to measure intentional change among the respondents. This model focuses on individual's adaptability; when change occurs, an individual passes through a number of stages of intentional changes. The first stage being pre-contemplation in which an individual is not ready to perform the behavior. The next stage being of contemplation in which an individual recognizes some benefits in adopting and is ambivalent about the change. The next stage is "Preparation" in which he or she is planning to change. In our study, respondents were tested based on chain of statements- where each statement represented a particular stage of the model. Respondents were further asked to mark the statement which best described them. These statements were: "Never thought of bicycling as a commuting option"(pre-contemplation); "Knowing the benefits, I consider commuting by bicycle sometimes" (contemplation); "I commute by bicycle sometimes" and " I am thinking of using a bicycle for commuting at least for short trips"(preparing for action).

4 Data Collection

4.1 Questionnaire Survey

A well-structured questionnaire based on the theory of planned behavior was developed. It begins with enquiring about the socio-demographic information like gender, age and educational qualification of the respondents. Data was also collected on the type of accommodation, commute mode and trip purpose of the commuters. This was followed by 27 statements measured on 5 points Likert scale which ranged from 1(highly disagree) to 5(highly agree). The data for the current study is gathered through face to face interview. People were approached at their workplaces, universities and restaurants. The agenda of the survey was well explained before their participation. A total of 655 individuals as students and Government (GVT.) and Private (PVT.) employees took part in the survey. The summary of the socio-economic variables is given in Table 1

Table 1 Summary of socio-economic variables

Sample characteristics	%
a. Gender	
Male	78.5
Female	21.5
b. Age	
<18	3
18–24	47.1
25–33	41.5
34–44	6.3
>44	2.1
c. Accommodation	
Family	71.5
Hostel	28.5
d. Commute mode	
Public transport	35.1
Personal car	26.2
Bike	16.3
Walking	10.8
Bicycle	11.6
e. Trip purpose	
College	38.7
Work	45.8
Recreation	8.7
Health	6.8
f. Occupation	
Student	53.3
Govt. employee	23.5
Pvt. employee	23.2

4.2 Sample Characteristics

Our sample had a high proportion of males (78.5%) with fairly young respondents with the majority of them belonging to the age group of 18–33 (88.6%). The most common preferred mode of transportation was public transport (35.1%), of which 51.6% were female commuters. This was followed by personal cars (25.2%). Out of 11.6% who chose bicycles as commute mode, 90.9% were males and only 9.1% were females, with 80% of trips made for health reasons, 17.2% for recreation, and only 3.8% for work.

5 Analysis and Results

The data were analyzed in two stages. Firstly, the principal component analysis was used to reduce the dimensionality and remove the potential Multicollinearity between the variables. Before proceeding for factor analysis suitability of data was checked on the basis of sample size and sample type. This was measured by the sample to variable ratio N: p, where N is a number of participants and p is the number of variables. N: p was recommended to be in the range of 3:1–20:1 by Williams et al. [34]) and in our case, it was 10.5, which meets the recommendation. This was followed by factor extraction. PCA constructs a new set of uncorrelated principal components by gathering highly correlated independent variables. Only the factors that have eigenvalues greater than 1 are considered significant [35] and the variables with greater load are considered important than others. After deriving latent factors from PCA, logistic regression is further used in the second stage. Logistic regression is used to predict the factors influencing the propensity of commuters toward bicycle commuting.

5.1 Principal Component Analysis

PCA was carried out in Statistical Package for the Social Sciences (SPSS) with the aim to reduce large sets of variables into smaller sets called principal components. The sample adequacy was given by Kaiser–Meyer–Olkin and was found out to be 0.858 and Bartlett's Test of Sphericity was found significant at $p = 0.000$. Five-factor with Eigen value >1, were extracted and explained a total variance of 64.488%. Varimax rotational approach was further used to achieve orthogonal rotation of factors. Rotation makes the interpretation of the analysis much more lucid and cogent. The extracted factors then were labeled logically based on the variables loading. The variables loading high on factor one are related to leisure, environmental awareness, affordability, physical fitness, congestion, and pollution reduction, and therefore is labeled as perceived benefits. The variables loading high on factor two are related to normative beliefs and therefore are labeled as social norms. The variables loading high on factor three are related to the physical barriers. Similarly, factor four is related to safety concerns and finally, factor five is named as "comfort and convenience". After deriving factors from PCA, we then check the reliability of each latent factor which tells us how strong inter-correlation each of the factors has. This is given by the Cronbach alpha and each factor being, greater than 0.7, indicated high reliability. Table 2 shows the variable mean scores (V.M) and factor mean scores (F.M) with alpha and Eigen values of each of the latent factors.

Based on the mean scores, the most positively influencing factors (motivators) and major deterrents to bicycle commuting (barriers), were identified. From the mean scores, it is evident that the respondents hold great environmental values (4.48) and identified the presence of dust (3.55) as one of the major barriers for making bicycle mode choice. Respondents seem to believe that cycling can bring down air pollution (4.41) as well as traffic congestion (4.215). Overall respondents did not seem to care about social norms (2.3367), but were concerned about safety, mainly due to high-speed vehicles (4.42). The absence of signals for cyclists (3.324) was another safety impediment for the respondents. Although the respondents did seem to recognize that commuting by bicycle demands too much exertion (3.23), but at the same time, agreed that it would help in improving their physical fitness (4.32). The presence of steep slopes (2.79) was not perceived as an issue and most of them did admit it to be affordable means of transport (4.22). People disagreed that their bicycle mode choice has any association with the cold weather but also acknowledged snow and rain as a pivotal obstacle for the same. This is because they find it unsafe and inconvenient to cycle in the absence of bicycle infrastructure (3.789) and in mixed traffic conditions (3.22), respectively. Table 2 also demonstrates the summary of factor loadings, variable mean, factor mean, standard deviation and Cronbach alpha (α).

5.2 *Binary Logistic Regression*

The factors achieved from PCA, along with two demographic variables- age and gender, were further used in binary logistic regression with behavioral intention being the predicted variable. The behavioral intent was redefined into two groups. The first group included people with little intent to commute by bicycle (pre-contemplation phase), while the second group included respondents with medium to high intention to change their mode choice behavior and shift toward bicycling (contemplation and preparation for action). Two logistic models were built, (1) Personal car commuters and (2) public transport commuters.

The first model is for those who commute by their personal car and allowed to classify 77.3% of the total cases correctly. This model accounted for 39.6% of the variance, where (R-square = 0.396, chi-square (8) = 72.562, $p < 0.001$) (Table 3). The significant predictors found in this model are safety concerns and age. The odds ratio indicates that greater is the safety concerns among the car commuters less likely

Table 2 Summary of factor loadings, variable mean, factor mean, standard deviation and Cronbach alpha

Factors (Eigenvalue)	Attitudinal variables	F.L	V.M	F.M	S.D	α
Perceived benefits (5.744)	By cycling, I can better enjoy the landscapes around me	0.791	3.91			
	Cycling should be promoted to save our environment	0.782	4.48			
	Transportation by cycle is affordable	0.771	4.22			
	Commuting by cycle can help in physical fitness	0.754	4.32	4.31	0.806	0.89
	Cycling will reduce traffic congestion	0.742	4.21			
	Cycling may result in less pollution	0.724	4.41			
Social Norms (3.389)	People will look down on me	0.794	2.11			
	People will not find cycling cool option to commute	0.794	2.10	2.33	0.800	0.85
	Everyone will make fun of me	0.793	2.42			
	Friends and family will not be supportive	0.789	2.33			
Physical barriers (2.368)	I do not cycle in winters because I feel cold	0.756	2.4			
	It requires too much exertion	0.723	3.23			
	I do not cycle due to the presence of dust on roads	0.709	3.55	3.66	0.950	0.89
	Cannot cycle in Snow and rain	0.709	3.33			
Safety concerns (1.482)	Unsafe due to high-speed vehicles	0.785	4.42			
	Unsafe to cycle with no separate bicycle lanes	0.722	3.78	3.91	986	0.82
	Unsafe to cycle with no signals for cyclists	0.652	3.32			
Comfort and convenience (1.204)	inconvenient to cycle on steep slopes	0.738	2.75			
	Inconvenient to cycle due to the clothes we wear	0.736	2.77	2.92	750	0.87
	Inconvenient to cycle in mixed traffic conditions	0.628	3.22			

Table 3 Binary logistic regression results on car and public transport commuters

	Car		Public transport	
	Sig	Exp (B)	Sig	Exp (B)
Perceived benefits	0.861	1.031	0.005	2.703
Social norm	0.561	0.905	0.100	0.481
Physical barriers	0.252	0.810	0.131	0.622
Safety concerns	0.016	0.648	0.019	0.835
Comfort and convenience	0.109	0.752	0.240	0.535
Gender (1)	0.158	0.478	0.000	0.401
Age	0.000		0.510	
Age (1)	0.019	2.886	0.247	0.207
Age (2)	0.000	0.049	0.419	0.217
Constant	0.024	0.605	0.008	45.816

are they to have any intention of commuting by bicycle, and age indicating that people in their old age are less likely to have an intention of bicycling compared to younger respondents.

The second model is for those who commute by public transport and allowed to classify 88.3% of the total cases correctly. This model accounted for 60.8% of the variance, where (R-square = 0.608, chi-square (8) = 37.991, $p < 0.001$). The public transport commuters seem to recognize the benefits of commuting by bicycle but the major deterrent for them seem to be safety concerns. More safety concerns they have less they are going to opt for bicycle mode choice. It further shows that male public transport commuters are 2.49 times more likely to have the intention of bicycle commuting than females.

6 Conclusion

This study identifies psychological factors and their relative importance in influencing bicycle mode choice. Our results indicated that people are less concerned with social norms and do not consider it as an impediment to bicycle commuting. Rather, safety and security were found to be of prime importance and a major obstacle. People felt unsafe with high-speed vehicles and considered it inconvenient to peddle in mixed traffic conditions. They perceived bicycle commuting to be dangerous with no separate bicycle lanes as was concluded by Daley and Rissel [9]. Overall, comfort and convenience were perceived to be of lesser importance and interestingly the route topography did not seem to affect their decision to bicycle. This was in line with the perception of the general respondents of [20] and contrary to his experts. Bicycle commuting was in fact, positively related to physical fitness and respondents believed that bicycling would allow them to enjoy the surrounding landscapes.

This is understood as SMA is known for its scenic beauty, as the city is surrounded by lush green mountains. Respondents seemed to hold a great value for the environment as environmental consciousness was found to have the strongest influence on bicycle mode choice. Dust on roads also affected their decision to bicycle and was also perceived as a deterrent. People believed that commuting by cycling would significantly reduce air pollution as well as address traffic congestion. Therefore, fostering bicycle commuting as cost-effective and eco-friendly mode of transport can positively influence commuters as also recommended by Verma et al. [19]. People seemed least bothered of cold weather and considered bicycling as an affordable means of transport. They, however, were not sure of how snow and rain would affect their bicycle mode choice, by causing safety concerns. One of the reasons for this could be the lack of adequate bicycle infrastructure and related facilities. This was similar to the findings of [30]. Further, male public transport commuters were much more willing for bicycle commuting than female public transport users. However, the major deterrents for them were safety concerns. It was found that the young car commuters were more likely to have an intention of bicycle commuting; they seemed to recognize the benefits associated with it, provided that safety is ensured.

In order to address the safety issue, cyclists should be segregated from the motorized vehicle fleet all across the network. Priority should be given to cyclists at the intersections which will also improve their social status. Road commuters will develop a positive attitude toward cyclists and the cycle commuters will no longer be side-lined or looked down upon. Improving policies related to cyclists will not be enough as other counter measures also need to be taken to ensure sustainable mobility. Levying a tax on private cars should be encouraged to reduce the ownership of cars. This, in turn, will drive people toward alternate modes of transport and thereby address the emission problem. Bicycling should be promoted and health benefits should be communicated. Celebrities can play an important role of influencers in promoting cycling, as people usually have the tendency to copy TV and sports stars. Bike- sharing would also be an effective way of increasing bicycle commuting, particularly at tourist places like in Srinagar Metropolitan Area (SMA).

References

1. Liu J, Santos G (2015) Plug-in hybrid electric vehicles' potential for urban transport in China: the role of energy sources and utility factors. Int J Sustain Transp 9:145–157. https://doi.org/10.1080/15568318.2012.738776
2. Litman T (2004) Quantifying the benefits of nonmotorized transportation for achieving mobility management objectives. J Transp Res Board 1441:134–140
3. Rastogi R (2010) Willingness to shift to walking or bicycling to access suburban rail: Case study of Mumbai, India. J Urban Plan Dev 136:3–10. https://doi.org/10.1061/(asce)0733-9488(2010)136:1(3)
4. Rietveld P (2016) Biking and walking: the position of non-motorized transport modes in transport systems. In: Handbook of transport systems and traffic control. Barcelona and Seville, pp. 299–319
5. NTDCP (2013) Regional stakeholder's meeting, Interim Report. Mumbai

6. Kim S, Ulfarsson GF (2008) Curbing automobile use for sustainable transportation: analysis of mode choice on short home-based trips. Transp (Amst). 35:723–737. https://doi.org/10.1007/s11116-008-9177-5
7. Lumsdona L, Tolley R (2001) The national cycle strategy in the UK: To what extent have local authorities adopted its model strategy approach? J Transp Geogr 9:293–301. https://doi.org/10.1016/S0966-6923(01)00022-9
8. Mohan D, Tiwari G (2014) Mobility, environment and safety in megacities. IATSS Res 24:39–46. https://doi.org/10.1016/s0386-1112(14)60016-9
9. Daley M, Rissel C (2011) Perspectives and images of cycling as a barrier or facilitator of cycling. Transp Policy 18:211–216. https://doi.org/10.1016/j.tranpol.2010.08.004
10. Ajzen I (1991) The theory of planned behavior. Orgnizational Behav Hum Decis Process 50:179–211. https://doi.org/10.1016/0749-5978(91)90020-T
11. Handy SL, Xing Y (2011) Factors correlated with bicycle commuting: a study in six small U.S. cities. Int J Sustain Transp 5:91–110. https://doi.org/10.1080/15568310903514789
12. Heinen E, Maat K, Van Wee B (2011) The role of attitudes toward characteristics of bicycle commuting on the choice to cycle to work over various distances. Transp Res Part D Transp Environ 16:102–109. https://doi.org/10.1016/j.trd.2010.08.010
13. Andrade V, Jensen OB, Harder H, Madsen JCO (2011) Bike infrastructures and design qualities: enhancing cycling. Danish J Geoinformatics L Manag 46:65–80
14. Dill J (2009) Bicycling for transportation and health: the role of infrastructure. J Publ Health Policy 30:95–110. https://doi.org/10.1057/jphp.2008.56
15. Moudon AV, Lee C, Cheadle AD, Collier CW, Johnson D, Schmid TL, Weather RD (2005) Cycling and the built environment, a US perspective. Transp Res Part D Transp Environ 10:245–261. https://doi.org/10.1016/j.trd.2005.04.001
16. Jain (2010) Evaluating bicyclists comfort and safety perception. In: 12th world conference for transportation research. Portugal, pp 1–19
17. Engbers LH, Hendriksen IJM (2010) Characteristics of a population of commuter cyclists in the Netherlands: perceived barriers and facilitators in the personal, social and physical environment. Int J Behav Nutr Phys Act 7:4–8. https://doi.org/10.1186/1479-5868-7-89
18. de Souza AA, Sanches SP, Ferreira MAG (2014) Influence of attitudes with respect to cycling on the perception of existing barriers for using this mode of transport for commuting. Procedia Soc Behav Sci 162:111–120. https://doi.org/10.1016/j.sbspro.2014.12.191
19. Verma M, Rahul TM, Reddy PV, Verma A (2016) The factors influencing bicycling in the Bangalore city. Transp Res Part A Policy Pract 89:29–40. https://doi.org/10.1016/j.tra.2016.04.006
20. Majumdar BB, Mitra S (2015) Identification of factors influencing bicycling in small sized cities: a case study of Kharagpur India. Case Stud Transp Policy 3:331–346. https://doi.org/10.1016/j.cstp.2014.09.002
21. Willis DP, Manaugh K, El-Geneidy A (2015) Cycling under influence: summarizing the influence of perceptions, attitudes, habits, and social environments on cycling for transportation. Int J Sustain Transp 9:565–579. https://doi.org/10.1080/15568318.2013.827285
22. Daley M, Rissel C, Lloyd B (2007) All dressed up and nowhere to go? A qualitative research study of the barriers and enablers to cycling in inner Sydney. Road Transp Res 16:42–52
23. Titze S, Stronegger WJ, Janschitz S, Oja P (2008) Association of built-environment, social-environment and personal factors with bicycling as a mode of transportation among Austrian city dwellers. Prev Med (Baltim) 47:252–259. https://doi.org/10.1016/j.ypmed.2008.02.019
24. De Geus B, De Bourdeaudhuij I, Jannes C, Meeusen R (2008) Psychosocial and environmental factors associated with cycling for transport among a working population. Health Educ Res 23:697–708. https://doi.org/10.1093/her/cym055
25. Panter J, Griffin S, Dalton AM, Ogilvie D (2013) Patterns and predictors of changes in active commuting over 12 months. Prev Med (Baltim) 57:776–784. https://doi.org/10.1016/j.ypmed.2013.07.020
26. Winters M, Davidson G, Kao D, Teschke K (2011) Motivators and deterrents of bicycling: comparing influences on decisions to ride. Transp (Amst). 38:153–168. https://doi.org/10.1007/s11116-010-9284-y

27. Stinson MA, Bhat CR (2005) A comparison of the rout preferences of experienced and Inexperienced bicycle commuters. In: Trb
28. Nankervis M (1999) The effect of weather and climate on bicycle commuting. Transp Res Part A Policy Pract 33:417–431. https://doi.org/10.1016/S0965-8564(98)00022-6
29. Flynn BS, Dana GS, Sears J, Aultman-Hall L (2012) Weather factor impacts on commuting to work by bicycle. Prev Med (Baltim) 54:122–124. https://doi.org/10.1016/j.ypmed.2011.11.002
30. Amiri M, Sadeghpour F (2015) Cycling characteristics in cities with cold weather. Sustain Cities Soc 14:397–403. https://doi.org/10.1016/j.scs.2013.11.009
31. Garrard J, Rose G, Lo SK (2008) Promoting transportation cycling for women: The role of bicycle infrastructure. Prev Med (Baltim) 46:55–59. https://doi.org/10.1016/j.ypmed.2007.07.010
32. Dill J (2007) Factors affecting bicycle demand: Initial survey findings from the Portland region. In: 86th annual meeting of the transportation research board. p 21
33. Prochaska JO, Velicer WF (1997) The transtheoretical model of health behavior change. Am J Heal Promot 12:38–48. https://doi.org/10.4278/0890-1171-12.1.38
34. Williams JH, DeBenedictis A, Ghanadan R, Mahone A, Moore J, Morrow WR, Price S, Torn MS (2012) The technology path to deep greenhouse gas emissions cuts by 2050: The pivotal role of electricity. Science (80-.):335, 53–59. https://doi.org/10.1126/science.1208365
35. Hair FJ (2012) Multivariate data analysis. Pearson, USA

Key Performance Indicators of Sustainable Urban Freight for Jaipur City in India

Pankaj Kant and Sanjay Gupta

1 Introduction

Sustainable transportation is an essential part of the overall sustainable development of cities. Externalities, economic, and social impacts are three dimensions that are often referred to as the three E's of sustainability [1]. Key performance indicators (KPIs) of urban freight transport provide a deeper insight into the urban freight sector in most situations. A mix of various indicators is always useful to cover the impacts and objectives of heterogenous stakeholders involved in the city logistics operations [2]. KPIs help in both aggregate and disaggregate level of analysis of urban freight stakeholders. KPIs complements the decision-making process by local policymakers for freight travel patterns of the goods vehicle, physical, social, economic, and externalities impacts in cities [3].

The typical objectives of the typical freight studies are the diagnosis of freight problems, simulation of possible solutions, and decisions based on results. KPIs are indeed always helpful and somewhat essential for achieving sustainability objectives [4]. There is always a possibility to compare the strategic position of cities for their productivity and level of service in the urban freight sector with help of KPIs freight efficiency [5]. KPIs of infrastructure efficiency show the optimum resource utilization, whereas KPIs of infrastructure effectiveness reflect the outputs required to achieve the freight objectives [6].

The logistics performance index (LPI) depicts the detailed logistics environment of a country. Domestic LPI covers the four aspects of freight data related

P. Kant (✉) · S. Gupta
Department of Transport Planning, School of Planning and Architecture, 4-Block-B Indraprastha Estate, New Delhi 110002, India
e-mail: pankaj179phd17@spa.ac.in

S. Gupta
e-mail: s.gupta@spa.ac.in

M. Parida et al. (eds.), *Proceedings of the Fifth International Conference of Transportation Research Group of India* , Lecture Notes in Civil Engineering 220,
https://doi.org/10.1007/978-981-16-9925-2_22

to time, distance, institution, and logistics process. LPI mainly focused on logistics constraints on the freight sector in the country [7]. Local policymakers have low awareness of the urban freight sector due to the involvement of multiple stakeholders and complex supply chains of commodities. Outcomes of freight policies and initiatives in city logistics are always difficult to predict due to the inherent complexities of the urban freight sector [8].

This research paper evaluates the KPIs weights for sustainable urban freight across commodities in Jaipur city of India by the analytical hierarchy process (AHP) method.

AHP is a multi-actor multicriteria decision-making method that can aid the local policymakers in setting the priorities and decision making in the urban freight sector [9]. There is a significant research gap in the urban freight sector especially in the Indian context regarding KPIs and LPIs in the purview of sustainability.

The rest of the research paper is organized in sequential sections which consist of literature review, case city profile, research methodology and data collection, AHP method, descriptive overview of dataset, and results with policy implication on urban freight.

2 Literature Review

Different types of KPIs are there in freight literature subject to the objectives of freight studies. There are some common indicators in the urban freight sector related to establishment profile, freight carriers, delivery patterns, and logistics decisions. KPIs can also be summarized in sustainable indicators which cover each stakeholder involved in the urban freight sector [10]. Data and evaluation methods with KPI are essential for freight policy measures, evaluation, and benchmarking. One of the main objectives of European Union countries is the sustainability of goods transport in their cities [11]. The impact of urban goods transport can be analyses in various categories like fleet size, urban deliveries, safety, externalities, and economic impacts. Each category of impact has different sets of KPIs and thus requires different freight surveys for quantification [12].

Important freight transport issues, gaps in knowledge, and potential future research can be addressed by different freight data and modeling methods [13]. Economic indicators of establishment and fleet operators were used to assess the changes in freight operation over 30 years for different cities of the U.K. [14, 15]. KPIs can also be used to assess the quantum of urban goods movement in cities for integration in local transport plans for better public decision support. KPIs are always diverse and always are subject to the objectives of freight initiatives and studies [16].

Social and externalities related to KPI were used in London for freight activity patterns, fleet operator's recognition scheme, and goods vehicle safety [17]. Economic indicators related to travel time and delivery reliability were used by Transport for London (TfL) authority to assess and demonstrate the potential benefits and impacts of the London consolidation construction center [18]. Variation in travel time as an indicator associated with a change in the location of the markets was used for the reconfiguration of London's five wholesale markets [19].

Variations in the evaluation of indicators by (GPS based and pencil paper-based) different survey methods were highlighted for the Peel region for urban goods distribution [20]. KPIs are essential for qualitative analysis of urban goods movement in the purview of sustainability. KPIs are also useful for post evaluation of freight initiatives and comparison of results [21]. Establishment-driver-related indicators answer the critical concerns of the stakeholders for the impacts goods movement at city level including sustainability aspect [22, 23].

Emission-related KPIs for freight transport are important for monitoring and reduction of greenhouse gas (GHG) emissions. A weight-based indicator like ton kilometer traveled (TKT) is not appropriate to assess the volume-based goods movement especially at the urban scale [24]. The logistics performance index (LPI) is a benchmarking tool suggested by the World Bank to assess the logistics performance at the country level. LPI helps in identifying the challenges involved in logistics performance with six key indicators. Currently, domestic LPI is not available due to the absence of freight data at the city scale, especially in developing countries [7].

AHP method has been widely used for complex decision making in various filed like planning, corporate policy, and resource management. The main advantage of the AHP method is that it is easy to use and the hierarchical structure of this method can be adjusted to various sizes of problems. The main problem with the AHP method is the inconsistency of ranking and judgment due to the interdependence of criteria and alternatives. AHP method has an advantage for its easy use and not being data intensive. The hierarchy structure of AHP can easily adjust to fit many sized problems. Problems in the AHP method arrive due to interdependence between criteria and alternatives, which can lead to inconsistencies between judgment and ranking criteria [25].

AHP method can also be applied in urban freight sectors for evaluation of freight schemes as this sector is driven by multiple stakeholders and institutions [26]. AHP method was successfully applied for the evaluation of transportation choices for a multimodal freight transport system at a regional scale. AHP method is a suitable approach for the urban freight sector due to multiple heterogeneous stakeholders. AHP method allows flexibility in arranging the alternatives according to different criteria [27].

Literature review suggests that various indicators are available to assess the sustainability of urban freight. Different commodities have a different supply chain, which makes urban goods distribution more challenging in the preview of sustainability. The heterogeneous nature of urban freight stakeholders may tend to reflect the variation in the weights of indicators. Assessment of KPI weights of different commodities will be helpful for local policymakers for urban freight initiatives.

3 Case City Profile

Jaipur is one of the oldest and capital of Rajasthan state in India. Jaipur city is famous for its architecture, gems, jewelry stone crafting, and local culture. The Jaipur city is 280 km from Delhi the capital of India. The total area of Jaipur is 2939 km^2 out of which 17 km^2 consists of a heritage walled city area (old area). As per the census of the year 2011, the population of Jaipur city is 3.05 million. The decadal growth rate of Jaipur city is +35%. The gross population density of Jaipur is 64 PPH, and the workforce participation rate is 34.7% [28]. The land use distribution of Jaipur city as per master plan 2011 consists of 44.8% residential, 6.7% commercial, 6% industrial, 10.5% public and semi-public, 15.4% circulation, and 11.3% recreational land use. There are various wholesale markets in Jaipur city mainly consist of electronics, building hardware, food grains, fruits and vegetables, pharmaceutical, and meat markets. Apart from the wholesale market, there are industrial areas like Satpura and Vishwakarma, etc., which acts as major freight handling areas in the city.

The nature of goods distribution of these wholesale markets varies from commodity to commodity. Some of these markets have goods distribution in urban areas only, like furniture, dairy, and meat products. Some of the markets have more regional distribution of goods like industrial products, chemicals, and fertilizers. Some markets have both urban and regional goods distribution like textiles and handicrafts [28].

Inferring from the literature review, this research paper purviews KPIs variations in the context of the perishability of goods/commodities. Perishability has both a dimension of time-bound delivery and storage (inventory) of goods. Three case markets were selected based on the perishability of goods. These case markets are the fruits and vegetable market at Muhana, Grain Market at NH-11, and building the hardware market at New Aatish Nagar in Jaipur. For evaluation of logistics performance indicators, medium-scale industries located at Vishkarma Industrial area (VKI) in Jaipur has been selected.

4 Methodology and Data Collection

The main objective of this research study is to assess the weights of KPIs across commodity distribution in Jaipur city. The criteria used for KPI selection was based on a literature review and are sustainable indicators of the economy, social and externalities additional to LPI given by the World Bank.

For shippers (wholesalers) related to KPI, three commodity markets of Jaipur city were selected for the study. These markets are the grain market (NH11), fruits, and vegetable market (Muhana Mandi), and building hardware market (New Aatish market) in the city of Jaipur. Distribution characteristics of the building hardware market are business-to-costumer (B-C), while grain and fruits and vegetable markets are more of business-to-business (B-B) distribution. Distribution characteristics of industrial products of industries situated at Vishkarma Industrial area (VKI) area are majorly regional.

For economic-related KPI, transport operators and carriers situated at Transport Nagar area in Jaipur has been selected. Both private carriers and for-hire carriers at wholesale markets and Transport Nagar have been surveyed. Social experts and academicians were consulted and surveyed for the evaluation and assessment of social and externalities KPI. The research results are based on primary surveys of wholesalers, transport carriers, social experts, and academicians. Table 1 shows the sample size for each stakeholder.

Figure 1 shows the KPIs selected from the literature for this research study.

The framework proposed for the assessment of KPIs weights and across commodities for Jaipur city is shown in Fig. 2.

Table 1 Stakeholders sample size

S. No.	Market	Sample size
1	Fruits and vegetable wholesalers	22
2	Grain market wholesalers	21
3	Building material and hardware wholesalers	21
4	Transport operators	27
5	VKI industrial area	20
6	Social experts and academicians	19

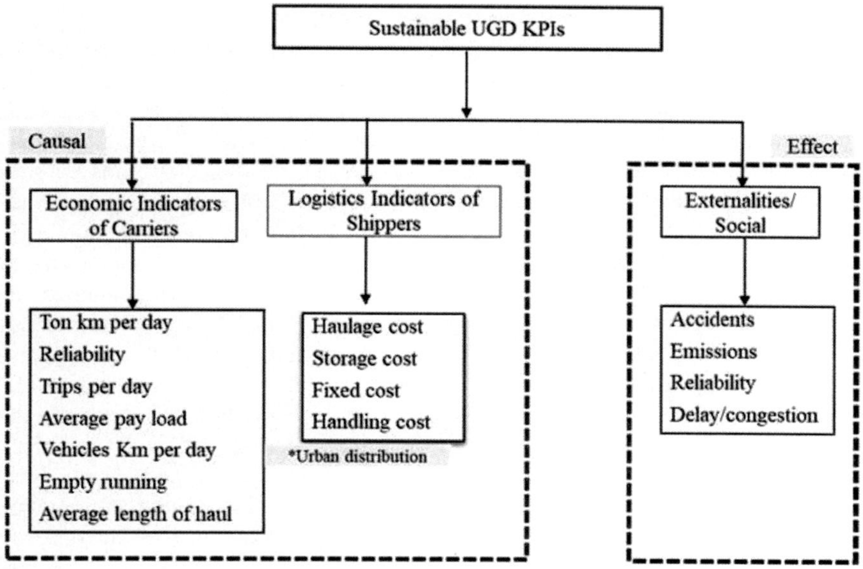

Fig. 1 Key performance indicators

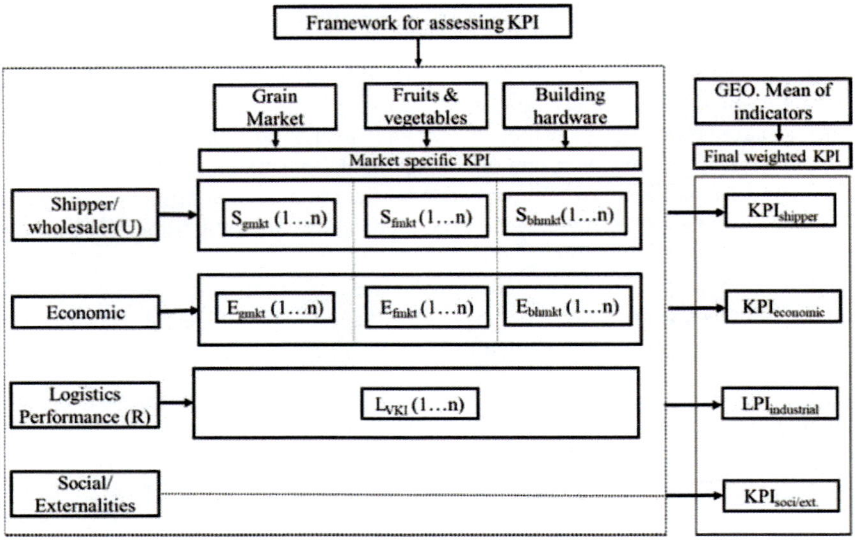

Fig. 2 Framework for assessment of KPIs weights

5 AHP Formulation

AHP is a multi-actor multicriteria decision-making technique used to solve complex decision operations. AHP solves a complicated problem by subdividing the problem in the hierarchical structure of objectives of stakeholders, available criteria to assess, and available alternatives. AHP method can combine quantitative and qualitative factors affecting multiple stakeholders [29]. AHP evaluates a pairwise comparison of criteria and alternatives in a hierarchical manner to arrive at the best decision among alternatives. AHP is a commonly used tool where subjectivity may affect the result of the decision-making process [30]. A nine-point scale is used for the pairwise comparison of alternatives. Each point of scale reflects the importance of alternatives in the choice set of paired comparisons. When both alternatives have equal importance, then scale 1 is assigned. Scale 3 is assigned to a latter alternative when it is slightly favorable than the former alternative. Similarly, scale 5 is assigned when it is strongly favorable, scale 7 when it is strongly favored, and scale 9 is assigned to the latter alternative when it is absolutely important in the highest order. Scales 2, 4, 6, 8 are assigned to the latter alternative when the importance lies in between odd scales of 1, 3, 5, 7, and 9. If the former alternative is more important than the latter alternative, then the reverse scale is of 1/3, 1/5, 1/7, and 1/9 which is assigned in the same manner [31–33]. AHP produces the final results on the ratio scale where the weight of each alternative is proportional to the sum of all alternatives [34].

The pairwise comparison of $\boldsymbol{n}$ alternatives and its results can be depicted in the form of a matrix (1).

$$A = (a_{ij})_{n\times m}\begin{vmatrix} a_{11} & a_{12} & * & a_{1m} \\ a_{21} & a_{22} & * & a_{2m} \\ * & * & * & * \\ a_{n1} & a_{n2} & * & a_{nm} \end{vmatrix} \quad (1)$$

The comparison matrix of alternatives is then normalized to produce relative weights of alternatives. The right eigenvector gives the relative weights of alternatives (w) concerning the largest eigenvalue $\lambda_{\max}$ (2),

$$A_w = \lambda_{\max} w \quad (2)$$

The consistency index (CI) for pairwise comparison of individual judgments is calculated by Eq. (3) [35].

$$\mathrm{CI} = \frac{\lambda_{\max} - n}{n-1} \quad (3)$$

Final results from the AHP method are evaluated by the consistency ratio (C.R.) check. To accept the results, C.R. should be less than 10%. The C.R. is computed as the ratio of CI and random index (RI) as per Eq. (4)

$$\mathrm{CR} = \frac{\mathrm{CI}}{\mathrm{RI}} \quad (4)$$

The value of the random index (RI) varies with the number of alternatives in the overall choice set, for example, RI value for five alternatives in the choice set is 0.90 and the RI value is 1.45 for nine alternatives.

6 A Descriptive Overview of Dataset

A descriptive overview of datasets for selected freight handling areas in the case of Jaipur city is given in Table 4. There are a lot of variables associated with freight handling areas (wholesale markets), but only two variables, i.e., tonnage handled weekly and shipments per week, are selected for the descriptive overview. These two variables are important as they reflect the commodity lifted and moved within and outside the city. The importance of goods delivery mode and its selection is also covered in these two variables. Variables in the case of establishments at VKI represent regional distribution, whereas establishments in other markets representing intracity freight distribution.

Table 2 Descriptive statistics of freight handling areas

Freight generating areas		Tonnage handled weekly/establishment	Shipment frequency weekly/establishment
Aatish market (building hardware and material)	*N*	21	21
	Mean	17.05	8.43
	Median	11.00	7.00
	Std. deviation	14.041	5.912
Grain market (NH-11)	*N*	21	21
	Mean	14.76	13.24
	Median	16.00	13.00
	Std. deviation	5.61	4.51
Muhana Mandi (fruits and vegetables)	*N*	22	22
	Mean	23.41	18.09
	Median	17.50	19.50
	Std. deviation	11.76	4.57
VKI (industrial products)	*N*	20	20
	Mean	565.70	37.90
	Median	536.50	36.00
	Std. deviation	252.29	16.83

The maximum deviation in tonnage handled (252 tons) per week per establishment is observed in the VKI area and the minimum in the fruits and grain market (5.6 tons). The maximum and minimum deviation in the shipment frequency of goods is also observed in these two markets, respectively. Data collected and presented in Table 2 shows typical weekly behavior observed in shipment size and shipment frequency at wholesale markets.

Descriptive statistics for transport operators involved in goods distribution in the case city of Jaipur India are given in Table 3. Vehicle kilometers traveled (VKT) is used for descriptive statistics. VKT has a direct implication on transporter's efficiency and externalities due to urban freight distribution patterns. The average length of haul is observed maximum for HCV (360 km/day) and least for three-wheel

Table 3 Descriptive statistics of transport operators (VKT/day)

Vehicle type	*N*	The capacity of mode (tons)	Mean	Median	Std. deviation
3W	9	0.5	59.56	60.00	15.05
HCV	5	18	360.00	350.00	114.01
LCV	5	1.5	71.00	70.00	14.31
4W	8	1	62.75	65.00	18.20

Note 3W = 3-wheeled commercial vehicles, 4W = 4-wheeled commercial vehicles, LCV = light commercial vehicle, HCV = heavy commercial vehicle

Table 4 Expert group dataset

S. No.	Experts	Sample size
1	Academicians (transport field)	4
2	Environment professionals	3
3	Social experts	3
4	City planners	4
5	Traffic management/enforcement officials	2
6	Market association people	3
	Total	19

commercial vehicles (60 km/day). The minimum deviation in VKT is observed for LCV (14.3 km). HCV is representing the regional distribution, i.e., tier-1 only. This research study has not considered non-motorized transport (NMT) in goods distribution.

Table 4 shows the sample size to assess the weights of social and externalities KPIs.

7 AHP Results

AHP results for key performance indicators for goods distribution in the case city are presented under four categories and are as following.

7.1 Performance Indicators

Logistics performance indicators associated with regional goods distribution by small- and medium-scale industries in Jaipur are given in Table 5. Quality of logistics services and reliability have a significant impact (23–25%) for regional distribution of

Table 5 Logistic performance indicators (LPIs)

LPI	AHP rating	
	Geometric mean of weights	In (%)
Border efficiency	0.12	12
Infrastructure	0.15	15
Ease of arranging shipment	0.16	16
Quality of logistics service	0.25	25
Reliability	0.23	23
Live tracking	0.09	9

Table 6 Wholesalers specific KPIs weights

KPI	Grain market	Fruits and vegetable market	Building hardware market	Overall AHP rating	
				Avg. weights	In (%)
Haulage cost	0.37	0.30	0.35	0.34	34
Handling cost	0.10	0.05	0.03	0.06	6
Fixed cost	0.22	0.21	0.16	0.20	20
Storage cost	0.31	0.44	0.46	0.40	40

goods, which indicates the lack of value addition facilities for finished goods and lack of transporter reliability in delivering the goods on the stipulated time frame. Ease of arranging shipment is a 16% impact factor in regional goods distribution, which reflects the load consolidation issues, (un)loading issues, and variation in demand orders. Infrastructure impact weightage is 15%, which reflects the inadequacy which can range from parking facilities, docking facilities for maneuvering of trucks, and adequate storage facilities. Border efficiency (12%) and live tracking (9%) have low impact weightage, which reflects the effects of goods and service tax implementation by the government of India and penetration of the Internet and mobiles.

7.2 Wholesaler (Shippers) Key Performance Indicators

Shipper-specific performance indicators associated with urban goods distribution by three wholesale markets in Jaipur are given in Table 6. Storage cost (40%) is the highest for shippers among all wholesale markets, followed by haulage cost (33%). Storage cost (46%) is highest in the case of building hardware markets due to the high cost of building hardware materials. Similarly, storage cost (44%) for fruits and vegetable markets is high due perishable nature of fruits and vegetables and the additional cost of cold storage. Fixed cost is urban local bodies fix almost the same among all wholesale markets under study as market rents and cost of other facilities. Handling cost has the least impact of 6% among all other costs for shippers. The highest variation is observed in the storage cost indicator (15%) among the three wholesale markets.

7.3 Transport Operators Key Performance Indicators

Economic indicators associated with transporters and carriers involved in urban goods distribution for three wholesale markets in Jaipur are given in Table 7. The reliability indicator has the highest impact of 24% among all economic indicators. Fruits and vegetable markets have the highest impact of reliability by 35% in distribution as its value keeps decreasing over time, and it is a daily consumption commodity. The

Table 7 Weights of economic KPIs of transport operators

KPI	Grain market	Fruits and vegetable market	Building hardware market	AHP rating	
				Avg. weights	In (%)
VKT	0.08	0.13	0.11	0.11	11
TKT	0.12	0.26	0.29	0.22	22
Reliability	0.25	0.35	0.12	0.24	24
Avg. length of haul	0.08	0.04	0.07	0.06	6
Avg. payload	0.07	0.11	0.14	0.11	11
Empty running	0.28	0.04	0.12	0.15	15
Trips per day	0.12	0.07	0.15	0.11	11

same is the case with grain market commodities by 25% as it also consists of daily consumption products, especially edible oils and spices, along with other products. There is frequent/stable ordering of these products from retailers due to predictable consumption patterns. Reliability is the least important factor (12%) in the case of building hardware products as these products are ordered in advance anticipation as per the construction schedule. Ton kilometer traveled (TKT) is the next important factor that has an impact on urban goods distribution. It is least (12%) in the case of fruits and vegetable market due to the volumetric nature of goods and less on weight based. In the case of building hardware material, TKT is most important (29%), which suggests that haulage cost is mostly based on weight, and the same is the case with grain market products. It is observed that vehicle kilometer traveled (VKT), average payload, and trips per day have almost similar weightage in urban distribution with an average variation of 5%. Empty running has a weightage of 15%, but the least variation of 4% is observed for fruits and vegetable products, which reflects the more involvement of hired carriers and further scope for load consolidation in urban distribution.

7.4 Social Key Performance Indicators

Social and externalities indicators associated with urban goods distribution for the case city of Jaipur are given in Table 8. Accidents have the highest impact factor of 46% and most important concerns associated with urban goods distribution, followed by emissions by 24%. Urban goods carriers are perceived to contributes 17% of congestion on urban roads and cause a 13% delay in travel time in the urban area of Jaipur city.

Table 8 Social and externalities indicators

KPI	AHP weights	
	Geometric mean of weights	In (%)
Accidents	0.46	46
Emissions	0.24	24
Delay	0.13	13
Congestion	0.17	17

8 Conclusions

This research paper presents the weights of KPIs across the commodity distribution at the city level in the context of sustainability. The results show that there is a variation among KPIs according to commodities due to their supply chains. LPI of industrial goods distribution is also explored to make a case for LPI at the city level for freight policy interventions.

It is observed that economic indicators important for fleet operators are reliability in goods delivery by 24%, followed by TKT by 22%, while the average haulage length indicator is least important (6%). For shipper-specific KPI in urban goods distribution, it is that storage cost and haulage costs of goods have the highest weights of 40% and 34%, respectively, while handling cost has a minimum weightage of 6%. Policy intervention which can enhance the load consolidation and haulage cost of goods has a direct implication on the efficiency of goods distribution. In the case of logistics performance indicators, it is observed that the quality of logistics services and reliability of delivering goods have similar weights of 23–25% each, and they are the critical decision factors in efficiency, while live tracking of goods movement indicator is least important with 9% weight only. It is observed that accidents and fatalities are the primary cause of concern, which affects 46% of issues in goods distribution, followed by externalities by 24%. Urban freight transport policies in India need to be more oriented to minimize externalities and accidents.

Extensive freight studies need to be carried out with more commodities across varying city sizes in India for robustness in the weights of KPIs in the purview of sustainable development of Indian cities. The clustering of KPI for sustainable goods distribution needs to be explored more scientifically in future studies.

References

1. Gudmundsson H, Hall RP, Marsden G, Zietsman J (2015) Sustainable transportation—indicators, frameworks, and performance management, 1st edn. Samfundslitteratur, Frederiksberg
2. U.S. EPA (2008) EPA's report on the environment (ROE), EPA/600/R-07/045F (NTIS PB2008-112484). U.S. Environmental Protection Agency, Washington, DC
3. Litman T. Developing indicators for sustainable and livable transport planning. http://www.vtpi.org/wellmeas.pdf. Accessed 18 Mar 2019

4. Patier D, Routhier J-L (2008) How to improve the capacity of urban goods movement data. In: 8th international conference on survey methods in transport, Annecy
5. Chu X, Fielding GJ, Lamar BW (1992) Measuring transit performance using data envelopment analysis. Transp Res Part A Policy Pract 26(3):223–230
6. Carvalho M, Syguiy T, Nithack e Silva D (2015) Efficiency and effectiveness analysis of public transport of Brazilian cities. J Transp Lit 9(3):40–44
7. World Bank. worldbank.org. https://lpi.worldbank.org/international, https://lpi.worldbank.org/domestic. Accessed 16 Apr 2019
8. Lindholm M (2010) A sustainable perspective on urban freight transport: factors affecting local authorities in the planning procedures. Procedia Soc Behav Sci 2:6205–6216
9. Saaty TL (2008) Decision making with the analytic hierarchy process. Int J Serv Sci 1(1):83–98
10. European Commission. Indicators and data collection methods on urban freight distribution. https://repository.uantwerpen.be/docman/irua/c338e3/149143.pdf. Accessed 15 May 2019
11. OECD. Delivering the goods: 21st century challenges to urban goods transport. https://www.itf-oecd.org/delivering-goods-21st-century-challenges-urban-goods-transport. Accessed 10 Feb 2019
12. BESTUFS II. Best urban freight solutions II. http://www.bestufs.net/download/BESTUFS_II/key_issuesII/BESTUFS_II_results_datacollection/BESTUFS_II_data_collection_synthesis_report.pdf. Accessed 19 Nov 2018
13. NCHRP Synthesis 410. Freight transportation surveys. https://www.nap.edu/catalog/13627/freight-transportation-surveys. Accessed 16 Mar 2019
14. Browne M, Allen J (2019) Analyzing the results of U.K. urban freight studies. https://www.researchgate.net/publication/232415414_Analysing_the_results_of_UK_urban_freight_studies
15. Schoemaker J, Allen J, Huschebeck M, Monigl J. Quantification of urban freight transport effects I. http://www.bestufs.net/download/BESTUFS_II/key_issuesII/BESTUF_Quantification_of_effects.pdf. Accessed 13 May 2018
16. Gonzalez-Feliu J, Gardrat M, Pluvinet P, Ambrosini C. https://halshs.archives-ouvertes.fr/halshs-00778480/document. Accessed 05 Nov 2019
17. Allen J, Browne M, Woodburn A. London freight data report: 2014 update. http://content.tfl.gov.uk/london-freight-data-report-2014.pdf. Accessed 19 Jan 2019
18. Transport for London (TfL). London construction consolidation centre. http://www.ndslogistik.se/files/reports/1425975813_9.pdf. Accessed 7 Jan 2019
19. Greater London Authority. London wholesale markets review. https://www.london.gov.uk/file/12723/download?token=j4mEZDd6. Accessed 02 Aug 2019
20. McCabe S, Roorda M, Kwan H (2013) Comparing GPS and non-GPS survey methods for collecting urban goods and service movements. Int J Transp Econ 40(2)
21. Wolpert S, Reuter C (2012) Status quo of city logistics in scientific literature: a systematic literature. Transp Res Rec J 2269:110–116
22. Gonzalez-Feliu J, Toilier F, Routhier J-L (2010) End consumer goods movement generation in French. Procedia Soc Behav Sci 2(3):6189–6204
23. Ambrosinia C, Patiera D, Routhier J-L (2010) Urban freight establishment and tour-based surveys for policy-oriented modelling. Procedia Soc Behav Sci 2(3):6013–6026
24. Holden R, Xu B, Greening P, Piecyk M, Dadhich P (2016) Towards a common measure of greenhouse gas related logistics activity using data envelopment analysis. Transp Res Part A Policy Pract 91:105–119
25. Velasque M, Hester PT (2013) An analysis of multi-criteria decision-making methods. Int J Oper Res 10(2):56–66
26. De Brucker K, Macharis C, Verbeke A (2011) Multicriteria analysis in transport project evaluation: an institutional approach. Eur Transp J 47:3–24
27. Kopytov E, Abramov D (2012) Multiple-criteria analysis and choice of transportation alternatives in multimodal freight transport system. Transp Telecommun J 13(2):148–148
28. Jaipur Development Authority. http://jda.urban.rajasthan.gov.in/. Accessed 13 May 2019
29. Sharma M, Moon I, Bae H (2008) Analytical hierarchical process & optimize distribution network. J Appl Math Comput 202:256–265

30. Semih T, Seyhan S (2011) A multi-criteria factor evaluation model for gas station site selection. J Glob Manag 2(1):12–21
31. Saaty RW (1987) The analytic hierarchy process—what it is and how it is used. J Math Model 9(3):161–176
32. Saaty TL (2006) Rank from comparisons and from ratings in the analytic hierarchy/network processes. Eur J Oper Res 168(2):557–570
33. Saaty TL (2008) Decision making with the analytic hierarchy process. Int J Serv Sci 1(1)
34. Schoner B, Wedley WC (1989) Ambiguous criteria weights in AHP: consequences and solutions. Decis Sci 20:462–475
35. Saaty TL (1980) The analytic hierarchy process: planning, priority setting, resource allocation. McGraw-Hill, New York

Data Envelopment Analysis (DEA) Approach to Assess the Efficiency of Wholesale Markets in Various Cities of India

Check for updates

Pankaj Kant and Sanjay Gupta

1 Introduction

The rising level of urbanisation and consumption of goods have increased the logistics and transport activities in urban areas. Continuous improvements are required in the transport sector in an integrated manner to pass on the expected benefits to all stakeholders including the local community [1]. Lack of knowledge and awareness of urban freight generates a low level of interest among policymakers. Freight transport operations represent a substantial proportion of externalities in urban areas [2]. Urban goods carriers in the case of Delhi city India contributes 20% of total emission and 50% contribution in PM10 emissions while having a modal share of urban goods vehicle is 3.7% only [3]. Trade volumes have increased the logistics activities in urban areas which require the larger need of land parcels in cities. The rising cost of land parcel in cities have increased the logistics sprawl phenomenon and pushed the distribution centres on the periphery of the urban areas. Global consumption of goods has increased the movement of goods from core areas of the cities to ports [4]. Various mitigations options are available to policymakers to minimise the negative impacts with specific policy interventions in the urban freight sector as adopted by various cities globally [1, 5]. Several indicators in literature are described for evaluating the freight efficiency in general like tonnage handled, population (density), urban area, road network density, freight, (un)loading time, vehicle kilometres travelled (VKT) and tonnes kilometres travelled (TKT) [6].

Freight handling areas within and across cities can be compared with the indicators of efficiency and effectiveness for their productivity and level of service.

P. Kant (✉) · S. Gupta
Department of Transport Planning, School of Planning and Architecture, 4-Block-B Indraprastha Estate, New Delhi 110002, India
e-mail: pankaj179phd17@spa.ac.in

S. Gupta
e-mail: s.gupta@spa.ac.in

M. Parida et al. (eds.), *Proceedings of the Fifth International Conference of Transportation Research Group of India* , Lecture Notes in Civil Engineering 220,
https://doi.org/10.1007/978-981-16-9925-2_23

Efficiency and effectiveness are inversely related thus required due diligence in their balance to maximise the benefits [7]. Infrastructure efficiency indicates the level of resource utilisation, whereas effectiveness indicates the use of outputs to maximise freight transport interests [8]. DEA is a nonparametric benchmarking method widely employed in the efficiency analysis in the transport sector, ports and airports sectors [9].

Measurement of the performance and efficiency of freight handling areas like wholesale markets are not much focused in India but have a significant role in sustainable goods distribution in cities. This research paper evaluates the efficiency of wholesale markets in five Indian cities through data envelopment analysis method (DEA). The next section reviews literature related to DEA applications; Sect. 3: methodology; Sect. 4: DEA model formulation; Sect. 5: case city profile and selection of DMU; Sect. 6: DEA efficiency analysis of wholesale markets areas in case cities and policy implications.

2 Literature Review

DEA technique is mostly used for comparing the performance of decision-making units (DMU) with multiple variables. Most DEA studies covered airports, ports, public transport and railways. The input inputs variables in the DEA method varies from 3 or 4, and the number of outputs variables remains 1 or 2 [9]. The bootstrap-based DEA technique was used to analyse the efficiencies of interchanges subway stations in Seoul city [10]. DEA study reveals that privatisation urban rail transit system has a positive impact on its efficiency [11]. DEA method was used to benchmark road freight companies with different operational characteristics in the purview of emission reduction targets [12].

DEA method with three different models was used to analyse the congestion management. CCR model assessed the congestion but was unable to differentiate technical efficiency with the congestion effect [13]. DEA was used for optimising delivery routes for a significant retailer in Australia. The study results confirm that the results are similar to conventional route optimising methods [14]. DEA method was used to analyse the energy efficiency practice of freight hauliers to minimise greenhouse gases (GHG) emissions. Study results reveal that due to the lack of incentives for improvements, energy efficiency is not crucial for hauliers [15]. DEA method was to benchmark the KPIs in freight transport in the U.K. for potential freight efficiency gains. Study results reveal that similar distribution operations can have variations in efficiency across the supply chain of industrial products [16].

DEA method was used to assess the efficiency of trucks involved in the Norwegian road construction sector. Study results reveal that inputs saving potential can vary from 24 to 26%, whereas output increasing potential varies from 23 to 59%. Make of trucks has no role in the overall performance [17]. DEA method was used to analyse the efficiency of shipping firms in Korea to identify the best practice firms. The study suggests that optimising on non-shipping costs may lead to a smaller size of firms [18].

DEA method was used to evaluate the technical efficiency of four container ports of Australia with 12 international container ports. Results revealed that inefficiency is due to slack in container birth labour issues [19]. DEA method was used for efficiency analysis of 30 container terminals within the ports of China and Korea for efficiency improvement [20]. DEA was used to analyse the relative efficiencies of 50 ASEAN container ports to improve the ASEAN trade competitiveness. The traditional output-oriented DEA-CCR method was applied for efficiency evaluation [21].

DEA method was used for productivity analysis of 45 U.S. commercial airports. The study reveals that small airport hubs outperform the airport hubs in [22]. DEA method was used to analyse the cost efficiency of Italian and Norwegian airports over a fixed period. The study suggested that the level of competition impacts the airport's efficiency. Military use of airports has a positive impact on airport efficiency [23]. DEA method with rough set theory was used to evaluate the supply chain efficiency of the furniture market in China [24]. DEA method was used to evaluate the efficiency of petroleum distribution centres facilities in the USA [25]. DEA was used to determine the performance levels of departments in Dokuz Eylul University (Turkey) with the technical scores and scale scores of departments to investigate the leading cause of inefficiency [26].

Sustainability and efficiency in freight transport are gaining momentum at regional and urban levels. DEA method is an appropriate evaluation method for efficiency and benchmarking studies. The DEA application in multiple sectors is continuously growing including the freight transport sector [27]. DEA method in the passenger transport sector is widely applied in the Indian context. This research study attempts to apply DEA in the urban freight sector for the efficiency of wholesale markets.

3 Methodology

This research study proposes a sequential approach of DEA methodology from the selection of DMUs, indicators, DEA model to sensitivity analysis [28]. The application of several DEA models is demonstrated in the literature to determine relative efficiencies within the compared DMUs. This research paper assessed the performance and efficiency measurement of the wholesale grain market in five different Indian cities of varying sizes. The DEA analysis was done using secondary and primary data on infrastructure efficiency and operational service effectiveness. Efficiency is important for local policy planners, whereas service effectiveness is important for wholesalers and transport operators. The wholesale markets (DMU) were selected by considering the availability of market and city data from the secondary source. The selected DMU (wholesale grain markets in the present study) have similar objectives and working environments. The data was collected and analysed for the year 2017.

Table 1 Efficiency variables

S. No.	Infrastructure efficiency	Selected variables	Operational effectiveness	Selected variables
	Input/output variable		Input/output variable	
1	Population	√	Area of market	√
2	Area	√	Avg. size of establishment	
3	Road density		Avg. tonnage handled/day	√
4	commercial vehicles		Avg. shipment size	√
5	Average speed		Avg. shipment frequency	
6	Length of the road network	√	Dwell time	√
7	Land use distribution		Loading/unloading charges	√
8	Nos. of commercial centres		Parking turnover	
9	Freight generation/person		Usage of storage area	
10	Intracity freight moved	√	The average length of haul	
11	Intercity freight moved		Employment/establishment	
12			Freight trips att./gen. rate	

The list of variables for efficiency analysis from literature and variables selected in this research for DEA analysis are given in Table 1. Selected variables are very generic due to the absence of standardised data from secondary sources. Still, this proposed research framework will be useful for rapid efficiency analysis to policymakers as a starting point before detailed benchmarking studies. The framework for DEA analysis for wholesale market efficiency used in this study is shown in Fig. 1.

Sensitivity analysis of DEA results is important for stability and robustness [29]. Sensitivity analysis is also important to gauge the impact of a particular variable on the overall efficiency of DMUs. Two methods are widely applied in the sensitivity analysis of DEA results. In one method, sensitivity is performed with the addition or deletion of DMU in the DEA model, and the other method for sensitivity analysis is performed by increasing or decreasing the input/output variables [30]. This research study adopts the latter method for sensitivity analysis. The outputs of efficient DMUs will be decreased, and the inputs will all be increased by a stipulated amount of 5%. For inefficient DMUs, all outputs will be increased, and all inputs will be decreased by 5%.

4 DEA Model

The basic DEA model can be categorised into input-oriented and output-oriented models. Input-oriented model is important when input is to be minimised without

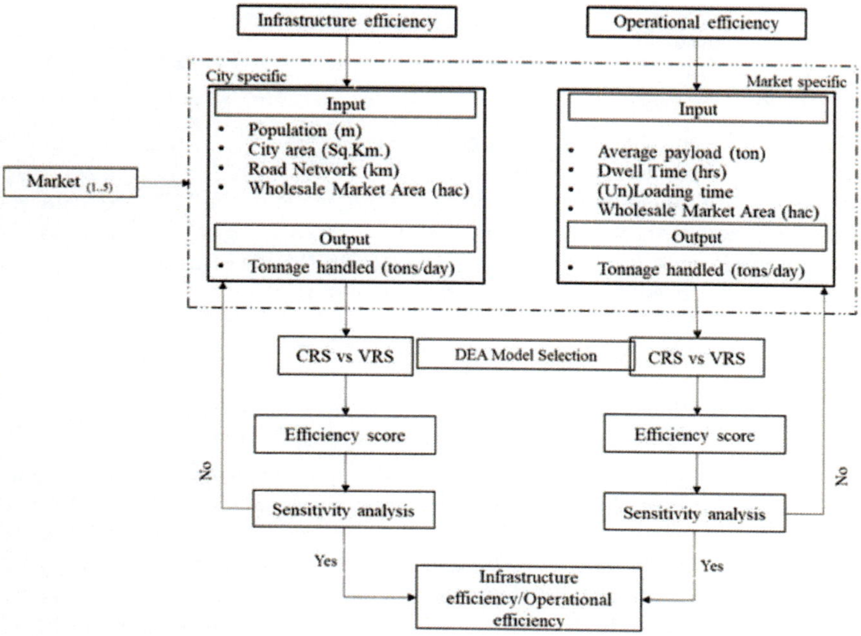

Fig. 1 DEA framework for wholesale market efficiency analysis

affecting the outputs, whereas an output-oriented model is used when output has to be maximised without altering inputs. This research study proposed to use the CCR (or CRS) output-oriented models for efficiency analysis of wholesale markets.

There are procedural issues relating to variables that need to be examined before applying DEA. Some of these issues are homogeneity of units, selection of input/output variables, measurement of variables and their weights. Each of these issues can present difficulties in DEA analysis [31].

In DEA method, productivity or efficiency of DMU is calculated by comparing the amount of outputs concerning the amount of inputs. DEA is widely used for the evaluation of DMUs in multiple sectors including transport due to its ability to evaluate the efficiency of multiple input–output units. DMU is a generic term used in DEA analysis and application for a wider spectrum of activities [32, 33].

4.1 *Efficiency Ratio*

The efficiency ratio is a ratio of the outputs produced with the number of inputs by a DMU. The weighted ratio is used (Eq. 1) for the calculation of the efficiency of a DMU

$$\text{performance} = \frac{\text{virtual output}}{\text{virtual input}} = \frac{u_1 y_{10} + \cdots u_s y_{so}}{v_1 x_{10} + \cdots + v_m x_{mo}} \tag{1}$$

where

x and y are resp., the input and output vectors and

u_s output weight, v_m inputs weight.

The two basic models of DEA are the CCR-DEA model and the BCC-DEA model. In the CCR model, there is a constant rate of returns to scale, whereas in the BCC model, there is a variable rate of return to scale [33, 34]. This research study used the CCR-DEA model for the efficiency evaluation of DMUs.

4.2 CCR Model

The CCR model converts the multiple input and outputs into single input and output combinations for each DMUs in the model. Linear programming used the CCR-DEA model to calculate the efficiency of a DMU. The performance of a single DMUs can be analysed in the form of rankings for better comparisons. The model gives the outputs in the form of relative efficiency from low to high scores. The highest score of 100% efficient is assigned to DMU having the highest relative efficiency. The CCR-DEA model involves both maximisation and minimisation problems in efficiency evaluation [33].

4.3 BCC Model

BCC-DEA model is similar to the CCR-DEA model having the same mathematical formula, but it evaluates the variable returns to scale instead of constant returns to scale. Compared to the CCR-DEA model, there are proportional changes in the input/output variables in BCC-DEA. The returns to scale is either decreasing or in increasing order [34].

4.4 Outputs from the DEA Model

This research study uses the frontier analyst software@4 for DEA analysis. Outcome results of the DEA model are efficiency score of DMU, lambda value, lambda peer indicator and return to scale (RTS). The lambda values are raw weights of peer units and attached to the peer indicator. RTS values 0 indicates the constant return, +1 indicates increasing returns, and −1 indicates decreasing returns to scale. The efficiency scores of DMU is always displayed in % in DEA evaluation. Lambda peer indicator value is either zero or a cell reference where it relates to [35].

5 Case City Profile and Market Selection

A total of five major wholesale grain markets are selected as DMUs in the Indian cities of Jaipur, Udaipur, Kanpur and Amritsar, for performance and efficiency analysis. The brief profile of these cities is mentioned in Table 2.

Table 2 Cities profile

Grain market	City	State	Brief profile
Suraj Pole Mandi	Jaipur	Rajasthan	Jaipur city population is 30.13 lac (2011) with an additional 10% floating population. The population growth rate city is 23.10%. The total area of Jaipur city is 467 km^2. The city is divided into 91 municipal wards. The length of the transport network is around 1500 km. Commercial land use in Jaipur is around 4% as per the master plan [36]
Dhaan Mandi	Udaipur	Rajasthan	Udaipur is known for its handicrafts and rich mineral and tourism sector. The total area of the city is 109 km^2 and having a population of 4.5 lac (2011). The city attracts 15–18 thousand daily tourist arrivals. The total road network of the city is 426 km. Commercial land use share is 2% [37, 38]
Galla Mandi	Kanpur	Uttar Pradesh	Kanpur is the biggest city in Uttar Pradesh state. According to the 2011 census, the city had a population of 25.5 lac. The population growth rate of Kanpur city is 23.10%, and commercial land use is 3% Kanpur has developed linearly along River Ganga and G.T road. Kanpur city has a linear spatial urban form. Kanpur is known for its small-scale industries [39]
SSCMY Market	Pune	Maharashtra	The Pune city area is 244 km^2 with a population of over 30 lacs. The city is the second-largest metropolitan city in Maharashtra state. Pune city is delineated into four main zones and further sub-divided into 14 administrative wards. The population growth rate of the city is 23.10%, and the share of commercial land use is 2% [40]
Bhagtawala Mandi	Amritsar	Punjab	According to the 2011 census, the population of Amritsar was 12.7 lac. The population growth rate is 15.7%. The area of Amritsar city is 170 km^2. Road network length is 612 km. The share of commercial land use is 5% of the city (CMP Amritsar) [41]

Note SSCMY = Shri Shri Chatrapati Shivaji Market Yard

Table 3 Input/output parameters for infrastructure efficiency DMUs

DMU (grain market)	City	Input (1)				Output (2)
		Pop (m)	City area (km^2)	Road net. (km)	Area of market (ha)	Tonnage handled per day
SSCMY	Pune	3.13	243	950	73	3750
Galla Mandi	Kanpur	1.53	296	360	50	12,135
Bhagtawala	Amritsar	1.27	142	612	83	63,262
Dhaan Mandi	Udaipur	0.45	109	426	26	6579
Suraj Pole	Jaipur	3.04	467	1500	16	4704

Source [36–41]

5.1 Input/Output Parameters for Infrastructure Efficiency

Selected input and output parameters for infrastructure effectiveness of grain markets in case cities are given in Table 3. It is observed that daily tonnage handled in grain markets varies from 3750 tonnes in Pune to 63,262 tonnes in Amritsar which suggests that Amritsar city grain market is more involved in intercity (regional) distribution compared to intracity distribution, whereas the Pune market is more of intracity distribution.

5.2 Input/Output Parameters for Operational Effectiveness

Selected input and output parameters for the operational efficiency of grain markets in case cities are given in Table 4. There are only one output and five input variables. All input/output variable is very generic, as detailed operational variables are not

Table 4 Input/output parameters for operational effectiveness DMUs

DMU (grain markets)	Input					Output	
	Avg. payload (ton)	Dwell time (min)	Loading time (min)	Unloading time (min)	Area of market (ha)	Tonnage per day (ton)	Tonnage per day/ha
SSCMY	1.3	450	90	70	73	3750	51
Galla Mandi	2.2	420	60	30	50	12,135	243
Bhagtawala	2.5	480	90	90	83	63,262	762
Dhaan Mandi	1.2	420	90	150	26	6579	253
Suraj Pole	1.8	240	90	240	16	4704	294

Source Primary surveys

available in the secondary source. It is observed that tonnage generated per hectare by Bhagtawala Market Amritsar is highest, and SSCMY Pune is lowest among all wholesale markets.

6 Performance Analysis Using DEA

Performance analysis of wholesale markets by DEA method is presented in the following three parts, i.e., infrastructure efficiency, operational effectiveness and comparison of efficiency versus effectiveness.

6.1 Infrastructure Efficiency of the Wholesale Markets

CRS efficiency score of wholesale markets (DMUs) for their infrastructure efficiency is given in Table 5. Shri Chatrapati Market Yard (SSCMY) Pune has the least efficiency score (15%), followed by Dhaan Mandi Udaipur (33%) and Suraj Pole Mandi Jaipur (39%). Bhagtawala Grain Mandi Amritsar got the most efficient score (100%).

Sensitivity analysis of infrastructure efficiency for the wholesale market is given in Table 6. The efficiency score of Bhagtawala Grain Mandi remains stable, whereas the efficiency of other markets increased. Suraj Pole Market Jaipur observed a maximum increase in efficiency score from 18 to 43% by 5% decrease in selected input variables in DEA.

Table 5 Infrastructure efficiency score of the wholesale markets

DMU No.	DMU name	CRS efficiency (%)	Sum of lambdas	Return of scale (RTS)
1	SSCMY	04	0.059	Increasing
2	Galla Mandi	33	0.192	Increasing
3	Bhagtawala	100	1.000	Constant
4	Dhaan Mandi	29	0.104	Increasing
5	Suraj Pole	18	0.074	Increasing

Table 6 Sensitivity analysis of infrastructure efficiency

DMU No.	DMU name	CRS efficiency (%)	Sum of lambdas	Return of scale (RTS)
1	SSCMY	7	0.059	Increasing
2	Galla Mandi	36	0.192	Increasing
3	Bhagtawala	100	1.000	Constant
4	Dhaan Mandi	37	0.104	Increasing
5	Suraj Pole	43	0.074	Increasing

Table 7 Operational effectiveness score of wholesale market

DMU No.	DMU name	CRS efficiency (%)	Sum of lambdas	Return of scale (RTS)
1	SSCMY	11	0.059	Increasing
2	Galla Mandi	32	0.192	Increasing
3	Bhagtawala	100	1.000	Constant
4	Dhaan Mandi	33	0.104	Increasing
5	Suraj Pole	39	0.074	Increasing

Table 8 Sensitivity analysis of operational effectiveness

DMU No.	DMU name	CRS efficiency (%)	Sum of lambdas	Return of scale (RTS)
1	SSCMY	14	0.066	Increasing
2	Galla Mandi	39	0.212	Increasing
3	Bhagtawala	100	1.000	Constant
4	Dhaan Mandi	41	0.115	Increasing
5	Suraj Pole	47	0.082	Increasing

6.2 Operational Effectiveness of the Wholesale Markets

CRS efficiency score of wholesale markets (DMUs) for their operational effectiveness is given in Table 7. Shri Chatrapati Market Yard (SSCMY) Pune has the least efficiency score (11%), followed by Dhan Mandi Udaipur (33%), Suraj Pole Mandi Jaipur (39%) and Galla Mandi Kanpur (58%). Bhagtawala Grain Mandi Amritsar market got the most efficient score of (100%) ineffectiveness.

Sensitivity analysis of operational efficiency for the wholesale markets is given in Table 8. The relative efficiency score of the Bhagtawala Grain Market remains stable, whereas the relative efficiency scores of other markets increased. Dhaan Mandi Udaipur observed a maximum increase in efficiency score by 8%.

6.3 Comparison of Infrastructure Efficiency with Oerational Effectiveness

A comparison of infrastructure efficiency and operational effectiveness is given in Table 9. Maximum variation (lowest score) is observed in the case of SSCMY Pune (0.36) followed by Suraj Pole Mandi Jaipur. Both markets need to increase their infrastructure efficiency. Galla Mandi Kanpur and Bhagtawala Mandi Amritsar have a maximum ratio of both infrastructure efficiency and operational efficiency.

Table 9 Infrastructure efficiency versus operational effectiveness

DMU No.	DMU name	City	Infrastructure efficiency (1) (%)	Operational effectiveness (2) (%)	Ratio (1)/(2)
1	SSCMY	Pune	04	11	0.36
2	Galla Mandi	Kanpur	33	32	1.03
3	Bhagtawala	Amritsar	100	100	1.00
4	Dhaan Mandi	Udaipur	29	33	0.88
5	Suraj Pole	Jaipur	18	39	0.46

7 Conclusion

This research study employed the DEA method to assess the efficiency of wholesale markets in Indian cities. Ranking of wholesale markets can be done based on relative efficiency scores predicted by the DEA method. This research study considered infrastructure efficiency and operational effectiveness of wholesale markets as two different cases. Both cases have different input variables but similar output variables for relative efficiency calculation. Both yields a similar kind of performance result for the wholesale grain markets in terms of efficiency and effectiveness ranking.

It is observed that the Amritsar wholesale market is both effective and efficient with an efficiency score of 100% in comparison with all other wholesale markets. SSCMY Pune has the least efficiency score of 4% and an effectiveness score of 11%. It is observed that Galla Mandi Kanpur has higher operational effectiveness (33%) than infrastructure efficiency (32%). Results observed from sensitivity analysis confirm improvement in infrastructure effectiveness and operational efficiency score with marginal perturbation of in-input variable concerning most efficient DMU. The maximum increase in efficiency during sensitivity analysis is observed in the Suraj Pole Market by 25%. The observed improvement in the effectiveness score of wholesale markets remains in a range of 7–8%, except SSCMY Pune (3%) in the sensitivity analysis.

More relevant data specific to the operational effectiveness of wholesale markets such as parking turnover, goods characteristics, storage and operational characteristics will need to be considered in future studies. Another area of research is to assess the performance levels of different wholesale markets within a city.

References

1. Paweł W, Lorenc A. Integrated, sustainable approach to the management of urban freight transport—review of the world's solutions. www.ejournals.eu/pliki/art/3/394/. Accessed 11 June 2018
2. Lindholm M (2010) A sustainable perspective on urban freight transport: factors affecting local authorities in the planning procedures. Procedia Soc Behav Sci 2:6205–6216
3. DPCC (2017) https://www.dpcc.delhigovt.nic.in. Accessed 5 May 2017

4. Dablanc L, Geneviève G, Holliday K, Obrien T (2014) Best practices in urban freight management: lessons from an international survey. Transp Res Rec (TRR) 29–38
5. Browne M, Allen J, Nemoto T, Patier D, Visser J (2012) Reducing social and environmental impacts of urban freight transport: a review of some major cities. Procedia Soc Behav Sci 39:19–33
6. OECD. Delivering the goods: 21st century challenges to urban goods transport. https://www.itf-oecd.org/sites/default/files/docs/03deliveringgoods.pdf. Accessed 12 Apr 2019
7. Chu X, Fielding L, Gordon J, Bruce W (1992) Measuring transit performance using data envelopment analysis. Transp Res Part A Policy Pract 26(3):223–230
8. Carvalho M, Syguiy T, Nithack Silva D (2015) Efficiency and effectiveness analysis of public transport of Brazilian cities. J Transp Lit 9(3):40–44
9. Somogyi RM. Measuring efficiency in transport: the state of the art of applying data envelopment analysis. https://journals.vgtu.lt/index.php/Transport/article/download/5545/4811. Accessed 16 May 2019
10. Kim C, Kim SW, Kang HJ, Song S-M. What makes urban transportation efficient? Evidence from subway transfer stations in Korea. https://www.mdpi.com/2071-1050/9/11/2054/pdf. Accessed 05 May 2019
11. Jain P, Cullinane S, Cullinane K (2008) The impact of governance development models on urban rail efficiency. Transp Res Part A Policy Pract 42(9):1238–1250
12. Holden R, Xu B, Greening P, Piecyk M, Dadhich P (2016) Towards a common measure of greenhouse gas related logistics activity using data envelopment analysis. Transp Res Part A Policy Pract 91:105–119
13. Kao C (2010) Congestion measurement and elimination under the framework of data envelopment analysis. Int J Prod Econ 123(2):257–265
14. Lau K-H (2013) Measuring distribution efficiency of a retail network through data envelopment analysis. Int J Prod Econ 146(2):598–611
15. Liimatainen H, Stenholm P, Tapio P, McKinnon A (2012) Energy efficiency practices among road freight hauliers. Energy Policy 50:833–842
16. McKinnon AC (2009) Benchmarking road freight transport: review of a government-sponsored program. Benchmarking Int J 16(5):640–656
17. Odeck J, Hjalmarsson L (1996) Performance of trucks—an evaluation using data envelopment analysis. Transp Plan Technol 20(1):49–66
18. Pattanamekar P, Ki C, Park D, Lee K (2011) Technical efficiency analysis of shippers using DEA. J Adv Transp 45:161–172
19. Tongzon J (2009) Efficiency measurement of selected Australian and other international ports using data envelopment analysis. Transp Res Part A Policy Pract 35(2):628–638
20. Zheng XB, Park NK (2016) A study on the efficiency of container terminals in Korea and China. Asian J Ship Logist 32(4):2013–2020
21. Kutin N, Nguyen TT, Valléec T (2017) Relative efficiencies of ASEAN container ports based on data envelopment analysis. Asian J Ship Logist 33(2):67–77
22. Bazargan M, Vasigh B (2003) Size versus efficiency: a case study of U.S. commercial airports. J Air Transp Manag 9(3):187–193
23. Merkert R, Mangia L (2014) Efficiency of Italian and Norwegian airports: a matter of management or the level of competition in remote regions. Transp Res Part A Policy Pract 62:30–33
24. Xu J, Li B, Wu D (2009) Rough data envelopment analysis and its application to supply chain performance evaluation. Int J Prod Econ 122(2):628–638
25. Ross AD, Droge C (2004) An analysis of operations efficiency in large-scale distribution systems. J Oper Manag 21(6):673–688
26. Goksen Y, Dogan O, Ozkarabacak B. A data envelopment analysis application for measuring efficiency of university departments. https://www.sciencedirect.com/science/article/pii/S2212567115000246. Accessed 13 May 2019
27. Zhou H, Yang Y, Chen Y, Zhu J (2018) Data envelopment analysis application in sustainability: the origins, development and future directions. Eur J Oper Res 264:1–16

28. Golany B, Roll Y (1989) An application procedure for DEA. Omega 17(3):237–250
29. Cooper WW, Seiford LM, Tone K (2006) Introduction to data envelopment analysis and its uses. Springer, New York
30. Seiford LM, Zhu J (1998) Sensitivity analysis of DEA models for simultaneous changes in all the data. J Oper Res Soc 49(10):1060–1071
31. Dyson RG, Allena R, Camanho AS, Podinovski VV, Sarrico CS, Shalea EA (2001) Pitfalls and protocols in DEA. Eur J Oper Res 132(2):245–259
32. Farrell MJ (1957) The measurement of productive efficiency. J R Stat Soc Ser A (Gen) 120(3):253–281
33. Charnes A, Cooper WW, Rhodes E (1978) Measuring the efficiency of decision-making units. Eur J Oper Res 2(6):429–444
34. Banker RD, Charnes A, Cooper WW (1984) Some models for estimating technical and scale inefficiencies in data envelopment analysis. Manag Sci 30(9):1078–1092
35. Frontier Analyst. https://banxia.com/frontier/. Accessed 24 Sept 2019
36. Comprehensive mobility plan Jaipur. https://transport.rajasthan.gov.in. Accessed 15 July 2019
37. Comprehensive mobility plan Udaipur. https://wedocs.unep.org/bitstream/handle/Pub-licTransport_Udaipur. Accessed 11 June 2019
38. City development plan Udaipur. http://www.udaipurmc.org/PDF/Others/Ud_CDP_Interim_Report.pdf. 11 June 2019
39. Comprehensive mobility plan Kanpur. http://kmc.up.nic.in/PDF_Files/others/CMP%20Kanpur-%20Draft.pdf. Accessed 09 May 2019
40. Comprehensive mobility plan Pune. http://www.pmrda.gov.in/Marathi/images/PDF. Accessed 13 May 2019
41. Comprehensive mobility plan Amritsar. https://wricitieshub.org/sites/default/files/Comprehensive%20Mobility%20Plan%20for%20Amritsar%20City.pdf. Accessed 19 Apr 2019

Emission Modelling of Passenger Cars in India: A Case of Hyderabad City

Saladi S. V. Subbarao, A. Deepti, H. Amit, A. Akshay, and S. K. Charan

1 Introduction

Due to the rapid growth in population and economic development in India, the number of personalized vehicles on the roads is increasing at a rapid pace. This is leading to an escalation in emissions and contributing to global warming, which is one of the most severe concerns faced by the world today. Traffic emissions, being a sizeable contributor to the global greenhouse gases (GHG) (14% of the entire GHG), grabs focus on the ongoing efforts toward reducing GHG. The emissions from road transport are considered as serious threats to urban air quality and global warming [1]. On the other side, the transport sector is indefinitely a significant contributor to the economy of any country and plays a crucial role in the daily activities of people in India like many other developing countries. At the same time, transportation also becomes a significant source of air and noise pollution and is presently one of the largest emission sources in India. Major cities in India are experiencing an emission growth rate of 6% per year over the past decade, which is quite alarming. Hence, the detailed understanding and accurate case-specific quantification of vehicular emissions are very much necessary to design policies to minimize traffic-associated emissions. Emission models are to be used for the estimation of emissions. Fuel consumption and exhaust emission can be modelled based on the driving cycle (DC).

A driving cycle provides an estimate of speed–time profile, and it represents on-road driving conditions for a particular region or country. Driving cycles are developed to take into account the varying on-road driving conditions (acceleration, deceleration, cruise, idle) and serve as an important input for estimating emissions

S. S. V. Subbarao (✉) · A. Deepti · H. Amit · A. Akshay · S. K. Charan
Department of Civil Engineering, Mahindra University, Hyderabad, India

A. Deepti
e-mail: deepti.avirneni@mechyd.ac.in

A. Akshay
e-mail: akshay15014@mechyd.ac.in

M. Parida et al. (eds.), *Proceedings of the Fifth International Conference of Transportation Research Group of India* , Lecture Notes in Civil Engineering 220,
https://doi.org/10.1007/978-981-16-9925-2_24

and fuel consumption of different categories of the vehicles. Two types of DC's are generally used: Legislative DC's and actual/real-world DC's. Legislative DCs are constructed from several constant acceleration and speed phases. Examples of legislative driving cycles include Japanese DC, California DC, Indian DC, etc. Many studies reported that these driving cycles are unsuitable for assessing fuel consumption due to their gentle acceleration, braking, and longer idle period [2]. In India, Indian driving cycle (IDC) and modified Indian driving cycle (MIDC) are used for the legislative purpose [3].

Real-world driving cycles (RWDCs) are the profiles of test vehicles derived from their movement on the road under real-world traffic conditions. RWDC differs from city to city as each city has its unique infrastructure such as type of facility, composition, and type of vehicle, topography, and time of the day [4–7]. For heterogeneous traffic conditions where sharp acceleration and decelerations are the main modes of operation, the RWDCs are the best suitable way of representing traffic conditions. Many researchers have successfully developed real-world driving cycles for different cities from different parts of the world. Some of them include American DC [8], European DC [9], Australian DC [10], Hong-Kong DC [11], etc. Further, Tong and Hung reviewed more than 100 such real-world driving cycles [12]. Developing countries like India where the traffic is heterogeneous require class-specific (passenger cars, buses, motorcycles, etc.) and case-specific (different regions) driving cycles. Researchers have made some attempts to construct driving cycles in India for different cities include Pune [5], Chennai [13], Delhi [14], Dhanbad [15], Hyderabad [16], and Bangalore [17]. As most of the above popular studies have used "micro-trip" as a basic construction unit for the development of the driving cycle, the present study also uses the same method for developing the driving cycle by giving more attention to local characteristics (geography of the region, driver and vehicular characteristics), and the same is used for estimation of emissions.

Vehicular emissions can be estimated by using driving cycles. Emission models serve this purpose, and they are developed by using average travel speed or by speed–time profiles of every second. Though average speed emission models are available in the literature, they do not account for speed change on a small scale due to traffic congestion. Further, they are unable to recognize the driving modes like acceleration and deceleration, which results in inaccurate estimation of the emissions (e.g., MOBILE). Modal emission models, on the other hand, focus on intricate details of dynamic conditions of traffic and could able to estimate the variations in emission rates due to various driving modes. Some of the modal emission models available in the literature are the comprehensive modal emission model (CMEM) [18] and VT-Micro [19] which requires second to second vehicle speed data to estimate instantaneous emission rates. Further, these models are used to derive acceleration characteristics and per cent time spent in each driving mode.

Based on the above literature, it is quite evident that various studies focussed on developing RWDCs by following different construction methods and estimated vehicular emissions based on those RWDC. Considering the existing heterogeneous traffic conditions in Indian roads and different vehicle classes with different driving behaviours makes the research project unique in various ways. As a part of the research project, the present study deals with the development of an RWDC for passenger cars for Hyderabad city and the estimation of emission factors by using CMEM.

2 Study Area and Data Collection

Hyderabad city, the capital of Telangana state, is selected as a study area. Hyderabad is the sixth-largest city in India with a population of 70 lakh and a total area spread over 625 km^2. Currently, the city has over 5.3 million vehicles, and nearly a thousand vehicles get added every day to the existing volume in the form of registrations, which has led to widespread traffic congestion. Though it is necessary to collect activity data throughout the city for a comprehensive representation of the city's driving patterns, due to budget constraints, the study is limited to selected major corridors in the city. A total of three corridors were selected for the study, i.e., financial district to Kukatpally region of around 14 km stretch, Kukatpally to Secunderabad of around 16 km stretch, and Secunderabad to Hitech city of around 18 km stretch. All three selected corridors have mixed land use along with a dense commercial area. The study covers all three corridors during morning peak and evening peak hours. The total distance covered in the study area is 96 km, and it takes 5.5 h (~19,800 s) to travel in those stretches. Though the number of stretches covered in the study is less, a total of 19,800 data points are considered for the study.

Two types of data collection methods are used for obtaining vehicle speed profiles, viz. on-board measurement method and chase-car method. The chase-car method involves an instrumented car that follows randomly chosen vehicles and imitates their driving patterns. In the on-board measurement method, a vehicle with an instrument collects trip data and speed data. The on-board measurement method is preferred over the chase-car method for avoiding difficulties in following a randomly chosen vehicle. Racelogic performance box (PBox) is used for recording speed data of passenger cars for this study. Data collected on all three corridors during morning peak and evening peak hours. The data from the performance box is extracted by using the software, VBOX tools, which are customized for the PBox instrument. The software will directly extract the speed time profiles of a selected stretch.

3 Methodology

Broadly, the methodology followed for this study can be divided into two parts: Generation of representative driving cycle and estimation of emissions using CMEM.

3.1 Generation of Real-World Driving Cycle (RWDC)

The procedure for the generation of representative DC involves the following steps: Selection of route, data collection, micro-trip generation, and construction of the driving cycle.

Selection of routes: The selection of routes is one of the important processes in the development of the driving cycle. Representative routes are selected from the total study area by considering the type of road, land use, topography, average traffic, population density, etc. The authors used their experience and knowledge regarding the local traffic conditions for the selection of study routes. If the selected routes are failed to represent the actual network and the traffic behaviour in that area, the representative sample will be biased, and hence, the results obtained from the driving cycle will deviate from its actual situation. Hence, in the present study, the authors selected travelling routes to represent the actual network at its best and the typical traffic flow conditions by considering major road links, public transport hubs, etc. The details of these locations are shown in Fig. 1.

Data collection: Data collection process always plays a pivotal role in any type of research. The quality of the output was greatly influenced by the reliability, representativeness, and consistency of the data which was collected from the field. The on-board measurement method is adopted due to the chaotic nature of mixed traffic streams in Indian conditions. Racelogic performance box is installed in the selected vehicle and the equipment logs data points, which include speed, latitude, longitude, etc. In total, data amounting to about 50 km is collected on three study stretches

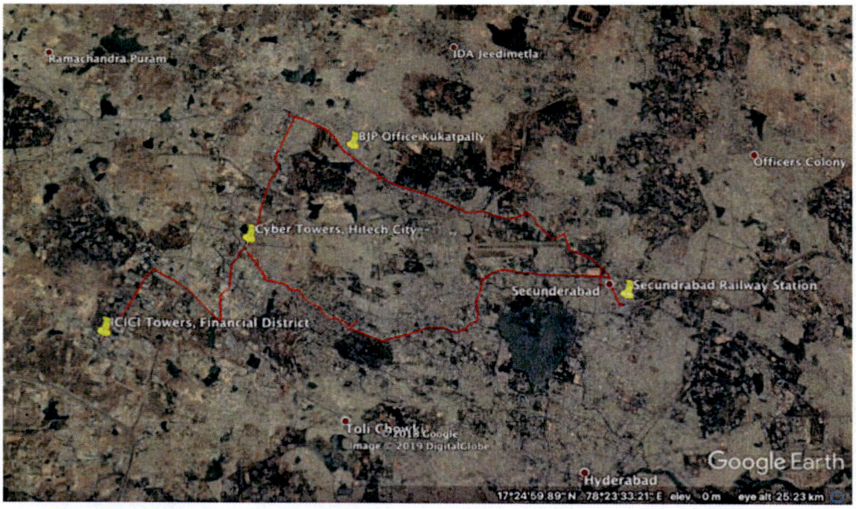

Fig. 1 Illustration of selected study routes taken from Google Earth

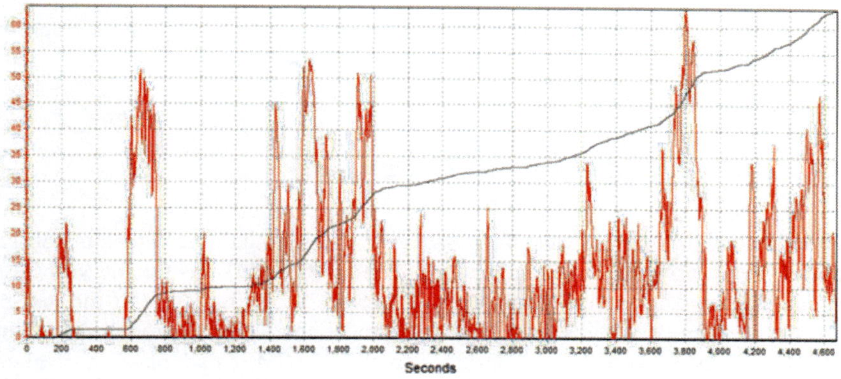

Fig. 2 Speed time profile of trip from financial district to Kukatpally

during morning peak and evening peak hours. The second by second speed data is recorded on all three routes. The sample plot of the speed versus time profile of one of the selected stretches is shown in Fig. 2.

Generation of micro-trips: The driving cycle is developed by considering micro-trips, which are illustrative of field traffic conditions. Micro-trip indicates a time gap between two successive points at which the vehicle is stopped. A micro-trip may consist of acceleration, cruise, and deceleration modes. The uniqueness of this study methodology is that the driving cycle is constructed considering the important target parameters of the speed–time profile, namely the per cent time in acceleration, deceleration, cruise, and idle speed. The entire data was separated into several micro-trips. A computer program was written to enable ease of the above process.

Construction of driving cycle: From the base data, parameters like % time spent by the vehicle in acceleration, deceleration, cruise, and idle need to be calculated along with the average speed. These parameters are generally called target parameters. The percentages of target parameters obtained from the base data are % time in acceleration—43, % time in deceleration—41, and % time in idle—16. The same kind of analysis is repeated on all identified micro-trips from the base data and calculated the parameters corresponding to micro-trip data. Later, parameters of the first micro-trip are needed to be compared with the parameters of all other micro-trips for reducing the repeated number of micro-trips of the same duration. A similar procedure is to be repeated to all other micro-trips. This procedure is necessary to obtain a feasible driving cycle duration. Then, an RWDC is to be built by selecting a sub-set of micro-trips to match the target parameters of the base data. If the overall parameters of selected micro-trips do not match with the target parameters, then some of the micro-trips are repeated according to the frequency of each micro-trip. Once the target parameters match with the parameters of micro-trips, the final RWDC is obtained by concatenating the individual micro-trips. The duration of RWDC should be long enough to represent all types of traffic situations and estimating the emissions accurately.

3.2 Estimation of Emissions Using CMEM

To estimate the emissions from the passenger cars using their corresponding driving cycles, the comprehensive modal emission model (CMEM) was used. CMEM estimates the vehicular emissions by considering the operating mode of the vehicle. For a wide range of vehicle categories, the model could able to predict second-by-second tailpipe emissions and consumption of fuel. CMEM adopts a physical and power-demand approach to analytically represent emission production. The model divides the entire emission estimation process into various components that correspond to the physical process associated with operating conditions of the vehicle and emissions production. Each component in the process is then modelled as an analytical representation by considering various parameters. These parameters vary with the type of vehicle, engine model, emission technology, and operating conditions of a vehicle. The model development phase is the essence of this modelling approach rather than its application. Data requirements for applications and updating the model to include new vehicles are easy once the model forms are established. The limited data requirement for future applications is the biggest advantage of this model. Further, this modelling approach provides a detailed understanding and explanation about the difference of variations in emissions among the different types of vehicles with various driving modes and other conditions [20].

The CMEM takes a wide variety of input variables which include vehicle type, maximum engine power, engine speed at maximum engine power, number of gears, vehicle mass, etc., along with the operating conditions of the vehicle. The results obtained from the CMEM include second-by-second tailpipe emissions and fuel consumption, which are used to arrive at the fuel consumption ratio and also the emission factor of the vehicle during the driving period.

4 Results and Discussion

Data collected from three selected study stretches and speed–time profiles of the trips were determined for all three stretches. Racelogic performance box was used for collecting data by administering the on-board measurement technique, and it can log values with a minimum velocity of 0.1 kmph and a maximum velocity of 1600 kmph. The target parameters from the base data are: per cent time in acceleration is 43, per cent time in deceleration is 41, and per cent time in idle is 16. The driving cycle is obtained from micro-trips, and the total number of micro-trips identified in the study are 141. Similar to the above analysis, the parameters of all micro-trips were obtained by developing a computer program. The computer program compares all

Table 1 Selected micro-trips and their parameters

S. No.	Micro-trip No.	% time in acceleration	% time in deceleration	% time in idle
1	45	55	43	17
2	48	37	45	18
3	55	44	43	13
4	59	42	49	9
5	62	33	47	20
6	66	44	44	12
7	72	37	50	13
8	77	45	45	10
9	94	49	48	3
10	138	48	43	9
11	107	49	41	10
12	90	50	33	17
13	63	50	40	10
14	58	43	49	8
15	48	36	45	19
16	21	50	33	17
17	25	50	44	6
18	17	49	39	12

micro-trips with one another by considering all targeted parameters of the individual micro-trips within a tolerance limit ranging from 5 to 15%, where 5% is considered as a lower limit and 15% is considered as an upper limit. After comparison, out of identified 141 micro-trips, 18 micro-trips are selected. The details of the selected micro-trips and their parameters are given in Table 1.

Based on the procedure explained in the methodology, the overall parameters of the selected micro-trips are matched with the target parameters of the base data. This procedure enables the identification of the duration of the driving cycle. The parameters obtained after the 18th micro-trip are exactly matching with the target parameters of the base data. The final duration of RWDC obtained from the study is 2277 s. The RWDC developed for Hyderabad city is as shown in Fig. 3.

A comprehensive modal emission model (CMEM) provides a second-wise emission from a vehicle. The input details provided to CMEM are second-wise changes in velocity, vehicle details, and road surface details. The output from CMEM is in the form of total distance travelled, fuel consumption, and the weight of gases like CO_2, CO, HC, and NO_x. Average emission values per kilometre are obtained for the Hyderabad city as CO_2—344 g/km, CO—1.5 g/km, HC—0.06 g/km, NO_x—0.4 g/km. Average fuel consumption was obtained as 110 g/km, which was verified by the inbuilt car equipment as it shows a mileage of 11 km/l in city driving conditions.

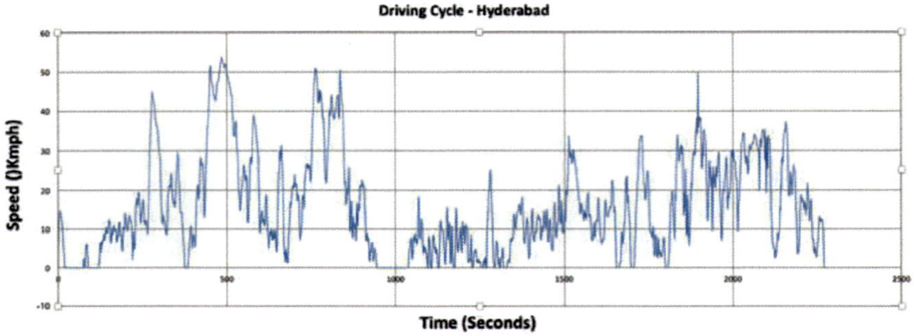

Fig. 3 Developed real-world driving cycle for Hyderabad city

5 Conclusions and Future Directions

In heterogeneous traffic conditions, every city or region needs a real-world driving cycle for emission estimation. The present study focus on that aspect and the study attempted to develop an RWDC for passenger cars in Hyderabad city. The developed RWDC is relatively much more complex than the modified Indian driving cycle. As the study incorporated local traffic conditions in the form of a time–space profile, acceleration and deceleration characteristics, etc., the approach is expected to be a better representation of traffic behaviour in the study area. The developed RWDC is used to estimate emissions released from passenger cars. The per kilometre emission values obtained for the study area are CO_2—344 g/km, CO—1.5 g/km, HC—0.06 g/km, and NO_x—0.4 g/km, which are much higher than the BSIV standards of light-duty petrol vehicles like cars. Hence, it can be concluded that the number of pollutants released into the atmosphere in the study area is at an alarming level. The study needs to be extended to other vehicle classes for accurately estimating the emissions in a city or region. Further, the life cycle assessment study can be conducted by comparing conventional vehicles with electric vehicles.

References

1. Saija S, Romano D (2002) A methodology for the estimation of road transport air emissions in urban areas of Italy. Atmos Environ 36(34):5377–5383
2. Bata R, Yasser Y, Wang W, Lyons D, Gambino M, Rideout G (1994) Heavy duty testing cycles: survey and comparison. SAE technical paper series no. 942263, Washington, DC
3. MoRTH (2010) Details of standards for emission of gaseous pollutants from petrol engine vehicles and test procedures effective from 1st April 1991. MoRTH/CMVR/TAP-115/116, 2(4)
4. Yu L, Zhang X, Qiao F, Qi Y (2010) Genetic algorithm-based approach to develop driving schedules to evaluate greenhouse gas emissions from light-duty vehicles. Transp Res Rec 2191:166–173

5. Kamble SH, Mathew TV, Sharma GK (2009) Development of real-world driving cycle: a case study of Pune, India. Transp Res Part D Transp Environ 14(2):132–140
6. Saleh W, Kumar R, Kirby H, Kumar P (2009) Real-world driving cycle for motorcycles in Edinburgh. Transp Res Part D Transp Environ 14(5):326–333
7. Hung WT, Tong HY, Lee CP (2007) Development of a practical driving cycle construction methodology: a case study in Hong Kong. Transp Res Part D Transp Environ 12(2):115–128
8. Esteve-Booth A, Muneer T, Kubie J, Kirby H (2002) A review of vehicular emission models and driving cycles. Proc Inst Mech Eng Part C J Mech Eng Sci 216:777–797
9. Tzirakis E, Pitsas K, Zannikos F, Stournas S (2006) Vehicle emissions and driving cycles: comparison of the Athens driving cycle (Adc) with Ece-15 and European driving cycle (EDC). Global NEST J 8:282–290
10. Kenworthy J, Newman P (1982) A driving cycle for Perth: methodology and preliminary results. In: Proceedings of the joint SAE-A/ARRB 2nd conference on traffic, energy, and emissions, Melbourne
11. Tong HY, Hung WT, Cheung CS (1999) Development of a driving cycle for Hong Kong. Atmos Environ 33:2323–2335
12. Tong HY, Hung WT (2010) A framework for developing driving cycles with on-road driving data. Transp Rev 30(5):589–615
13. Nesamani KS, Subramanian KP (2011) Development of a driving cycle for intra-city buses in Chennai, India. Atmos Environ 45:5469–5476
14. Chugh S, Kumar P, Muralidharan M, Mukesh Kumar B, Sitanathan M, Gupta A, Basu B, Malhotra RK (2012) Development of Delhi driving cycle: a tool for realistic assessment of exhaust emissions from passenger cars in Delhi. SAE technical paper 2012-01-0877
15. Adak P, Sahu R, Elumalai SP (2016) Development of emission factors for motorcycles and shared auto-rickshaws using real-world driving cycle for a typical Indian city. Sci Total Environ 544:299–308
16. Perugu H (2018) Emission modelling of light-duty vehicles in India using the revamped VSP-based MOVES model: the case study of Hyderabad. Transp Res Part D Transp Environ 68:150–163
17. Saikiran M, Verma A (2018) A novel methodology for construction of driving cycles for Indian cities. Transp Res Part D Transp Environ 65:725–735
18. Barth M, An F, Younglove T, Scora G, Levine C (2000) Development of a comprehensive modal emissions model. Final report for NCHRP project 25-11. Transportation Research Board
19. Ahn K, Trani AA, Rakha H, Aerde MV (1999) Microscopic fuel consumption and emission models. In: Proceedings of the 78th annual meeting of the transportation research board, Washington, DC
20. Scora G, Berth M (2006) Comprehensive modal emission model (CMEM): user's guide. University of California, Riverside, Centre for Environmental Research and Technology, NCHRP, Project 25-11